Advances in biocontrol of crop insect pests

It is widely recognised that agriculture is a significant contributor to global warming and climate change. Agriculture needs to reduce its environmental impact and adapt to current climate change whilst still feeding a growing population, i.e. become more 'climate-smart'. Burleigh Dodds Science Publishing is playing its part in achieving this by bringing together key research on making the production of the world's most important crops and livestock products more sustainable.

Based on extensive research, our publications specifically target the challenge of climate-smart agriculture. In this way we are using 'smart publishing' to help achieve climate-smart agriculture.

Burleigh Dodds Science Publishing is an independent and innovative publisher delivering high quality customer-focused agricultural science content in both print and online formats for the academic and research communities. Our aim is to build a foundation of knowledge on which researchers can build to meet the challenge of climate-smart agriculture.

For more information about Burleigh Dodds Science Publishing simply call us on +44 (0) 1223 839365, email info@bdspublishing.com or alternatively please visit our website at www.bdspublishing.com.

Related titles:

Biopesticides for sustainable agriculture
Print (ISBN 978-1-78676-356-3); Online (ISBN 978-1-78676-358-7, 978-1-78676-359-4)

Improving integrated pest management in horticulture
Print (ISBN 978-1-78676-753-0); Online (ISBN 978-1-78676-755-4, 978-1-78676-756-1)

Integrated management of diseases and insect pests of tree fruit
Print (ISBN 978-1-78676-256-6); Online (ISBN 978-1-78676-258-0, 978-1-78676-259-7)

Integrated management of insect pests: Current and future developments
Print (ISBN 978-1-78676-260-3); Online (ISBN 978-1-78676-262-7, 978-1-78676-263-4)

Microbial bioprotectants for plant disease management
Print (ISBN 978-1-78676-813-1); Online (ISBN 978-1-78676-815-5, 978-1-78676-816-2)

Chapters are available individually from our online bookshop:
https://shop.bdspublishing.com

climate•SMART•publishing
IN AGRICULTURAL SCIENCE

BURLEIGH DODDS SERIES IN AGRICULTURAL SCIENCE
NUMBER 164

Advances in biocontrol of crop insect pests

Edited by Professor Travis Glare, Lincoln Agritech Ltd and Lincoln University, New Zealand and Professor Johannes A. Jehle, Institute of Biological Control – Federal Research Centre for Cultivated Plants (JKI), Germany

burleigh dodds
SCIENCE PUBLISHING

Published by Burleigh Dodds Science Publishing Limited
82 High Street, Sawston, Cambridge CB22 3HJ, UK
www.bdspublishing.com

Burleigh Dodds Science Publishing, 1518 Walnut Street, Suite 900, Philadelphia, PA 19102-3406, USA

First published 2025 by Burleigh Dodds Science Publishing Limited

Library of Congress Control Number: 2025931713

British Library Cataloguing in Publication Data
A catalogue record for this book is available from the British Library

ISBN 978-1-80146-839-8 (print)
ISBN 978-1-80146-840-4 (ePub)
ISBN 978-1-80146-841-1 (PDF)
ISSN 2059-6936 (print)
ISSN 2059-6944 (online)

DOI: 10.19103/AS.2024.0149

Typeset by Deanta Global Publishing Services, Dublin, Ireland

Contents

Part 2 Improving biocontrol product development and use

Series list

Title	Series number
Achieving sustainable cultivation of maize - Vol 1 From improved varieties to local applications Edited by: Dr Dave Watson, CGIAR Maize Research Program Manager, CIMMYT, Mexico	001
Achieving sustainable cultivation of maize - Vol 2 Cultivation techniques, pest and disease control Edited by: Dr Dave Watson, CGIAR Maize Research Program Manager, CIMMYT, Mexico	002
Achieving sustainable cultivation of rice - Vol 1 Breeding for higher yield and quality Edited by: Prof. Takuji Sasaki, Tokyo University of Agriculture, Japan	003
Achieving sustainable cultivation of rice - Vol 2 Cultivation, pest and disease management Edited by: Prof. Takuji Sasaki, Tokyo University of Agriculture, Japan	004
Achieving sustainable cultivation of wheat - Vol 1 Breeding, quality traits, pests and diseases Edited by: Prof. Peter Langridge, The University of Adelaide, Australia	005
Achieving sustainable cultivation of wheat - Vol 2 Cultivation techniques Edited by: Prof. Peter Langridge, The University of Adelaide, Australia	006
Achieving sustainable cultivation of tomatoes Edited by: Dr Autar Mattoo, USDA-ARS, USA and Prof. Avtar Handa, Purdue University, USA	007
Achieving sustainable production of milk - Vol 1 Milk composition, genetics and breeding Edited by: Dr Nico van Belzen, International Dairy Federation (IDF), Belgium	008
Achieving sustainable production of milk - Vol 2 Safety, quality and sustainability Edited by: Dr Nico van Belzen, International Dairy Federation (IDF), Belgium	009
Achieving sustainable production of milk - Vol 3 Dairy herd management and welfare Edited by: Prof. John Webster, University of Bristol, UK	010
Ensuring safety and quality in the production of beef - Vol 1 Safety Edited by: Prof. Gary Acuff, Texas A&M University, USA and Prof. James Dickson, Iowa State University, USA	011
Ensuring safety and quality in the production of beef - Vol 2 Quality Edited by: Prof. Michael Dikeman, Kansas State University, USA	012
Achieving sustainable production of poultry meat - Vol 1 Safety, quality and sustainability Edited by: Prof. Steven C. Ricke, University of Arkansas, USA	013
Achieving sustainable production of poultry meat - Vol 2 Breeding and nutrition Edited by: Prof. Todd Applegate, University of Georgia, USA	014

Introduction

With an ever-growing number of products, the biopesticides sector is expanding and diversifying rapidly in supplying more sustainable solutions to the management of crop insect pests. This collection reviews some of the key developments in this area. It builds on our successful 2020 title: *Biopesticides for sustainable agriculture* (ed. Birch and Glare).

Part 1 Understanding and disrupting insect pests

The first group of chapters reviews a wide range of novel, more environmentally-friendly approaches to managing crop insect pests. As Chapter 1 points out, stimuli associated with host plant finding, mostly vision and olfaction, provide the basis not just for pest monitoring but also ways of exploiting these stimuli to manipulate pest and natural enemy dynamics for improved pest management. This is particularly important for small insect pests such as thrips, aphids, psyllids and whiteflies which can be particularly difficult to detect and manage. The chapter examines the visual and olfactory systems associated with the period before physical contact with the plant and how stimuli could be exploited for improved biological control.

Chapter 2 reviews key methods of genetic biocontrol focusing on those systems that have been developed for crop insect pests. These include "sterile release systems" where e.g. sterilized males produce sperm which carries dominant lethal mutations that result in their offspring failing to develop or prevents functional sperm from being produced. These are contrasted with "fertile release systems" which involve releasing insects that are capable of producing some offspring but with genetic defects that e.g. mean that fewer females survive. The chapter discusses the varied strengths and weaknesses of these different systems.

Chapter 3 discusses challenges and opportunities in using botanical pesticides to replace traditional synthetic products. These pesticides act throughout varied modes of action against insects which causes neurotoxic effects ranging from repellency to toxicity. The chapter focuses on how these biocontrol techniques can be deployed using the example of insect pest control in black truffle orchards. The chapter discusses the experience of using essential oils (a key botanical biocontrol agent) in field conditions as a solution against *Leiodes cinnamomeus*, the most damaging pest in truffle cultivation.

Chapter 4 discusses developments in neuropeptide-based biocontrol agents. The chapter addresses the key opportunities and challenges in using neuropeptides as bioinsecticides. Rational design of neuropeptide analogues could provide a new generation of eco-friendly, pollinator-safe bioinsecticides.

The chapter also focuses on manufacture and improving sustainability of peptide bioinsecticides. Continued optimization of appropriate adjuvants and formulation for peptides along with green improvements would help in increasing their use as a viable alternative to conventional insecticides.

Chapter 5 focuses on using gene silencing (RNA interference - RNAi) techniques to produce safe insecticidal compounds. This is a rapidly-developing area with the first RNAi-based insecticide having received commercial approval in the USA in 2023. A key advantage of this biotechnology lies in its high specificity, as it targets unique sequences in the mRNA of the intended pest organism. The chapter also addresses how scientists are addressing challenges such as variation in effectiveness between species due to factors such as dsRNA degradation, cellular uptake and genomic variations.

As Chapter 6 indicates, plant resistance to herbivores is a phenotypically flexible, and effective use of resistance in future pest management programs will be a matter of learning how best to manage this phenotypic plasticity in crops. This includes a better understanding of how best to deploy substances with the potential to induce plant defense mechanisms, including chemical elicitors (such as jasmonates, salicylic acid and its analogues, and aminobutyric acids), soil microorganisms (with biostimulant and bioprotective effects) and soil (e.g. silicon) amendments. As the chapter shows, greater understanding of the mechanisms by which these substances increase resistance and of the factors that influence their effectiveness will be needed to leverage phenotypic plasticity for improved pest management.

Part 2 Improving biocontrol product development and use

The second group of chapters addresses ways of improving the development of new biocontrol products. Chapter 7 outlines key steps in the development of biocontrol agents (BCAs) for plant protection, particularly in addressing the key challenge of product instability in the face of environmental variables such as the effects of UV light, humidity and temperature. This challenge can be addressed by good practice in formulation and production, including encapsulation materials, control of water activity, drying and storage.

Chapter 8 addresses ongoing challenges in commercialising microbial-based biopesticides for controlling insect pests. Issues holding back new biocontrol products include: regulation and policy, achieving economic mass production, improving formulation for long shelf life and survival after application, the reality (and perceptions) of product field efficacy, and the lack of novel actives and approaches. The chapter outlines ways forward in addressing these challenges.

Chapter 9 addresses developments in application technologies for biocontrol agents for insect pest control. The chapter considers the particular challenges associated with the use of biocontrol agents in foliar spray applications, including the effects of equipment choice, spray parameters and formulation on treatment efficacy. It discusses the various factors that affect effective foliar treatment and provides examples of how they can be optimised. These factors include the persistence of biocontrol agents, the impact of spray systems on surface coverage, the potential of equipment and adjuvants to develop more targeted foliar spray applications and, finally, the rainfastness of biocontrol agents.

As Chapter 10 points out, while most biopesticides are generally considered to be of low risk, robust risk assessments are still needed to support their registration. The chapter identifies some of the common issues faced by regulators in assessing risks of microbial and biochemical pesticides as well as dsRNA pesticides. The chapter discusses problems related to identification, exposure analysis and hazard testing, especially in nontarget organism testing with microbial pesticides. Advancement in clarifying these issues will strengthen biopesticide risk assessments by increasing their resolution and decreasing the need for conservative assumptions.

Part 1

Understanding and disrupting insect pests

Chapter 1

Understanding the use of vision and olfaction by thrips and other small flying insects to enhance biological control

D.A.J. Teulon and M-C. Nielsen, The New Zealand Institute for Plant and Food Research Limited, New Zealand; K. Lopez-Reyes, Lund University, Sweden; R. Meyhöfer, Leibniz Universität Hannover, Germany; and R.W.H.M. Van Tol, BugResearch Consultancy, The Netherlands

1 Introduction

As biological control programmes of insect pests evolve and seek to become more effective, they increasingly require a greater understanding of the genetic, neurobiological, morphological, behavioural, ecological and environmental parameters associated with the targeted species. For thrips and other small insects, a thorough understanding of the components, processes and behaviour in host finding, including the constraints imposed by miniaturisation, will provide the basis for more informed research and application for pest management including biological control (Bernays and Chapman 1994; Döring 2014; Schnaitmann et al. 2020; Ge et al. 2022; Hu et al. 2023; Liu et al. 2023).

Miniaturisation is a widespread trend in animal evolution and one of the principal directions of evolution in insects (Hanken and Wake 1993; Polilov

http://dx.doi.org/10.19103/AS.2024.0149.01

2015). The benefits of being small are likely to be many but include the ability to avoid natural enemies and to occupy niches that are inaccessible to larger organisms (Palavalli-Nettimi et al. 2019). Miniaturisation plays a crucial role in morphological and physiological design, the performance of sensory and information-processing systems, and therefore the lifestyle of organisms (Palavalli-Nettimi et al. 2019). A simple example of an adaptation associated with miniaturisation is the feather-like ciliated wings (Hanken and Wake 1993) characteristic of terebrantian thrips (Lewis 1997).

It is widely recognised that vision and olfaction are important for host finding in small insects such as thrips (Kirk et al. 2021; Lopez-Reyes et al. 2022), aphids (Pettersson et al. 2007; Webster 2012), psyllids (Civolani et al. 2023; Silva et al. 2023) and whiteflies (Schlaeger et al. 2018). However, much of what we understand about host finding in insects is based on relatively few species (Bernays and Chapman 1994) and particularly larger species which have the ability to fly upwind in response to host plant stimuli (Cardé 2021). Very little is known about how small, winged insects use vision and olfaction to locate their hosts (Bernays and Chapman 1994). In this chapter, we examine host finding in thrips (Thysanoptera) (especially flower-inhabiting thrips) and related small, winged insects, aphids (Aphididae), psyllids (Psyllidae) and whiteflies (Aleyrodidae), in the context of behaviours that could be exploited to improve biological control. Most winged insects are very small (Johnson 1957) and include many of the hemipteroid pest species that are the target of biological control programmes. Small and weak flying insects pose unique and scientifically challenging questions for the study and application of host finding in pest management, because of their limited ability to fly upwind (Compton 2002), their space constrained neurological, morphological and physiological systems (Palavalli-Nettimi et al. 2019) and the challenges of studying them both in the laboratory and in the field (Bernays and Chapman 1994; Döring and Chittka 2007).

Thrips are small insects about 1–2 mm in length (Moritz 1997). Of the approximately 6000 known species, over 100 are plant pests, especially species from the genera *Frankliniella* and *Thrips* (Thripidae: Thripinae) (Mound et al. 2022), causing damage through feeding and through transmission of plant viruses (Lewis 1997, Rotenberg and Whitfield 2018). Several are invasive pests, and a few are of international quarantine concern (Vierbergen 1995; Kirk and Terry 2003). Western flower thrips (WFT) (*Frankliniella occidentalis*) is one of the most economically important pest insects of many crops worldwide (Mouden et al. 2017). Thrips are difficult to control because of their life history characteristics (Mound and Teulon 1995; Morse and Hoddle 2006) and because of their ability to rapidly develop insecticide resistance (Gao et al. 2012; Adesanya et al. 2020). Flower-inhabiting thrips (Thripidae), in particular, provide a useful model to study host finding in small insects because they have strong responses to both visual and olfactory stimuli in flight, which they use for host plant finding (Kirk et al. 2021; Lopez-Reyes et al. 2022), and because some are significant plant pests that are targeted by biological control agents (see Section 5).

In this chapter, we examine the visual and olfactory systems associated with host finding, that is the period before physical contact with the plant (habitat and host location and not host acceptance), and how responses to these stimuli might be exploited to enhance biological control. Our focus is on thrips and other small, winged insects and particularly those which are plant pests and the targets of biological control programmes. Pests also include those species where most research in this area has been undertaken. We do not consider host finding in wingless or immature insects.

Our approach is unlike most other reviews in this area in that it is integrative (e.g. across genomics, neurobiology, morphology and application) and comparative (across several small, winged insect taxa). It consolidates a large quantity of information into a single comprehensive resource for the first time. Our focus on host finding by small, winged insect pests excludes herbivore-induced plant volatiles, and sex and aggregation pheromones, although the analysis may be equally relevant to them. We also refrain from reporting on the numerous studies detailing the optimal visual or olfactory stimuli for specific species, as such summaries are readily available elsewhere.

Our overall objective is to better understand the components, mechanisms and behaviour of host finding in thrips and other small insect pests to improve their application in pest management in general and biological control in particular. To achieve this, we:

- review the genetic, neurobiological and morphological components for vision and olfaction;
- summarise the different methods used to study vision and olfaction and compare their advantages and disadvantages;
- examine how vision and olfaction are integrated to facilitate effective host plant finding;
- document where an understanding of host-finding stimuli can enhance biological control; and
- identify critical gaps for future research.

2 Ways small flying insects locate plant hosts

Insects utilise a range of environmental, ecological and biological parameters in host selection, including visual, olfactory, gustatory and tactile stimuli, as well as humidity and light intensity (Bernays and Chapman 1994). Body size plays a crucial role in the morphological and physiological design and performance of the sensory systems (Palavalli-Nettimi et al. 2019), with the smallest insects retaining the general structure, functionality and diversity of all their sense organs but with a reduction in the number of structural elements (Makarova et al. 2022a). We examine the two main modalities of insect host finding in flight (vision and olfaction) not only for thrips (especially) but also for aphids, psyllids and whitefly, in relation to the genetic, neurobiological and morphological components of these systems.

2.1 Vision

The two compound eyes are the principal visual receptors for insects. Ocelli are not thought to be involved in an insect's interaction with plants (Bernays and Chapman 1994). The eyes are composed of numerous, individual, self-contained optical units called ommatidia, each made up of a cornea, a crystalline cone, and between six and nine photoreceptor cells, which make up a cylindrical microvilli-rich structure called the rhabdom. Light passes through the cornea and crystalline cone and is focused into the rhabdom, where the photoreceptor cells absorb light and convert it into electrical signals. These signals are then sent via the axons of each photoreceptor cell through to the optic neuropils for processing (Land and Nilsson 2012; Warrant 2019; Schnaitmann 2020).

The spectral sensitivity of a photoreceptor is not solely determined by that of its photopigment, owing to a range of screening effects and the presence of sensitising pigments, leading to a richness in photoreceptor types (van der Kooi et al. 2021; Santer and Allan 2024). Thrips (Stukenberg et al. 2020; Lopez-Reyes et al. 2022), aphids (Döring and Chittka 2007) and whiteflies (Stukenberg and Poehling 2019) also appear to utilise colour opponency mechanisms (Kelber 2016) which allows an organism to discriminate between a range of colours, based on the integration of the signals from photoreceptors with different spectral sensitivities.

The response to colour, or more accurately the light wavelengths reflected from objects (e.g. traps) or sourced directly from light emitters (e.g. incandescent or LED light bulbs), comprises by far most of the research examining visual responses of small insects. However, visual stimuli are not restricted to wavelength but also brightness, hue, reflectance intensity, intensity contrast and saturation (van der Kooi et al. 2019; Schnaitmann et al. 2020). To date, genes associated with colour vision only (i.e. opsin genes) have been identified for insect vision (see below) and not for other visual stimuli.

Although colour vision is widespread among insects, colour perception varies among species depending on the spectral sensitivities of the participating photoreceptors and their interplay (van de Kooi et al. 2021). True colour vision is defined as the ability to discriminate colour (i.e. wavelength composition) independently from intensity (Kelber 2006) and has been reported to occur in aphids (Kelber et al. 2003), psyllids (Mauck et al. 2024) and whitefly (Stukenberg and Poehling 2019), but it is still unclear whether it occurs in thrips (Lopez-Reyes et al. 2022).

Three opsin (photosensitive molecules) subfamilies (long wavelength sensitive (LWS), short wavelength sensitive (SWS)-blue, short wavelength sensitive (SWS)-UV) are commonly associated with insect vision and are expressed in the photoreceptor's microvilli and the ocelli (Lopez-Reyes et al. 2022; Friedrich 2023). These opsin subfamilies are found in most hemipteroid groups, including thrips (Lopez-Reyes et al. 2022, Friedrich 2023) and the number of identified opsin genes (five to seven), including those associated with ocelli, is relatively consistent across different insect groups (e.g. thrips, aphids and psyllids) (Li et al. 2019; Lopez-Reyes et al. 2022; Gao et al. 2024). For thrips at least, there is very little understanding of the expression of these genes. In addition, there are numerous examples of the opsin proteins having been lost, duplicated or modified such that there is great variation across different insect species. The SWS-Blue-opsin subfamily is missing from the Heteroptera, aphids (Aphidoidea) and planthoppers (Fulgoroidea), with multiple independent expansions of the LW-opsin subfamily in the Heteroptera and the presence of tandem-duplicated UV opsins in aphids and planthoppers thought to effectively restore blue sensitivity in these groups (Friedrich 2023).

Reduction in body size places significant constraints on insect vision for a range of structural, physiological and metabolic reasons (Land 1997; Fischer et al. 2011; Warrant 2019). Reduction in the size of compound eyes is accompanied by a decrease in the number and size of the ommatidia facets, strong compaction of cellular elements and various morphological adaptations unique to different insect taxa (Makarova et al. 2022a). The number of ommatidia in the compound eye is positively correlated with body size across a wide range of insect species and taxa (Fig. 1) (Makarova et al. 2022a) although not within a group of 20 aphid species (Döring and Spaethe 2009). Compound eyes can be composed of tens to tens of thousands of ommatidia per eye (Warrant 2019) but small insects such as thrips (60–70, Lopez-Reyes et al. 2022), aphids (127–268, Döring and Spaethe 2009), psyllids (160–320, Farnier et al. 2015) and whitefly (83–86, *Trialeurodes vaporariorum*, Mellor et al. 1997) have comparatively few.

For an eye of a given size, there is a trade-off between light sensitivity (amount of light entering and being absorbed in the eye) and spatial resolution (the distance at which objects of a given size can be resolved) (Fischer et al. 2011; Farnier et al. 2015). A measure for acuity that is directly related to spatial resolution is the interommatidial angle: the smaller the interommatidial angle the greater the distance at which objects can be resolved (Land 1997) and the better insects can detect relatively small stimuli located at greater distance (Farnier et al. 2015). Measures of interommatidial angles in thrips (11–15°, Mazokhin-Porshnyakov and Kazyakina 1983), aphids (5.2–8.5°, Döring and

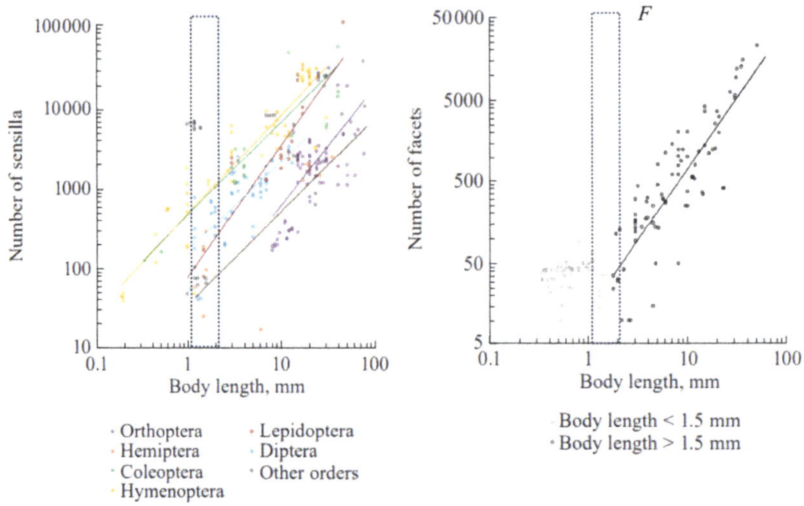

Figure 1 Relationships between body length and number of antennal sensilla and body length and number of facets (ommatidia), indicating the body length of insect (1–2 mm) which is the focus of this chapter (adapted from Makarova et al. 2022a,b).

Spaethe 2009) and psyllids (6.3–8.7°, Farnier et al. 2015) are relatively large compared to larger insects such as beetles, moths and flies (Table 1 in Land 1997) and would suggest these small insects possess relatively poor visual acuity.

Just how small insects cope with the demands for light sensitivity and spatial acuity in host finding is still to be determined. 'Acute zones' or specialised eye regions, such as UV receptors localised in the divided compound eyes of whiteflies (Mellor et al. 1997), could be one adaptation to the limitations of small eyes and are found in many insects (Warrant 2019), including possibly thrips (Lopez-Reyes et al. 2022). Nevertheless, there remains much to explore about the eyes of small insects and their functionality, including the opsin expression pattern in the eyes, photoreceptor fine structure and arrangement, absorptive properties of the putative spectral filters and the electrophysiological responses of individual photoreceptors (Lopez-Reyes et al. 2022).

2.2 Olfaction

Olfactory appendages of insects consist of antennae and – to a lesser extent – maxillary palps (Montagné et al. 2015). The antennae are the most important multimodal sensory organs for the insects, bearing not only the sensilla for olfaction but also those for taste, mechano-, hygro- and thermoreception and sometimes sensors for CO_2 (Keil 1999). They carry arrays of innervated hair structures or sensilla (Keil 1999) which are overwhelmingly distributed on the

Table 1 Numbers and types of antennal sensilla in selected Thysanoptera

Thysanoptera	Böhm Bristle	Campaniform	Trichoidea	Basoconica	Chaetica	Coeloconica	Cavity	Other	Reference
Dendrothrips minowai				8					Lu et al. 2015 (in Zhu et al. 2017)
Echinothrips americanus	6	1	2	**7**	51	**2**		6	Hu et al. 2020
Frankliniella intonsa	6	1	2	7	46–48	**2**	2	2	Li and Feng 2013
Frankliniella occidentalis	6	1	1	**7**	46–48	**2**	2	2	Li and Feng 2013; Zhang and Lei 2022
Frankliniella tenuicornis	6	1	2	7	47–49	**2**	2	2	Li and Feng 2013
Liothrips jatrophae			+	**+**					González-Orellana et al. 2022
Megalurothrips distalis		1	2	**7**	47	**2**	2		Liu et al. 2021
Megalurothrips usitatus	5	1	2	**7**	48		1	1	Wang et al. 2019a
Odontothrips loti		1	2	**6–7**	47	**2**	1		Liu et al. 2021
Scirtothrips dorsalis	6	2	3	**7**	47	**2**	1	1	Zhu et al. 2017
Sericothrips kaszabi		1	2	**7**	44	**2**	2		Liu et al. 2021
Thrips palmi	5	1	4	**5**	44		1	1	Wang et al. 2019a

+, recorded as present but number not quantified. Bold indicates the sensilla implicated as olfactory receptors or chemoreceptors in the respective publications.

antennae (Bernays and Chapman 1994; Qian et al. 2024). Odour molecules (such as green leaf, flower or fruit volatiles) pass from the external environment into the sensilla through surface pores and diffuse to the dendrites of the odour receptor neurons (ORNs), facilitated by odour-binding proteins (OBPs) (Wicher and Miazzi 2021). Subsequent excitation of the olfactory neurons occurs, converting chemical signals into electrical signals, which are transmitted as action potentials to higher nerve centres (i.e. antennal lobes and mushroom bodies) (Qian et al. 2024).

Thrips usually have seven to eight antennal segments (Moritz 1997), aphids five to six (Blackman and Eastop 2000), psyllids ten (Hollis 2004) and whiteflies seven (Anonymous 2004). A decrease in body size in insects is accompanied by a decrease in the number of sensilla (Fig. 1) and to a lesser degree, the size of the sensilla (Polilov 2015; Makarova et al. 2022b), and there is a linear relationship between antennal length and the number of sensilla for most insects although Chapman (1982) noted that hemipteroid species tend to have fewer sensilla than would be expected for the size of the antenna. The numbers, types and positioning of sensory sensilla show a high degree of similarity in thrips (Table 1), suggesting that they may have similar functional responses across these species (Liu et al. 2021). In comparison, there are relatively few findings (despite numerous electroantennogram (EAG) studies) describing the full array of sensory sensilla for aphids (see references in Kanturski et al. 2017; Song et al. 2020), psyllids (Onagbola et al. 2008) or whiteflies (Zhang et al. 2015; Mrisho et al. 2021). Different terminology for different sensilla (Keil 1999) makes comparison between insect groups difficult, and it is not yet possible to accurately identify the function and quantity of all olfactory sensilla, including those associated with host finding. Nevertheless, sensilla with possible or likely olfactory functions (usually evidenced by porous sensilla walls) have been found in thrips (up to nine) (Table 1), aphids (up to 14 placoid sensilla in 1°/2° rhinarium, Song et al. 2020), psyllids (between 5 and 28 with up to five sensilla types, Onagbola et al. 2008; Ge et al. 2022) and whiteflies (up to 15 basiconic and coeloconic sensilla, Mrisho et al. 2021). In comparison, the number of antennal sensilla in some insect species can reach the thousands or tens of thousands (Bernays and Chapman 1994).

Underpinning insect chemoreception are the odorant receptor (OR), gustatory receptor (GR) and ionotropic receptor (IR) gene families (Robertson 2019). ORs are responsible for the ability to detect most odours by a given insect (Montagné et al. 2015) and are almost exclusively found in the antennae (Karpe et al. 2021). While ORs solely detect volatile chemosignals, IRs are multimodal receptive entities (Wicher and Miazzi 2021), and GRs are primarily associated with insect taste (King and Gunathunga 2023). IRs and GRs are not considered here to any degree.

There is no apparent relationship between body size and genome size (Alfsnes et al. 2017), in contrast to body size and sensilla number. For insects, genome size is clearly phylogeny dependent, reflecting their life history and mode of development (Alfsnes et al. 2017). Of relevance to this chapter is the finding that hemimetabolous insects (presumably including species from a range of sizes) tend to have larger genomes compared to holometabolous insects, apparently related to their developmental complexity (Gregory 2005).

Table 2 details published information on the chemosensory genes and associated proteins for thrips, aphids, psyllids and whiteflies, illustrating the capacity for olfaction in these groups. Knowledge of olfactory protein families in thrips is scarce (Liu et al. 2023) but there appears to be even less for other hemipteroid groups. The number of OR genes in insects ranges from 3 to 503 (Robertson 2019; Yan et al. 2020) but values like the number found in hemipteroid species (Table 2) are not unusual in other insect groups irrespective of size (Montagné et al. 2015; Robertson 2019; Yan et al. 2020).

Comparison between the number of OR genes across different insect species and different studies should be carried out with caution. There may be varying ratios between pheromone receptor and general OR genes in different species (Nicholson et al. 2015), different criteria for categorising genes and pseudogenes (Robertson 2019), differential targeting of genomes and transcriptomes (e.g. complete genetic potential versus tissue-specific expression) and variation in the methods used for gene annotation with different degrees of accuracy (Montagné et al. 2015; Karpe et al. 2021). The size of an insect's OR gene family may be related to the complexity of the chemical environment in which the insect lives – it is thought that generalists have the largest repertoires (Robertson 2019). Liu et al. (2023) conjected that the larger number of OR genes for *F. occidentalis*, compared to other thrips species, may be related to this species' higher environmental adaptability, especially its broad host diversity. However, the reasons for the higher number of odorant genes for *Aphis pisum* (Smadja et al. 2009; Robertson et al. 2019), compared to other aphids, are not immediately obvious as this species is not polyphagous. The low number of odorant genes for *Odontothrips loti* (Liu et al. 2023), *Cacopsylla chinensis* (Xu et al. 2019) and *Bemisia tabaci* (Mrisho et al. 2021) reflect the analysis of head and antennal transcriptomes.

3 Methods used to study vision and olfaction in thrips and other small insects

Different methods are used to study the responses of small insects to the visual and olfactory stimuli used for host plant finding. Here we summarise the main ones and critically assess their relevance and limitations. Each method has its advantages and disadvantages, but all need to be understood in terms

Table 2 Numbers of odorant, gustatory and ionotropic receptor genes, and odorant binding and chemosensory protein genes in selected Thysanoptera and Hemiptera

		Odorant receptor genes	Gustatory receptor genes	Ionotropic receptor genes	Odorant-binding proteins (OBP)	Chemosensory proteins (CSP)	Reference
Thysanoptera							
Stenchaetothrips biformis	G*	34	30	36	17	9	Hu et al. 2023
Thrips palmi	G	53	73	42	22	11	Hu et al. 2023; Liu et al. 2023
	T	15	–	6	8	–	
Frankliniella occidentalis	G	49, 84	–, 102	–, 167	–, 11	–, 7	Tian et al. 2022; Hu et al. 2023;
	T	–, 84	–, –	–, >167	7, 12	8, 3	Li et al. 2022; Liu et al. 2023
Frankliniella intonsa	T				6	6	Li et al. 2022
Odontothrips loti	T	7		14	6	9	Liu et al. 2023
Aleyrodidae							
Bemisia tabaci	T	–, 10, 9–13	–, 9	–, 11	8, 3	13, –	Wang et al. 2017; Mrisho et al. 2021; Tian et al. 2022
Trialeurodes vaporariorum	G	20					Tian et al. 2022
Aphididae							
Aphis glycines	G	47	61	19	10		Robertson et al. 2019
Aphis gossypii	G	45, 47	–, –	14, –	–, –	–, –	Cao et al. 2014; Tian et al. 2022;
	T	–	–	–	9	9	Gu et al. 2013
Acyrthosiphon pisum	G	79, –, 87, 100	77, –, 78, –	19, –, 19, –	–, 15, 18, –	–, 13, –, –	Smadja et al. 2009; Zhou et al. 2010; Robertson et al. 2019; Tian et al. 2022

Diuraphis noxia	G	21	30		9		Nicholson et al. 2015
Lipaphis erysimi	T				11	4	Kuang et al. 2023
Megoura crassicauda	T	20	4	9			Huang 2023
Megoura viciae	T				10		Bruno et al. 2018
Myzus persicae	G				9	9	Wang et al. 2019b
	T				9	9	
Rhopalosiphum padi	T	15	19	16	14	9	Kang et al. 2018
Sitobion avenae	T				13	5	Xue et al. 2016
Psyllidae							
Cacopsylla chinensis	T	7	2	4	12	11	Xu et al. 2019
Diaphorina citri	G	–, 102	–, –	–, –	9, –	12, –	Zhang et al. 2020; Tian et al. 2022; Wu et al. 2016
	T	46	20	35	9	12	

*G, genome, T, transcriptome. Dashes are used to indicate lack of data from a particular reference where multiple references are cited.

of the responses they measure. Indeed, certain methods may have been developed to target a certain modality (e.g. olfaction in olfactometers), and when used with other modalities (e.g. vision), they may bias results towards that modality for which they were designed. Interpretation of the results from a single method can lead to misinterpretation (Kennedy 1977), with the most informative studies using several methods to understand biological responses.

There are challenges in measuring the responses of small insects to visual and olfactory stimuli, including the requirement of smaller components for the specialist apparatus (e.g. probes), the difficulty of observing or tracking small flying insects in open and confined spaces, and their ability to hide in small crevasses. In addition, in many bioassays, non-responders can make up a considerable part of a population (see Section 4.5).

3.1 Electrophysiology

Electroantennograms (EAG) (for olfaction) and electroretinograms (ERG) (for vision) measure neurological responses to stimuli. EAGs combined with gas chromatography–mass spectrometry can also help identify the chemicals involved.

Both EAGs and ERGs provide little indication of the associated insect behaviour, including whether the responses relate to host finding or not. This is especially pertinent for volatiles where responses might reflect behaviours related to gustation and mating (e.g. pheromones). For EAGs, a large response may indicate that many neurones are activated by the odour tested, whereas a small response may indicate that only a few receptor cells are activated (Bernays and Chapman 1994) – but the smaller response may be more important.

Targeting single sensilla or single photoreceptors may provide evidence that the responses are related to host finding, due to the assumed function of particular sensilla in detecting volatile compounds (see Section 2.2, Olfaction). Despite their small body size, single sensillum measurements have been successfully undertaken with thrips (González-Orellana et al. 2022), aphids (Huang 2023) and psyllids (Anderbrant et al. 2020) but apparently not whiteflies (Schlaeger et al. 2018). Except for the pea aphid (Döring et al. 2011), intracellular recordings of single photoreceptor cells do not seem to have been undertaken for these groups. Additionally, Mellor et al. (1997) examined ERG responses in dorsal and ventral parts of the whitefly eye. There appear to be few instances where EAGs and ERGs have been used in combination to measure visual and olfactory responses at the same time, and this may be very difficult for such small and fragile insects.

3.2 Olfactometers

Olfactometers are used to examine the behavioural responses of (mostly) walking insects to odours. They include still-air olfactometers (usually linear) and moving-air olfactometers, including two-arm (Y-tube, T-tube) and four- and six-arm constructs. Each olfactometer has its own advantages and disadvantages (Roberts et al. 2023). The Y-tube has been commonly used for thrips (Díaz et al. 2023) and the four-arm for aphids (Schlaeger et al. 2018; Roberts et al. 2023). Colour has sometimes been added as an additional stimulus, including for thrips in Y-tube studies (Mainali and Lim 2011) and for psyllids in Y-tube (Wenninger et al. 2009) and four-arm olfactometer (Volpe et al. 2020) studies.

The major limitation of olfactometers is that they measure the response of walking (or perhaps jumping) insects only (Kennedy 1977, Döring 2014, Roberts et al. 2023) and there appear to be no compelling analyses relating walking and flying responses. Therefore, data obtained with walking insects cannot be freely extrapolated to flying insects without additional experimentation.

3.3 Wind tunnels

Wind tunnels (sometimes called flight olfactometers) are laboratory devices primarily designed for studying the responses of flying insects to olfactory cues (Baker and Linn 1984; Kainoh 2011; Knudsen et al. 2018) and also visual stimuli, sometimes in combination (Knudsen et al. 2018). For small insects, wind tunnels must be capable of low air movement to allow for upwind flight: 0.3 m/s or less for thrips (Lopez-Reyes et al. 2023), aphids (Pettersson et al. 2007), psyllids (Davidson et al. 2014) and whitefly (Isaacs et al. 1999). An automated flight chamber has been used extensively for aphids (David and Hardie 1988) although it was developed primarily to research aphid flight duration. Wind tunnels have been used to study the responses of thrips to visual and olfactory cues separately and in combination (Teulon et al. 1999; Berry et al. 2006; Davidson et al. 2006, 2012; Lopez-Reyes et al. 2023). They have been used less frequently for psyllids (Davidson et al. 2014) and whiteflies (Isaacs et al. 1999).

One of the greatest challenges with small insects in wind tunnels is observing and measuring their flight activity (Kainoh 2011) and landing behaviour. Insects caught in traps placed within the observation chamber provide 'end-point' data from which it can be difficult to interpret the actual behaviour that led to that result (similar to traps in the field – see Section 3.4). More recently, real-time, 3D tracking and high-speed videography have enabled the collection of important flight behavioural data, including for thrips (Lopez-Reyes et al. 2023), helping to detect and categorise host-finding behaviour. The study of sensory responses in insect behaviour experiments has seen advancements in recent years by using neural networks and machine learning to develop new tools (e.g. DeepLabCut – Mathis et al. 2018) to quantitatively analyse insect behaviour through videography in 2D and 3D. Although not yet optimised or used for miniature insects, these types of tools could be useful in the future to answer more specific and targeted questions about how small insects respond to odours and colours and find suitable hosts.

3.4 Trapping

The response of flying insects to olfactory and visual stimuli has been extensively and successfully tested in greenhouses, semi-field and field experiments using different trapping systems (e.g. pan-traps, delta-traps and sticky-traps) and is widely reported in the literature. For example, see Terry (1997), Kirk et al. (2021) and Lopez-Reyes et al. (2022) for thrips; Döring and Chittka (2007), Pettersson et al. (2007) and Döring (2014) for aphids; Mauck et al. (2024) for psyllids and Byrne and Bellows (1991) and Schlaeger et al. (2018) for whiteflies.

In contrast to laboratory bioassays, trapping experiments more closely reflect the circumstances of host plant stimuli encountered by insects during host finding and can be used in applied settings. However, these experiments can be difficult to interpret because insect responses may be influenced both by extrinsic environmental factors and the physiological state of the insects themselves (Bernays and Chapman 1994; Nielsen 2013). In addition, the disparate conditions found in different experimental settings mean that results cannot be applied across all situations, and repeatability of experiments can be compromised, even though generalisations are required for application across a wide range of circumstances.

In trapping experiments, olfactory and visual stimuli can be tested separately and together in the same bioassay, although it can be difficult to completely limit light from these bioassays. However, the results usually reflect an 'end-point' response, making it difficult to interpret the behaviour that led to that response. In addition, odour from one trap may influence trap catch in an adjacent trap without odour (Chapman et al. 1981 for aphids, Teulon et al. 2007 for thrips), confounding the interpretation of relative trap catch and creating challenges for experimental design, especially in confined situations such as cage and small greenhouse tests.

3.5 Alternative bioassays

Various purpose-built methods have been developed to examine the responses of small insects to visual and olfactory cues (Mainali and Lim 2011; van Tol et al. 2012; Otieno et al. 2018; Ren et al. 2020; Stukenberg et al. 2020; Kalile et al. 2022; Grupe and Meyhöfer 2024) and it is not possible to consider them all here. Some methods have been developed to examine specific aspects of host finding including the combined response to visual and olfactory stimuli.

In contrast, other methods have only examined a limited aspect of host finding and/or measured 'end-point' response behaviour. 'Closed' units may make it more difficult to determine whether the behavioural response is due to walking or flying (or both). Due to the unique design of these methods, they may be more difficult to replicate across different laboratories and target species.

3.6 Genomics

The assembly and analysis of the genomes and transcriptomes associated with olfaction (especially) of thrips and other small insects (Table 2) has the potential to contribute to a greater understanding of host finding and its application for pest management and biological control through:

- identifying the physical location of expression of genes (e.g. in antennae) to inform their likely role in host finding or not (Kuang et al. 2023; Liu et al. 2023);
- sequence data comparison and phylogenetic analysis of chemosensory genes indicate whether targets are pheromones or non-pheromones (Yuvaraj et al. 2022 for Lepidoptera);
- comparing the structure and function of genes across taxa (e.g. family, genus and species) to determine phylogenetic affinities (Liu et al. 2023);
- a range of methods, including fluorescence binding, molecular dynamics, homologous modelling and molecular docking, to predict the binding affinity of potential agonists or antagonists (Wang et al. 2020; Liu et al. 2023);
- elucidating the mechanisms a particular insect may use to recognise plant volatiles (Ha and Smith 2022; Shih et al. 2023);
- high-throughput methods to screen for new volatiles of high activity (Wang et al. 2020, Silva et al. 2023; Qian et al. 2024). For some insects, this has included the expression of genes in other insect models and the use of machine learning to screen ligands (Caballero-Vidal et al. 2020); and
- the ability to identify olfactory (and visual) genes for silencing/RNA interference (RNAi) (Wang et al. 2020).

3.7 Biological control agents

Like their pest hosts, various predators and parasitoids are small flying insects. In many cases, methods like those described above can be used to study their responses to visual and olfactory stimuli of the host plants of their insect hosts, sometimes carried out in combination with that pest species. Limitations of each method described earlier would be similar for these natural enemies. Examples of these methods being used for natural enemies include EAGs for aphid parasitoids (olfaction) (Dong et al. 2008); ERGs for whitefly parasitoids (vision) (Mellor et al. 1997); olfactometers for aphid predators and parasitoids (Birkett et al. 2000) and thrips predators (Bennison et al. 2002) (both olfaction) and psyllid parasitoids (vision) (Kalile et al. 2022); closed cages for predators and parasitoids of various pests including thrips and aphids (vision) (Park et al. 2023); wind tunnels for aphid parasitoids (olfaction and vision) (Lo Pinto et al. 2004); and whitefly parasitoids (vision) (Sarig and Ribak 2021) and a range of trapping experiments for thrips predators (olfaction) (Broughton and Harrison 2012; Davidson et al. 2015), thrips parasitoids (olfaction) (Murai et al. 2000) and aphid parasitoids (vision) (Grupe et al. 2023). Kang et al. (2017) examined the genome of the aphid parasitoid *Aphidius gifuensis* to better understand its mechanisms of host detection. For the application of this information, see Section 5.

4 Integration of visual and olfactory stimuli for host finding

It is recognised that vision and olfaction are important for host finding in thrips (Kirk et al. 2021; Lopez-Reyes et al. 2022), aphids (Pettersson et al. 2007; Webster 2012), psyllids (Civolani et al. 2023; Silva et al. 2023) (in contrast see Mauck et al. (2024)) and whiteflies (Schlaeger et al. 2018), but research on each separate modality (i.e. vision or olfaction) has often occurred in isolation (with exceptions discussed in Section 4.2). Sensory stimuli never occur in isolation in the field for pests (Döring 2014) or their natural enemies (Traczyk et al. 2020).

4.1 Integration of sensory systems

The major structures containing sensory receptors for host finding, the compound eyes and antennae, are found on the head. However, there is surprisingly little reference to the integration of visual and olfactory inputs for small insects (Döring 2014; Ben-Yakir et al. 2020a). The brains of micro-insects such as parasitoid wasps (<0.3 mm in length) still retain the complete set of brain centres at almost the same proportions as larger insects (Polilov 2015) and the brain structure of *Heliothrips haemorrhoidalis* is consistent with these observations (Makarova and Polilov 2017). The axons from olfactory sense cells connect to the antennal lobe of the brain where the integration of olfactory sensory input occurs (Bernays and Chapman 1994). The axons from the compound eyes end up in the optic lobe, situated immediately behind the eye but continuous with the brain (Bernays and Chapman 1994).

4.2 Experimental evidence of integration

Positively directed walking responses towards odours have been recorded in several thrips species in many olfactometer studies (Kirk et al. 2021) where a white light is often used to illuminate the apparatus. However, few studies have tested different visual cues in olfactometers and even fewer have used combined visual and olfactory cues. One of the few, Mainali and Lim (2011), used an olfactometer to successfully measure the response of *F. occidentalis* to an artificial yellow flower shape combined with an odour. Similarly, olfactometers have been used to successfully investigate the walking responses for olfactory and occasionally visual, cues in aphids (Webster 2012), psyllids (Wenninger et al. 2009; Davidson et al. 2014; Volpe et al. 2020; Silva et al. 2023) and whiteflies (Morin et al. 2024).

Increased and orientated upwind flight responses towards some colour stimuli have been recorded in two thrips species (*F. occidentalis* and *Thrips obscuratus*) in several wind-tunnel bioassays (Teulon et al. 1999; Berry et al. 2006; Davidson et al. 2006, 2012; Smits et al. 2000; Lopez-Reyes et al. 2023). However, the addition of an 'attractive' odour reduced the proportion of thrips executing take-off in the presence and absence of a colour stimulus and reduced the proportion making upwind flight in the absence of a colour stimuli (Teulon et al. 1999, Berry et al. 2006). Odour concentration may influence this response (Davidson et al. 2006, 2012) but odour at any concentration did not increase the response of thrips in the presence or absence of a colour stimulus in almost all treatments (see also Smits et al. 2000). Döring and Chittka (2007) documented upwind aphid flight responses to colour stimuli in horizontal and vertical wind tunnels, but there are relatively few studies addressing the role of olfaction in flying aphids (Webster 2012). Using a vertical wind tunnel, Nottingham and Hardie (1993) suggested that host volatiles did not stimulate aphid flight towards a visually attractive stimulus, and that non-host volatiles may inhibit this behaviour. One wind tunnel study with the psyllid *Bactericera cockerelli* indicated an upwind flight response to certain colours but not odour (Davidson et al. 2014). The whitefly *Bemisia tabaci* made upwind orientated flights in the presence of a green stimulus (Isaacs et al. 1999) but there are no reports of its response to odour in a wind tunnel.

Numerous studies have reported increased capture of thrips in response to visual stimuli (especially colour) (Lopez-Reyes et al. 2022) and with visual and olfactory stimuli combined (Kirk et al. 2021). The few studies where odour was provided without a visual cue (i.e. black or transparent traps) also indicated that odour alone can increase thrips capture for some species, but to a lesser extent than in traps with colour alone (Teulon et al. 1999). Aphid trapping experiments relate mostly to increased trapping as a result of visual stimuli (especially colour) (Döring and Chittka 2007) but increased trap capture was reported for *Lipaphis erysimi* with an odour irrespective of trap colour (Dilawari and Atwal 1989) and Chapman et al. (1981) reported increased trap catches of *Cavariella aegopodii* with an odour and no visual stimulus. Increased capture of psyllids in traps with colour stimuli (Volpe et al. 2020; Civolani et al. 2023) and colour and odour stimuli have been reported in several field studies (Silva et al. 2023). A certain odour increased the trapping response of the Asian citrus psyllid towards a specific type of yellow (Khadka et al. 2020). Trapping has informed trap colour selection for whiteflies but not odour (Pinto-Zevallos and Vänninen 2013; Stukenberg et al. 2015).

4.3 The behavioural response

Given the large knowledge gaps and despite unequal research across taxa, the literature is remarkably consistent in the descriptions of small hemipteroid insect responses to visual and olfactory stimuli used in host finding. These include:

- a positive directional response of walking insects to both types of stimuli;
- a positive directional response of flying insects to visual stimuli;
- an uncertain response of flying insects to olfactory stimuli;
- a positive trapping response of flying insects to both types of stimuli; and
- generally, an additive trapping response to both types of stimuli when combined.

The behavioural response of thrips and other small insects to olfactory stimuli is still unclear. Several authors have suggested that olfactory stimuli may take the form of an arrestant rather than an attractant (see Kennedy 1977; Teulon et al. 1999; Finch and Collier 2000). But as Döring (2014) noted for aphids, the behavioural sequences of host finding in the field and the integration of different sensory stimuli remain mysterious. The degree to which small insects fly upwind within plant canopies or during periods of low wind speed, and their migratory inclination (Compton 2002; Ben-Yakir and Chen 2008; Sarig and Ribak 2021), requires further attention in relation to visual and olfactory stimuli.

Statements on the relative importance of vision or olfaction for host finding should be made with caution because of the inherent biases in the methods used for a particular stimulus and because these stimuli clearly interact with each other. Their relative contributions may be difficult to disentangle.

4.4 The question of distance

There is general agreement that vision and olfaction work interactively for host finding for thrips (Kirk et al. 2021), aphids (Döring 2014), psyllids (Patt et al. 2011; George et al. 2020) (although see Mauck et al. 2024) and whiteflies (Schlaeger et al. 2018). However, there is little agreement as to the relative importance of these stimuli at different stages of host finding and indeed for the different physiological stages of the insects involved. Such information is likely to be useful for informing how a particular combination of stimuli can be best used for pest management. This discussion needs to be considered in the context of what is meant by 'distance' – small insects can fly considerable distances (tens to hundreds of kilometres) as part of host finding, yet most experimental designs (olfactometers, wind tunnels) only measure responses over relatively short distances (<1 m), and field trapping experiments provide little information as to the distance from which the insects respond to a given stimulus. The small eyes of small insects are thought to have relatively poor acuity (see Section 2) and provide less specific information than larger eyes (Döring and Spathe 2009; Lopez-Reyes et al. 2022). One can presume that maximum acuity occurs at very close range and that perception of detail decreases as distance increases (Prokopy and Owens 1983), but it is unclear if and how small insects compensate for this (e.g. by relying more on olfaction).

There has been a general theme in the literature placing vision or olfaction individually as central to certain steps of the host-finding process (Finch and Collier 2000). For small insects, visual cues have been considered most important for long-distance host plant finding (Volpe et al. 2020; Mauck et al. 2024), but for other authors argue that olfaction is paramount (Khadka et al. 2020; Ren et al. 2020; Zhang et al. 2020; Silva et al. 2023; Grupe and Meyhöfer 2024). Yet as there appears to be very little evidence showing how either visual or olfactory stimuli work over larger distances, this seems largely to be conjecture. As Mauck et al. (2024) noted for psyllids, there is a temptation to extrapolate responses to volatile cues (and visual cues as well – our emphasis) that have been shown to operate over distances of only a few centimetres, sometimes only by walking insects, to field-level processes occurring over several metres or more. Also, as we have noted before, care needs to be taken in placing host finding in the context of the inability of small insects to fly upwind compared to other insects.

4.5 The non-responders

The lack of, and associated inconsistent response to, visual and olfactory stimuli by a substantial subset of insects in many studies (Teulon et al. 1999, Wenninger et al. 2009, Volpe et al. 2020) confounds our ability to understand host plant finding and apply this information for pest management and biological control. A range of biological parameters have been explored in some studies, such as insect age, sex and satiation (Davidson et al. 2006; Matteson and Terry 1992), but none have provided satisfactory and consistent answers for such large variations. An exception is the disregard for host plant (and mating) stimuli and the positive the response to UV by aphids in their migration phase, a phenomenon that is of variable duration for individuals in clonal colonies reared under similar conditions (Hardie 1994; Gatehouse 1997; Döring 2014).

Similar behaviour has been reported for whiteflies (Byrne 1999). Host finding in alate (winged) aphids is a complex behaviour that is closely linked to migration and the function of dispersal (Döring and Chittka 2007). Physiological and behavioural data show that thrips (Lopez-Reyes et al. 2022), psyllids (Kalile et al. 2022) and whiteflies (Mellor et al. 1997; Stukenberg and Poehling 2019), have UV receptors, are known to respond to UV (somewhat inconsistently), and are known to migrate over long distances (Lewis 1997; Gatehouse 1997; Isaacs et al. 1999; Silva et al. 2023). Flying whitefly parasitoids are also known to respond to UV (Sarig and Ribak 2021). The response of thrips, psyllids and whiteflies to UV, including in combination with other light wavelengths, is not understood. The influence of migratory behaviour on host plant finding may provide valuable insights into host plant selection for these insects.

5 Exploiting host finding stimuli for biological control

Biological control agents for pest thrips, especially as components of integrated pest management systems in protected crops, have received substantial attention.

- The number of documented predators (especially) and parasitoids attacking pest thrips is extensive (Loomans et al. 1997; Sabelis and van Rijn 1997; Castañé et al. 1999; Cox et al. 2006; Messelink et al. 2006; Mouden et al. 2017; Smagghe et al. 2023), and many (especially phytoseiid mites and anthocorid bugs) are commercially available in many parts of the world. Despite initial interest in thrips parasitoids, few are available from commercial sources (Mouden et al. 2017; van Lenteren et al. 2020; Lei et al. 2021; Smagghe et al. 2023).
- Several fungal pathogens have been identified as potential biological control agents for thrips pests (Wu et al. 2018) and some commercial products based on entomopathogenic fungi are available for thrips management (Mouden et al. 2017; van Lenteren et al. 2020; Smagghe et al. 2023).
- Additionally, various entomopathogenic nematode species are used against soil-inhabiting pupae of thrips (Shapiro-Ilan et al. 2017) although their use is perhaps not as widespread as other natural enemies (Smagghe et al. 2023).

Van Lenteren et al. (2020) and Smagghe et al. (2023) also list the wide range of commercially available biological control agents for aphids, psyllids and whiteflies.

Globally, the higher cost associated with biopesticides (and other biologicals – our emphasis) constitutes one of the major constraints to increasing their application (Smagghe et al. 2023). Approaches that optimise the use of biologicals can only help with their development and uptake such that natural enemies can be used in combination within a single growing system and integrated with other pest management tactics (Bennison et al. 2002).

Kirk et al. (2021) reported that research and application of semiochemicals for thrips pest management were still at a rudimentary stage, with current applications mainly used for thrips monitoring. This is largely true for the use of visual stimuli for thrips pest management as well. Recent reviews by Okosun et al. (2023) and Rodríguez and Coy-Barrera (2023) reinforce this view. Nevertheless, the use of visual and olfactory stimuli is increasingly considered to improve outcomes for pest management in general, and biological control in particular, for thrips and other small insects.

5.1 Response of natural enemies to host plant stimuli

There are numerous reports of winged natural predators and parasitoids responding to the visual and olfactory stimuli from plants used by their pest hosts (Ben-Yakir et al. 2020b, see below). For thrips, at least, the response is not consistent across natural enemy species or stimuli (Kirk et al. 2021) so that informed decisions for effective pest management may need to be made on a case-by-case basis. Trade-offs will need to be considered for intended and unintended consequences, such as information used from the trapping of parasitoids and their destruction because of trapping (e.g. Kalile et al. 2022 for psyllids).

5.2 Trapping for monitoring

Traps enhanced with visual and olfactory stimuli are commonly used to monitor the population density of thrips (including presence/absence) and can therefore provide information on the appropriate timing of application of natural enemies, especially at low pest densities (Teulon et al. 1993; Davidson et al. 2007). Muvea et al. (2014) (reflected colour plus odour) and Grupe and Meyhöfer (2024) (incident colour plus odour) found an improved correlation between thrips on traps and those found within the crop compared to standard traps. Visual and olfactory cues have also been used to enhance pest management and biological control for aphids and whiteflies (Pinto-Zevallos and Vänninen 2013; Döring 2014; Ben-Yakir et al. 2020b) but possibly to a lesser degree for psyllids (Civolani et al. 2023; Silva et al. 2023). Where pests and their natural enemies respond to the same stimuli, they can be monitored together on the same traps (Mellor et al. 1997; Böckmann and Meyhöfer 2017;

Kalile et al. 2022) with Grupe et al. (2023) indicating that natural enemy trap counts may be a useful proxy for pest density.

5.3 Mass trapping

Mass trapping could be an effective tool when used in conjunction with natural enemies and shows considerable promise. However, despite the many anecdotal references to mass trapping in the literature (e.g. see Teulon et al. 2017) there are very few proven examples of this technique being effective. One exception was the mass trapping of western flower thrips in a commercial strawberry crop in the United Kingdom, where blue roller sticky traps used in combination with the release of the predatory mite *Neoseiulus cucumeris* and insecticide treatments kept thrips numbers below the damage threshold (Sampson and Kirk 2013). George et al. (2020) described a multimodal (visual, olfactory, phagostimulant) attract and kill tactic for the management of psyllids, but it is unclear if this has been used in practice. Negative impacts of traps on natural enemy populations could be minimised by using wingless predators (e.g. predatory mites) or reducing trapping intensity when winged natural enemies are introduced.

5.4 Pest/natural enemy dynamics

There may be many opportunities to manipulate pest populations by strategic use of stimuli without trapping, to 'attract and retain' natural enemies and maximise the impact of biological control (Traczyk et al. 2020; Okosun et al. 2023; Park et al. 2023). Bennison et al. (2002) highlighted the potential use of thrips predators (*Orius*) in a 'push–pull' strategy where thrips and predators were 'attracted' to trap plants by a host plant volatile to enhance biological control. Attractive lures have been suggested for use with supplemental food resources such as flower strips to enhance the effectiveness of aphid predators in the context of an 'attract and reward' system (Simpson et al. 2011). Exploitation of host plant stimuli may also assist with the early arrival of ladybird predators associated with better control of aphid pests (Traczyk et al. 2020). Use of olfactory associative learning (e.g. with plant odours) during mass rearing before the release of natural enemies in the field is an exciting area of research (Ayelo et al. 2021) although we are unaware of any examples relating to the natural enemies of small insects such as thrips.

5.5 Targeted/spot treatments

Visual and olfactory stimuli can be used to congregate pest species in a particular location within the crop, and then natural enemies or biopesticides can be applied to that area rather than treating the whole crop. Mfuti et al. (2017) reported that a spot spray application of *Metarhizium anisopliae* for managing *Megalurothrips sjostedti* infesting cowpea could be a viable option for small-scale farmers in Kenya. Where the natural enemy also responds to the visual or olfactory stimuli, an added effect might be to keep the predator or parasitoid within this area (see Section 5.1). Semiochemicals could be used to enhance the effectiveness of physical immobilisation strategies via adhesive particles in selected areas of a crop, as described by van Zwieten et al. (2024).

5.6 Lure and Infect

The ability to attract insects through visual and olfactory stimuli to increase contact with entomopathogenic fungi (biopesticide), and hence increase the likelihood of their becoming infected and killed, has been explored for at least two species of thrips with some promise (Mfuti et al. 2021). The success of such a strategy depends on the compatibility of the semiochemical odour with the entomopathogenic fungus unless the lure and the fungus are effectively spatially separated (Mfuti et al. 2021). Recent attempts at developing effective attract-and-kill devices for the Asian citrus psyllid showed the importance of incorporating multiple sensory cues and three-dimensionality into the design of the devices for optimal efficiency (Silva et al. 2023).

5.7 Pest activation

Thrips are thigmotaxic and seek out confined spaces in their environment. This may limit their exposure to biological control agents, including biopesticides (Shipp and Zhang 1999). One synthetic thrips lure, based on a host plant volatile, blown over western flower thrips on leaf discs in the laboratory, increased thrips' walking and take-off behaviour (van Tol et al. 2012). More active and exposed thrips would probably be at greater risk from a range of natural enemies, but this approach requires further consideration. Pest activation may play a critical role in the greater exposure of pests to generalist predators and improve the efficacy of new generation technologies such as UV-C treatments (Montemayor et al. 2023).

6 Conclusion

Thrips and other small flying insects, such as aphids, psyllids and whitefly, are important plant pests. Reliance and overuse of insecticides have precipitated the need to develop knowledge-intensive integrated pest management systems, which could include the exploitation of visual and olfactory stimuli in various ways, and biological control, sometimes working in unison. A greater understanding of these tools and how they might be used together will enable us to develop more effective pest management.

Our approach in this chapter was to examine thrips and closely related small, winged hemipteroid insects to identify similarities and differences in host finding that may confirm observations and provide cross-taxa insight for future work. In addition, we contrasted the genetic, neurological and morphological components of vision and olfaction for these insects to identify areas where further work will help our understanding of host finding and exploit them for pest management.

Our findings show that thrips, aphids, psyllids and whiteflies have complex visual and olfactory sensory systems used for host finding although these appear to be constrained somewhat by the relatively small body size of these insects. Their small size also places practical challenges on the ability to undertake host-finding research, and information from more than one method is required to best understand insect responses to visual and olfactory stimuli.

Our work has illustrated that there are many areas where research would be fruitful – several of the more compelling ones are listed here:

- a better understanding of the behavioural responses of small, winged insects to visual and especially olfactory stimuli, and how these combine to affect host finding;
- clarification of how different modalities (i.e. vision, olfaction) are used in the sequence of host finding (e.g. temporal, distance);
- a greater understanding of the relevance of stimuli identified for walking insects compared to those stimuli used in flight;
- the degree to which small insects fly upwind within plant canopies or during periods of low wind speed, and their migratory inclination, requires further attention in relation to visual and olfactory stimuli; and
- the biological relevance of the substantial number of non-responders in behavioural bioassays and what this might mean for practical application.

There is a need for critical design of laboratory bioassays and field experiments to draw sound conclusions on insect behaviour and their application in response to visual and olfactory stimuli. In many cases, this may include confirmation of behaviour across several different methods. The use of methods that involve flight (not walking) should be prioritised over other methods, as flight is the most important activity during host plant finding for thrips, aphids, psyllids and whiteflies. Advances in the assembly and analysis of genomes, light (e.g. LED) and flight recording (e.g. video tracking) technology should underpin and assist these approaches. How visual and olfactory stimuli are integrated in different species is unclear, and different insect taxa may require a different emphasis on certain research aspects compared with other taxa.

Potential and actual applications for the use of host-finding stimuli for pest management and biological control appear to be numerous. However, despite recent advances in knowledge, there remains much to know to fully exploit host-finding stimuli for pest management and biological control. Currently, most applications involve pest and natural enemy monitoring, but there appear to be substantial opportunities to enhance biological control through mass trapping, attract and retain, lure and infect, targeted application, associative learning and pest activation.

7 Where to look for further information

The following articles provide a good background to the subject:

- Bernays, E.A. & Chapman, R.E. 1994. Host-Plant Selection by Phytophagous Insects. New York: Chapman & Hall.
- Compton, S.G. 2002. Sailing with the wind: dispersal by small flying insects. Dispersal Ecology: Pp. 113–133. In: Bullock, J.M. et al. (eds). The 42nd Symposium of the British Ecological Society Held at the University of Reading, 2–5 April 2001. Malden, MA: Blackwell Publishing.
- Kirk, W.D. et al. 2021. Semiochemicals for thrips and their use in pest management. Annual Review of Entomology 66: 101–119.
- Lopez-Reyes, K. et al. 2022. Colour vision in thrips (Thysanoptera). Philosophical Transactions of the Royal Society B 377 (1862): 20210282.
- Polilov, A.A. 2015. Small is beautiful: features of the smallest insects and limits to miniaturization. Annual Review of Entomology 60: 103–121.

8 Acknowledgements

We thank the students and colleagues in our laboratories who have supported our ongoing research in this field over many years and in many ways contributed to the ideas expressed in this chapter. This collaboration was supported by The Catalyst Seeding (New Zealand) (17-PAF-027-CSG) fund as part of the FARIS (Fundamental to Applied Research on Insect Sensing) project to which all authors belonged. We thank Melissa Jordan, Hayley Ridgway and Ruth Williams (Plant & Food Research) for reviewing an earlier draft of this manuscript. We apologise to colleagues whose articles were not cited due to space limitations.

9 References

Adesanya, A.W. et al. 2020. Multiple insecticide resistance in onion thrips populations from Western USA. *Pesticide Biochemistry and Physiology* 165: 104553.

Alfsnes, K., et al. 2017. Genome size in arthropods; different roles of phylogeny, habitat and life history in insects and crustaceans. *Ecology and Evolution* 7(15): 5939–5947.

Anderbrant, O. et al. 2020. Electrophysiological responses of carrot psyllids (*Trioza apicalis*), in different phases of their life cycle, to volatile carrot and conifer compounds. *Journal of Applied Entomology* 144(3): 236–240.

Anonymous. 2004. Whiteflies (Hemiptera: Aleyrodidae). Pp. 2548–2555. In: *Encyclopaedia of Entomology*. Dordrecht: Springer. https://doi.org/10.1007/0-306-48380-7_4585.

Ayelo, P.M. et al. 2021. Exploring the kairomone-based foraging behaviour of natural enemies to enhance biological control: A review. *Frontiers in Ecology and Evolution* 9: 641974.

Baker, T.C. & Linn Jr., C.E. 1984. Wind tunnels in pheromone research. Pp. 75–110. In: *Techniques in pheromone research*. New York, NY: Springer New York.

Bennison, J. et al. 2002. Towards the development of a push-pull strategy for improving biological control of western flower thrips on chrysanthemum. Pp 199–206. In: Marullo, R. & Mound, L.A. (eds). *Thrips and Tospoviruses: Proc. 7th International Symposium on Thysanoptera*. Canberra: Australian National Insect Collection.

Ben-Yakir, D. & Chen, M. 2008. Studies of thrips migratory flights in Israel. *Acta Phytopathologica et Entomologica Hungarica* 43(2): 243–248.

Ben-Yakir, D. et al. 2020a. Arthropod vision. Chapter 3. Pp: 19–48. In: Ben-Yakir, D. (ed.). *Optical Manipulation of Arthropod Pests and Beneficials*. Wallingford, UK: CAB International.

Ben-Yakir, D. et al. 2020b. Attraction of Pests. Chapter 7. Pp: 101–140. In: Ben-Yakir, D. (ed.). *Optical Manipulation of Arthropod Pests and Beneficials*. Wallingford, UK: CAB International.

Bernays, E.A. & Chapman, R.E. 1994. *Host-Plant Selection by Phytophagous Insects*. New York: Chapman & Hall.

Berry, N. et al. 2006. Reponses of New Zealand flower thrips (*Thrips obscuratus* (Crawford)) to synthetic and natural stimuli (odour and colour) in a wind tunnel. *New Zealand Journal of Horticulture & Crop Science* 34(2): 121–129.

Birkett, M.A. et al. 2000. New roles for cis-jasmone as an insect semiochemical and in plant defence. *Proceedings of the National Academy of Sciences* 97(16): 9329–9334.

Blackman, R.L. & Eastop, V.F. 2000. *Aphids on the world's crops: an identification and information guide*. John Wiley & Sons.

Böckmann, E. & Meyhöfer, R. 2017. Sticky trap monitoring of a pest–predator system in glasshouse tomato crops: are available trap colours sufficient? *Journal of Applied Entomology* 141(5): 339–351.

Broughton, S. & Harrison, J. 2012. Evaluation of monitoring methods for thrips and the effect of trap colour and semiochemicals on sticky trap capture of thrips (Thysanoptera) and beneficial insects (Syrphidae, Hemerobiidae) in deciduous fruit trees in Western Australia. *Crop Protection* 42: 156–163.

Bruno, D. et al. 2018. Sensilla morphology and complex expression pattern of odorant binding proteins in the vetch aphid *Megoura viciae* (Hemiptera: Aphididae). *Frontiers in Physiology* 9: 343351.

Byrne, D.N. 1999. Migration and dispersal by the sweet potato whitefly, *Bemisia tabaci*. *Agricultural and forest meteorology* 97(4): 309–316.

Byrne, D.N. & Bellows Jr., T.S. 1991. Whitefly biology. *Annual Review of Entomology* 36(1): 431–457.

Caballero-Vidal, G. et al. 2020. Machine learning decodes chemical features to identify novel agonists of a moth odorant receptor. *Scientific Reports* 10(1): 1655.

Cao, D. et al. 2014. Molecular characterization of the *Aphis gossypii* olfactory receptor gene families. *PLoS One* 9(6): e101187.

Cardé, R.T. 2021. Navigation along windborne plumes of pheromone and resource-linked odors. *Annual Review of Entomology* 66: 317–336.

Castañé, C. et al. 1999. Biological control of thrips. Pp. 244–253. In: *Integrated pest and disease management in greenhouse crops*. Dordrecht: Springer Netherlands.

Chapman, R.F. 1982. Chemoreception: the significance of receptor numbers. *Advances in Insect Physiology* 16: 247–356.

Chapman, R.F. et al. 1981. Attraction and repulsion of the aphid, *Cavariella aegopodii*, by plant odors. *Journal of Chemical Ecology* 7: 881–888.

Civolani, S. et al. 2023. Diversity, biology, and management of the pear psyllids: a global look. *Annals of the Entomological Society of America* 116(6): 331–357.

Compton, S.G. 2002. Sailing with the wind: dispersal by small flying insects. Dispersal Ecology: Pp. 113–133. In: Bullock, J.M. et al. (eds). *The 42nd Symposium of the British Ecological Society Held at the University of Reading, 2–5 April 2001*. Malden, MA: Blackwell Publishing.

Cox, P.D. et al. 2006. Potential for the use of biological agents for the control of *Thrips palmi* (Thysanoptera: Thripidae) outbreaks. *Biocontrol Science and Technology* 16(9): 871–891.

David, C.T. & Hardie, J.I. 1988. The visual responses of free-flying summer and autumn forms of the black bean aphid, *Aphis fabae*, in an automated flight chamber. *Physiological Entomology* 13(3): 277–284.

Davidson, M.M. et al. 2006. Starvation period and age affect the response of female *Frankliniella occidentalis* (Pergande) (Thysanoptera: Thripidae) to odor and visual cues. *Journal of Insect Physiology* 52(7): 729–736.

Davidson, M.M. et al. 2007. Pyridine compounds increase trap capture of *Frankliniella occidentalis* (Pergande) in a covered crop. *New Zealand Plant Protection* 60: 56–60.

Davidson, M.M. et al. 2012. Response of female *Frankliniella occidentalis* (Pergande) to visual cues and para-anisaldehyde in a flight chamber. *Journal of Insect Behavior* 25(3): 297–307.

Davidson, M.M. et al. 2014. A volatile compound 2undecanone increases walking but not flying tomato potato psyllid movement toward an odour source. *New Zealand Plant Protection* 67: 184–190.

Davidson, M.M. et al. 2015. Can semiochemicals attract both western flower thrips and their anthocorid predators? *Entomologia Experimentalis et Applicata* 155: 54–63.

Díaz, M.A. et al. 2023. Semiochemicals associated with the western flower thrips attraction: A systematic literature review and meta-analysis. *Insects* 14(3): 269.

Dilawari, V.K. & Atwal, A.S. 1989. Response of mustard aphid, *Lipaphis erysimi* (Kalt.) to allylisothiocyanate. *Journal of Insect Science* 2: 103–108.

Dong, W. et al. 2008. Electroantennogram responses of aphid parasitoid *Aphidius gifuensis* to aphid pheromones and host-plant volatiles. *Chinese Journal of Ecology* 27(4): 591–594. Abstract only.

Döring, T.F. 2014. How aphids find their host plants, and how they don't. *Annals of Applied Biology* 65(1): 3–26.

Döring, T.F. & Chittka, L. 2007. Visual ecology of aphids–a critical review on the role of colours in host finding. *Arthropod-plant Interactions* 1: 3–16.

Döring, T.F. & Spaethe, J. 2009. Measurements of eye size and acuity in aphids (Hemiptera: Aphididae). *Entomologia Generalis* 32: 77–84.

Döring, T.F. et al. 2011. Spectral sensitivity of the green photoreceptor of winged pea aphids. *Physiological Entomology* 36(4): 392–396.

Farnier, K. et al. 2015. Visual acuity trade-offs and microhabitat-driven adaptation of searching behaviour in psyllids (Hemiptera: Psylloidea: Aphalaridae). *The Journal of Experimental Biology* 218(10): 1564–1571.

Finch, S. & Collier, R.H. 2000. Host-plant selection by insects – a theory based on 'appropriate/inappropriate landings' by pest insects of cruciferous plants. *Entomologia Experimentalis et Applicata* 96: 91–102.

Fischer, S. et al. 2011. How small can small be: the compound eye of the parasitoid wasp *Trichogramma evanescens* (Westwood, 1833) (Hymenoptera, Hexapoda), an insect of 0.3-to 0.4-mm total body size. *Visual Neuroscience* 28(4): 295–308.

Friedrich, M. 2023. Parallel losses of blue opsin correlate with compensatory neofunctionalization of UV-Opsin gene duplicates in aphids and planthoppers. *Insects* 14(9): 774.

Gatehouse, A.G. 1997. Behavior and ecological genetics of wind-borne migration by insects. *Annual Review of Entomology* 42(1): 475–502.

Gao, Y.L. et al. 2012. Western flower thrips resistance to insecticides: detection, mechanisms and management strategies. *Pest Management Science* 68: 1111–21.

Gao, H. et al. 2024. Unravelling the G protein-coupled receptor superfamily in aphids: Contractions and duplications linked to phloem feeding. *General and Comparative Endocrinology* 347: 114435.

Ge, Y. et al. 2022. Morphological characterization and sexual dimorphism of the antennal sensilla in *Bactericera gobica* Loginova (Hemiptera: Psyllidae)–a scanning and transmission electron microscopic study. *PeerJ*.10: e12888.

George, J. et al. 2020. A multimodal attract-and-kill device for the Asian citrus psyllid *Diaphorina citri* (Hemiptera: Liviidae). *Insects* 11(12): 870.

González-Orellana, J. et al. 2022. Behavioural and electrophysiological responses of *Liothrips jatrophae* (Thysanoptera: Phlaeothripidae) to conspecific extracts and some of its identified compounds. *Physiological Entomology* 47(1): 11–19.

Gregory, T.R. 2005. Genome size evolution in animals. Pp. 3–87. Chapter 1. In: Gregory, R. (ed.). *The evolution of the genome*. Cambridge: Academic Press.

Grupe, B. & Meyhöfer, R. 2024. Blue LED trap and commercial lure improve western flower thrips (*Frankliniella occidentalis*) monitoring in cucumber crops. *Journal of Pest Science* 20: 1–2.

Grupe, B. et al. 2023. Keep an eye on natural enemies: what *Aphidius* on sticky traps tells us about aphid pest population dynamics. *Entomologia Experimentalis et Applicata* 171(10): 722–731.

Gu, S.H. et al. 2013. Identification and expression profiling of odorant binding proteins and chemosensory proteins between two wingless morphs and a winged morph of the cotton aphid *Aphis gossypii* Glover. *PLoS One* 8(9): e73524.

Ha, T.S. & Smith, D.P. 2022. Recent insights into insect olfactory receptors and odorant-binding proteins. *Insects* 13(10): 926.

Hanken, J. & Wake, D.B. 1993. Miniaturization of body size: organismal consequences and evolutionary significance. *Annual Review of Ecology and Systematics* 24(1): 501–519.

Hardie, J. 1994. Variation in behavioural migration in aphids. *European Journal of Entomology* 91(1): 115–120.
Hollis, D. 2004. *Australian psylloidea: Jumping plantlice and lerp insects*. Australian Government, Department of Environment and Heritage, Canberra.
Hu, Q. et al. 2020. Morphology, distribution, ultrastructure, and possible function of the external sensilla on antennae and mouthparts of *Echinothrips americanus* Morgan (Thysanoptera: Thripidae). *Journal of Asia-Pacific Entomology* 23(2): 599–605.
Hu, Q.L. et al. 2023. A chromosome-level genome assembly of *Stenchaetothrips biformis* and comparative genomic analysis highlights distinct host adaptations among thrips. *Communications Biology* 6(1): 813.
Huang, T. 2023. *Identification and functional studies of odorant receptors in aphids*. PhD dissertation. 162 pp. l'Université de Liège.
Isaacs, R. et al. 1999. Modulation of whitefly take-off and flight orientation by wind speed and visual cues. *Physiological Entomology* 24(4): 311–318.
Johnson, C.G. 1957. The distribution of insects in the air and the empirical relation of density to height. *Journal of Animal Ecology* 26: 479–94.
Kainoh, Y. 2011. Wind tunnel: A tool to test the flight response to semiochemicals. *Wind Tunnels and Experimental Fluid Dynamics Research* 27: 89–99.
Kalile, M.O. et al. 2022. UV light attracts *Diaphorina citri* and its parasitoid. *Biological Control* 170: 104928.
Kang, Z.W. et al. 2017. Identification and expression analysis of chemosensory receptor genes in an aphid endoparasitoid *Aphidius gifuensis*. *Scientific Reports* 7(1): 3939.
Kang, Z.W. et al. 2018. The identification and expression analysis of candidate chemosensory genes in the bird cherry-oat aphid *Rhopalosiphum padi* (L.). *Bulletin of Entomological Research* 108(5): 645–657.
Kanturski, M. et al. 2017. Morphology and sensilla of the enigmatic Bhutan pine aphid *Pseudessigella brachychaeta* Hille Ris Lambers (Hemiptera: Aphididae)–A SEM study. *Zoologischer Anzeiger* 266: 1–3.
Karpe, S.D. et al. 2021. InsectOR–Webserver for sensitive identification of insect olfactory receptor genes from non-model genomes. *PLoS One* 16(1): e0245324.
Kelber, A. 2006. Invertebrate colour vision. Pp. 250–290. In: Warrant, E. & Nilsson, D.E. (eds). *Invertebrate vision*. Cambridge, UK: Cambridge University Press.
Kelber, A. 2016. Colour in the eye of the beholder: receptor sensitivities and neural circuits underlying colour opponency and colour perception. *Current Opinion in Neurobiology* 41: 106–112.
Kelber, A. et al. 2003. Animal colour vision–behavioural tests and physiological concepts. *Biological Reviews* 78(1): 81–118.
Keil, T.A. 1999. Morphology and development of the peripheral olfactory organs. Pp. 5–47. Chapter 1. In: *Insect olfaction*. Berlin, Heidelberg: Springer Berlin Heidelberg.
Kennedy, J.S. 1977. Olfactory responses to distant plants and other odour sources. Pp. 67–91. Chapter 5. In: Shorey, H.H. & McKelvey, J.J. Jr. (eds). *Chemical Control of Insect Behavior, Theory and Application*. New York: Wiley, pp. 67–91.
Khadka, A. et al. 2020. Can the addition of odor and visual targets enhance attraction of the Asian citrus psyllid (Hemiptera: Liviidae) to sticky traps? *Journal of Economic Entomology* 113(5): 2563–2567.
King, B.H. & Gunathunga, P.B. 2023. Gustation in insects: taste qualities and types of evidence used to show taste function of specific body parts. *Journal of Insect Science* 23(2): 11.

Kirk, W.D.J. & Terry, L.I. 2003. The spread of the western flower thrips *Frankliniella occidentalis* (Pergande). *Agricultural and Forest Entomology* 5: 301–10.

Kirk, W.D. et al. 2021. Semiochemicals for thrips and their use in pest management. *Annual Review of Entomology* 66: 101–119.

Knudsen, G.K. et al. 2018. A wind tunnel for odor mediated insect behavioural assays. *Journal of Visualized Experiments* 30(141): e58385.

Kuang, Y. et al. 2023. Candidate odorant-binding protein and chemosensory protein genes in the turnip aphid *Lipaphis erysimi*. *Archives of Insect Biochemistry and Physiology* 113(4): e22022.

Land, M.F. 1997. Visual acuity in insects. *Annual Review of Entomology* 42(1): 147–177.

Land, M.F. & Nilsson, D.E. 2012 *Animal eyes*. Oxford, UK: Oxford University Press.

Lei, Y.A. et al. 2021. Advances on biological control of thrips pests. *Chinese Journal of Biological Control* 37(3): 393. Abstract only.

Lewis, T. 1997. Thrips as Crop Pests. Wallingford, UK: CAB International.

Li, W. & Feng, J. 2013. Ultrastructure of antennal sensilla in three species of *Frankliniella* Karny (Thysanoptera: Thripidae). *Acta Entomologica Sinica* 56(9): 1088–1100.

Li, C. et al. 2019. The expression and function of opsin genes related to the phototactic behavior of Asian citrus psyllid. *Pest Management Science* 76(4): 1578–1587.

Li, X. et al. 2022. Comparison and functional analysish of odorant-binding proteins and chemosensory proteins in two closely related thrips species, *Frankliniella occidentalis* and *Frankliniella intonsa* (Thysanoptera: Thripidae) based on antennal transcriptome analysis. *International Journal of Molecular Sciences* 23(22): 13900.

Liu, Y.Q. et al. 2021. Morphology and distribution of antennal sensilla in three species of Thripidae (Thysanoptera) infesting alfalfa *Medicago sativa*. *Insects* 12(1): 81.

Liu, Y. et al. 2023. Antennal transcriptome analysis of olfactory genes and characterization of odorant binding proteins in *Odontothrips loti* (Thysanoptera: Thripidae). *International Journal of Molecular Sciences* 24(6): 5284.

Lo Pinto, M. et al. 2004. Olfactory response of two aphid parasitoids, *Lysiphlebus testaceipes* and *Aphidius colemani*, to aphid-infested plants from a distance. *Entomologia Experimentalis et Applicata* 110(2): 159–164.

Loomans, A.J.M. et al. 1997. Interactions with hymenopterous parasitoids and parasitic nematodes. Pp. 355–398. In: Lewis, T. (ed.). *Thrips as crop pests*. Wallingford, UK: CAB International.

Lopez-Reyes, K. et al. 2022. Colour vision in thrips (Thysanoptera). *Philosophical Transactions of the Royal Society B* 377 (1862): 20210282.

Lopez-Reyes, K. et al. 2023. Tracking the flight and landing behaviour of western flower thrips in response to single and two-colour cues. *Scientific Reports* 13(1): 14178.

Mainali, B.P. & Lim, U.T. 2011. Behavioral response of western flower thrips to visual and olfactory cues. *Journal of Insect Behavior* 24: 436–446.

Makarova, A.A. & Polilov, A.A. 2017. Peculiarities of the brain organization and fine structure in small insects related to miniaturization. 4. Thrips (Thysanoptera, Thripidae). *Entomological Review* 97: 302–309.

Makarova, A.A. et al. 2022a. Scaling of the sense organs of insects. 1. Introduction. Compound Eyes. *Entomological Review* 102(2): 161–181.

Makarova, A.A. et al. 2022b. Scaling of the sense organs of insects. 2. Sensilla. Discussion. Conclusion. *Entomological Review* 102(3): 323–346.

Mathis, A. et al. 2018. DeepLabCut: markerless pose estimation of user-defined body parts with deep learning. *Nature Neuroscience* 21(9): 1281–1289.

Matteson, N. et al. 1992. Spectral efficiency of the western flower thrips, *Frankliniella occidentalis*. *Journal of Insect Physiology* 38(6): 453–459.

Mauck, K.E. et al. 2024. The hidden secrets of Psylloidea: Biology, behavior, symbionts, and ecology. *Annual Review of Entomology* 69: 277–302.

Mazokhin-Porshnyakov, G.A. & Kazyakina, V.I. 1983. Morphological description of the compound eyes and lateral ocelli of thrips (Thysanoptera). *Biologicheskie Nauki* 1: 57–60

Mellor, H.E. et al. 1997. Spectral efficiency of the glasshouse whitefly *Trialeurodes vaporariorum* and *Encarsia formosa* its hymenopteran parasitoid. *Entomologia Experimentalis et Applicata* 83(1): 11–20.

Messelink, G.J. et al. 2006. Evaluation of phytoseiid predators for control of western flower thrips on greenhouse cucumber. *BioControl* 51: 753–768.

Mfuti, D.K. et al. 2017. Lure and infect strategy for application of entomopathogenic fungus for the control of bean flower thrips in cowpea. *Biological Control* 107: 70–76.

Mfuti, D.K. et al. 2021. Exploiting thrips aggregation pheromones to develop a lure-and-kill strategy for the management of the bean flower thrips. *Agronomy* 11(7): 1269.

Montemayor, J.D. et al. 2023. Is UV-C light compatible with biological control of two-spotted spider mite? *Biological Control* 183:105269.

Montagné, N. et al. 2015. Advances in the identification and characterization of olfactory receptors in insects. *Progress in Molecular Biology and Translational Science* 130: 55–80.

Morin, S. et al. 2024. Whitefly–plant interactions: An integrated molecular perspective. *Annual Review of Entomology* 69: 503–525.

Moritz, G. 1997. Structure, growth and development. Pp. 15–64. In: Lewis, T. (ed.). *Thrips as crop pests*. Wallingford, UK: CAB International.

Morse, J.G. & Hoddle, M.S. 2006. Invasion biology of thrips. *Annual Review of Entomology* 51: 67–89.

Mouden, S. et al. 2017. Integrated pest management in western flower thrips: past, present and future. *Pest Management Science* 73(5): 813–822.

Mound, L.A. & Teulon, D.A.J. 1995. Thysanoptera as phytophagous opportunists. Pp. 3–19. In: Parker, B.L. et al. (eds.). *Thrips Biology and Management*. New York: Plenum Press.

Mound, L.A. et al. 2022. Problems with the concept of "pest" among the diversity of pestiferous thrips. *Insects* 13(1): 61.

Mrisho, L.M. et al. 2021. Influence of olfaction in host-selection behavior of the cassava whitefly *Bemisia tabaci*. *Frontiers in Ecology and Evolution* 9: 775778.

Murai, T. et al. 2000. Methyl anthranilate as an attractant for two thrips species and the thrips parasitoid *Ceranisus menes*. *Journal of Chemical Ecology* 26: 2557–2565.

Muvea, A.M. et al. 2014. Attraction of pest thrips (Thysanoptera: Thripidae) infesting French beans to coloured sticky traps with Lurem-TR and its utility for monitoring thrips populations. *International Journal of Tropical Insect Science* 34: 197–206.

Nicholson, S.J. et al. 2015. The genome of *Diuraphis noxia*, a global aphid pest of small grains. *BMC Genomics* 16: 429.

Nielsen, M.C. 2013. *Factors affecting the response of thrips to an olfactory cue*. PhD thesis. Lincoln University, New Zealand.

Nottingham, S.F. & Hardie, J. 1993. Flight behaviour of the black bean aphid, *Aphis fabae*, and the cabbage aphid, *Brevicoryne brassicae*, in host and non-host plant odour. *Physiological Entomology*, 18: 389–394.

Okosun, O.O. et al. 2023. Role of kairomones in biological control of pests: commercial potential. Chapter 4. Pp. 57–80. In: *Development and commercialization of biopesticides*. Cambridge: Academic Press.

Onagbola, E.O. et al. 2008. Morphological characterization of the antennal sensilla of the Asian citrus psyllid, *Diaphorina citri* Kuwayama (Hemiptera: Psyllidae), with reference to their probable functions. *Micron* 39(8): 1184–1191.

Otieno, J.A. et al. 2018. Efficacy of LED-enhanced blue sticky traps combined with the synthetic lure Lurem-TR for trapping of western flower thrips (*Frankliniella occidentalis*). *Journal of Pest Science* 91: 1301–1314.

Palavalli-Nettimi, R. et al. 2019. Miniaturisation reduces contrast sensitivity and spatial resolving power in ants. *Journal of Experimental Biology*: 222(12): jeb203018.

Park, Y.G. et al. 2023. Attractiveness of four wavelengths of LED light: UV (385 nm), violet (405 nm), blue (450 nm), and red (660 nm) for seven species of natural enemies. *Biological Control* 179: 105166.

Patt, J.M. et al. 2011. Multimodal cues drive host-plant assessment in Asian citrus psyllid (*Diaphorina citri*). *Environmental Entomology* 40(6): 1494–1502.

Pettersson, J. et al. 2007. Host-plant selection and feeding. Pp. 87–115. In: van Emden, H.F. & Harrington, R. (eds). *Aphids as Crops Pests*. Wallingford, UK: CAB International.

Pinto-Zevallos, D.M. & Vänninen, I. 2013. Yellow sticky traps for decision-making in whitefly management: What has been achieved? *Crop Protection* 47: 74–84.

Polilov, A.A. 2015. Small is beautiful: features of the smallest insects and limits to miniaturization. *Annual Review of Entomology* 60: 103–121.

Prokopy, R.J. & Owens, E.D. 1983. Visual detection of plants by herbivorous insects. *Annual Review of Entomology* 28: 337–364.

Qian, Q. et al. 2024. The plant volatile-sensing mechanism of insects and its utilization. *Plants* 13(2): 185.

Ren, X. et al. 2020. Behavioral responses of western flower thrips (*Frankliniella occidentalis*) to visual and olfactory cues at short distances. *Insects* 11(3): 177.

Roberts, J.M. et al. 2023. Scents and sensibility: Best practice in insect olfactometer bioassays. *Entomologia Experimentalis et Applicata* 171(11): 808–820.

Robertson, H.M. 2019. Molecular evolution of the major arthropod chemoreceptor gene families. *Annual Review of Entomology* 64: 227–242.

Robertson, H.M. et al. 2019. The chemoreceptors and odorant binding proteins of the soybean and pea aphids. Insect *Biochemistry and Molecular Biology* 105: 69–78.

Rodríguez, D. & Coy-Barrera, E. 2023. Overview of updated control tactics for western flower thrips. *Insects* 14(7): 649.

Rotenberg, D. & Whitfield, A.E. 2018. Molecular interactions between tospoviruses and thrips vectors. *Current Opinion in Virology* 33: 191–197.

Sabelis, M.W. & Van Rijn, P.C. 1997. Predation by insects and mites. Pp. 259–354. In: Lewis, T. (ed.). *Thrips as Crop Pests*. Wallingford, UK: CAB International.

Sampson, C. & Kirk, W.D.J. 2013. Can mass trapping reduce thrips damage and is it economically viable? Management of the western flower thrips in strawberry. *PLoS One* 8(11): e80787.

Santer, R.D. & Allen, W.L. 2024. Optimising the colour of traps requires an insect's eye view. *Pest Management Science* 80(3): 931–934.

Sarig, A. & Ribak, G. 2021. To what extent can the tiny parasitoid wasps, *Eretmocerus mundus*, fly upwind? *Journal of Applied Entomology* 145(7): 660–674.

Schlaeger, S. et al. 2018. Prospects for management of whitefly using plant semiochemicals, compared with related pests. *Pest Management Science* 74(11): 2405–2411.

Schnaitmann, C. et al. 2020. Color vision in insects: insights from *Drosophila*. *Journal of Comparative Physiology A* 206(2): 183–198.

Shapiro-Ilan, D. et al. 2017. Basic and applied research: entomopathogenic nematodes. Pp. 91–105. In: Lacey, L.A. (ed.). *Microbial Control of Insect and Mite Pests*. Cambridge: Academic Press.

Shih, P.Y. et al. 2023. Molecular mechanisms underlying host plant specificity in aphids. *Annual Review of Entomology* 68: 431–450.

Shipp, J.L. & Zhang, Y. 1999. Using greenhouse microclimate to improve the efficacy of insecticide application for *Frankliniella occidentalis* (Thysanoptera: Thripidae). *Journal of Economic Entomology* 92(1): 201–220.

Silva, M.S. et al. 2023. Asian citrus psyllid, *Diaphorina citri* (Hemiptera: Liviidae) responses to plant-associated volatile organic compounds: A mini-review. *Crop Protection* 106242.

Simpson, M. et al. 2011. Attract and reward: combining chemical ecology and habitat manipulation to enhance biological control in field crops. *Journal of Applied Ecology* 48(3): 580–590.

Smadja, C. et al. 2009. Large gene family expansions and adaptive evolution for odorant and gustatory receptors in the pea aphid, *Acyrthosiphon pisum*. *Molecular Biology and Evolution* 26(9): 2073–2086.

Smagghe, F. et al. 2023. Biological control of arthropod pests in protected cropping by employing entomopathogens: *Efficiency, production and safety*. *Biological Control* 186: 105337.

Smits, P.H. et al. 2000. Western flower thrips: reactions to odours and colours. *Proceedings of the section Experimental and Applied Entomology, N.E.V.* 11, 175–180.

Song, L. et al. 2020. Characterization of antennal sensilla and immunolocalization of odorant-binding proteins on spotted alfalfa aphid, *Therioaphis trifolii* (Monell). *Frontiers in Physiology* 11: 606575.

Stukenberg, N. & Poehling, H.M. 2019. Blue–green opponency and trichromatic vision in the greenhouse whitefly (*Trialeurodes vaporariorum*) explored using light emitting diodes. *Annals of Applied Biology* 175: 146–163.

Stukenberg, N. et al. 2015. Light emitting diode (LED)-based trapping of the greenhouse whitefly (*Trialeurodes vaporariorum*). *Journal of Applied Entomology* 139(4): 268–279.

Stukenberg, N. et al. 2020. Wavelength-specific behavior of the western flower thrips (*Frankliniella occidentalis*): evidence for a blue-green chromatic mechanism. *Insects* 11(7): 423.

Terry, L.I. 1997. Host selection, communication and reproductive behaviour. Pp. 65–118. In: Lewis, T. (ed.). *Thrips as Crop Pests*. Wallingford, UK: CAB International.

Teulon, D.A.J. et al. 1993. Volatile chemicals for thrips (Thysanoptera: Thripidae) host finding and applications for thrips pest management. *Journal of Economic Entomology* 86(5): 1405–1415.

Teulon, D.A.J. et al. 1999. Colour and odour responses of flying western flower thrips: wind tunnel and greenhouse experiments. *Entomologia Experimentalis et Applicata* 93(1): 9–19.

Teulon, D.A.J. et al. 2007. Odour-baited traps influence thrips capture in proximal unbaited traps in the field. *Entomologia Experimentalis et Applicata* 123: 253–262.

Teulon, D.A.J. et al. 2017. Methyl isonicotinate–a non-pheromone thrips semiochemical–and its potential for pest management. *International Journal of Tropical Insect Science* 37(2): 50–56.

Tian, J. et al. 2022. Diversity and molecular evolution of odorant receptor in Hemipteran insects. *Insects* 13(2): 214.

Traczyk, E. et al. 2020. Foraging behavior responses of *Orius insidiosus* to thrips cues. *Entomologia Experimentalis et Applicata* 168(10): 716–722.

Van der Kooi, C.J. et al. 2019. Functional significance of the optical properties of flowers for visual signalling. *Annals of Botany* 123(2): 263–276.

Van der Kooi, C.J. et al. 2021. Evolution of insect color vision: from spectral sensitivity to visual ecology. *Annual Review of Entomology* 66: 435–461.

Van Lenteren, J.C. et al. 2020. Biological control agents for control of pests in greenhouses. Pp. 409–439. In: Gullino, M.L. et al. (eds). *Integrated Pest and Disease Management in Greenhouse Crops*. New York: Springer International Publishing.

Van Tol, R.W. et al. 2012. Methyl isonicotinate induces increased walking and take-off behaviour in western flower thrips, *Frankliniella occidentalis*. *Entomologia Experimentalis et Applicata* 142(3): 181–190.

Van Zwieten, R. et al. 2024. Mimicking natural deterrent strategies in plants using adhesive spheres. *Proceedings of the National Academy of Sciences* 121(21): e2321565121.

Vierbergen, G. 1995. International movement, detection and quarantine of Thysanoptera pests. Pp. 119–32. In: Parker, B.L. et al. (eds.). *Thrips Biology and Management*. New York: Plenum Press.

Volpe, H.X.L. et al. 2020. Behavioral responses of *Diaphorina citri* to host plant volatiles in multiple-choice olfactometers are affected in interpretable ways by effects of background colors and airflows. *PLoS One* 15(7): e0235630.

Wang, R. et al. 2017. Identification and expression profile analysis of odorant binding protein and chemosensory protein genes in *Bemisia tabaci* MED by head transcriptome. *PLoS One* 12(2): e0171739.

Wang, X.S. et al. 2019a. Morphology and distribution of the antennal sensilla of two species, *Megalurothrips usitatus* and *Thrips palmi* (Thysanoptera: Thripidae). *Insects* 10(8): 251.

Wang, Q. et al. 2019b. Integrative transcriptomic and genomic analysis of odorant binding proteins and chemosensory proteins in aphids. *Insect Molecular Biology* 28(1): 1–22.

Wang, R. et al. 2020. Molecular and functional characterization of one odorant-binding protein gene OBP3 in *Bemisia tabaci* (Hemiptera: Aleyrodidae). *Journal of Economic Entomology* 113(1): 299–305.

Warrant, E. 2019. Invertebrate vision. Pp. 64–79. In: Choe, J.C. (ed.). *Encyclopedia of Animal Behavior*. Cambridge: Academic Press.

Webster, B.E. 2012. The role of olfaction in aphid host location. *Physiological Entomology* 37(1): 10–18.

Wenninger, E.J. et al. 2009. Roles of olfactory cues, visual cues, and mating status in orientation of *Diaphorina citri* Kuwayama (Hemiptera: Psyllidae) to four different host plants. *Environmental Entomology* 38(1): 225–234.

Wicher, D. & Miazzi, F. 2021. Functional properties of insect olfactory receptors: ionotropic receptors and odorant receptors. *Cell and Tissue Research* 383: 7–19.

Wu, Z. et al. 2016. Antennal and abdominal transcriptomes reveal chemosensory genes in the Asian citrus psyllid, *Diaphorina citri*. *PLoS One* 11(7): e0159372.

Wu, S. et al. 2018. Chapter 7 - Screening, efficacy and mechanisms of microbial control agents against sucking pest insects as thrips. *Advances in Insect Physiology* 55: 199–217.

Xu, J.W. et al. 2019. Chemosensory gene families in the oligophagous pear pest *Cacopsylla chinensis* (Hemiptera: Psyllidae). *Insects* 10(6): 175.

Xue, W. et al. 2016. Identification and expression analysis of candidate odorant-binding protein and chemosensory protein genes by antennal transcriptome of *Sitobion avenae*. *PLoS One* 11(8): e0161839.

Yan, H. et al. 2020. Evolution, developmental expression and function of odorant receptors in insects. *Journal of Experimental Biology* 223(Suppl 1): jeb208215.

Yuvaraj, J.K. et al. 2022. Sex pheromone receptors of the light brown apple moth, *Epiphyas postvittana*, support a second major pheromone receptor clade within the Lepidoptera. *Insect Biochemistry and Molecular Biology* 141: 103708.

Zhang, Z. & Lei, Z. 2022. Sensilla of the western flower thrips, *Frankliniella occidentalis* (Pergande)(Thysanoptera, Thripidae). *Revista Brasileira de Entomologia* 66: e20220023.

Zhang, X.M. et al. 2015. Comparison of the antennal sensilla ultrastructure of two cryptic species in *Bemisia tabaci*. *PLoS One* 10(3): e0121820.

Zhang, H. et al. 2020. Odorant-binding proteins and chemosensory proteins potentially involved in host plant recognition in the Asian citrus psyllid, *Diaphorina citri*. *Pest Management Science* 76(8): 2609–2018.

Zhou, J.J. et al. 2010. Genome annotation and comparative analyses of the odorant-binding proteins and chemosensory proteins in the pea aphid *Acyrthosiphon pisum*. *Insect Molecular Biology* 19: 113–122.

Zhu, W. et al. 2017. Ultrastructure and distribution of antennal sensilla of the chilli thrips *Scirtothrips dorsalis* Hood (Thysanoptera: Thripidae). *Microscopy Research and Technique* 80(12): 1283–1296.

Chapter 2

Genetic engineering of insects to inhibit insect pest reproduction

Alexis L. Kriete[1], Sarah E. Hudadoff[1], Ariel E. Tarrand[1], Amarish K. Yadav[1] and Maxwell J. Scott, North Carolina State University, USA

1 Introduction

Genetic biocontrol for suppresion of pest populations refers to a general approach in which sterile pests, or fertile pests with specific genetic factor(s), are released into wild populations to reduce reproduction and hinder population growth. Thus, the pest itself is the control agent. The type of genetic biocontrol is determined by the mechanism of the genetic factor, which can reduce population size. There are several schemas, some simple and some more complex.

1 These authors contributed equally to this work.

http://dx.doi.org/10.19103/AS.2024.0149.04

A primary advantage of genetic biocontrol is its specificity. Broad application of chemical pesticides affects other species in the area, reducing biodiversity and harming benign native organisms such as pollinators. While not chemically toxic, classical biocontrol involving the release of nonnative natural enemies can also have unintended consequences for the environment. Genetic biocontrol is, by its very nature, targeted to the species being released. Genetic changes are only induced in the pest species and are generally non-transmittable to other organisms. A gene drive targeting *Drosophila suzukii*, for example, will only impact the population of *D. suzukii* (Yadav et al., 2023). All other insect species will be unaffected, and the health of the ecosystem will be more stable.

An early successful application of genetic biocontrol was the development of the sterile insect technique (SIT), which led to the eradication of the New World screwworm from North and Central America (Knipling, 1955; Wyss, 2000). The SIT is discussed in detail later in the chapter. The success of the screwworm eradication campaign and other SIT programs led to the development of other genetic biocontrol approaches, which will be the main focus of this chapter. Researchers had observed natural genetic elements that altered the rates of their own inheritance in many species throughout the twentieth century. In his seminal paper in 2003, Burt described how 'selfish' homing endonuclease genes (HEGs), which drive their own proliferation through a population, might be engineered in the lab within organisms for genetic biocontrol (Burt, 2003). When CRISPR/Cas9 (Clustered Regularly Interspaced Short Palindromic Repeats/CRISPR-associated protein 9) gene editing became more widespread in the 2010s, researchers discovered new ways in which genetic elements for biocontrol could be created and implemented.

In this chapter, we will cover a few methods of genetic biocontrol, with emphasis on those systems that have been developed for crop insect pests. We have designated some of these methods as 'Sterile Release Systems,' though the released males may still produce sperm. For example, radiation-sterilized males produce sperm which carries dominant lethal mutations that result in their offspring failing to develop due to aneuploidy (Robinson, 2021). Another 'sterile' release system, the precision-guided SIT (pgSIT), does actually prevent functional sperm from being produced (Kandul et al., 2019). We contrast these systems with 'Fertile Release Systems.' The systems that we have designated as fertile release systems all involve releasing insects that are capable of producing some offspring that are expected to develop to adulthood and reproduce themselves. Living offspring which have a chance of passing on the genetic mechanisms of the system are a hallmark of these 'fertile' systems. The strengths and weaknesses of these systems vary, and these differences will be discussed.

2 Sterile release systems: the sterile insect technique

The SIT refers to the repeated release of radiation-sterilized insects into a wild population to lower its reproductive output (Klassen et al., 2021) (Fig. 1). For SIT to be effective, an excess of sterile males must be released relative to the number of fertile males in the area, in order for sterile males to have a high chance of mating with wild-type females (Knipling, 1960). This necessitates mass-rearing and mass-sterilization of the target species in large facilities designed to produce millions of insects per week. The scale of production

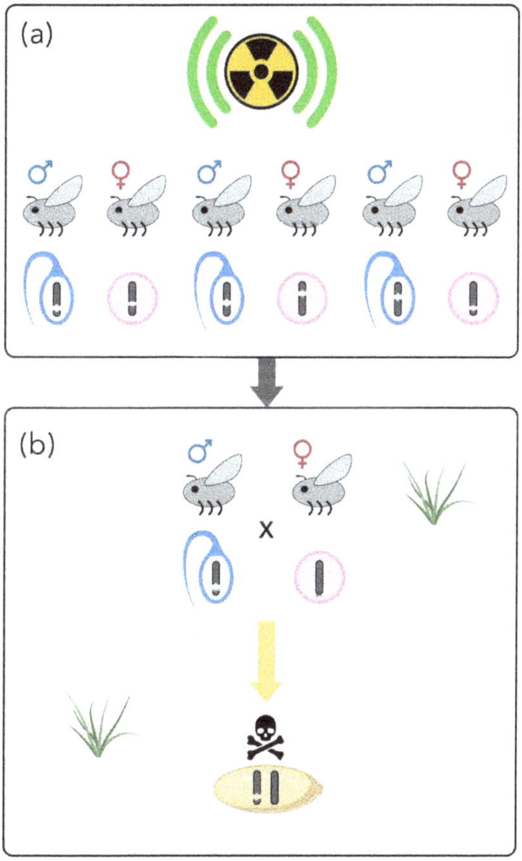

Figure 1 The sterile insect technique (SIT). (a) Insects are produced in a mass-rearing facility and irradiated. Radiation damages the chromosomes in developing gametes, causing dominant lethal mutations. In conventional SIT programs, both males and females are irradiated and released into the field. (b) Population suppression with SIT. Released males mate with wild females. Offspring inherit damaged chromosomes from their irradiated fathers, leading to failed mitosis and death, typically early in development. By repeatedly releasing irradiated males, the reproductive output of wild females is lowered and the wild population can be suppressed or, in some cases, eliminated.

requires optimization of diet, environmental conditions, and cage design to minimize costs and maximize fecundity (Parker et al., 2021). Sterilization is effected through exposure to ionizing radiation. The developmental stage at which insects are irradiated varies with species, but generally the late pupal stage is used. Pupae are ideal because they are easy to transport and cannot fly. Late-stage pupae have more developed germ tissues and are able to withstand higher doses of radiation. The pupae are exposed to gamma radiation from radioactive isotopes, commonly cesium-137 and cobalt-60 (Bakri et al., 2021). Insect pupae are placed into canisters, which are lowered into a large room or smaller container shielded with lead or water. Several source pencils (rods of radioactive isotope) are arrayed within the chamber to maximize uniformity of exposure. When the insect canisters are in position, the chamber is moved from a shielded position to an irradiation position, exposing the pupae to the gamma rays for a set time. Gamma radiation doses are experimentally tuned to cause chromosomal breaks but not to kill the insect. Enough DNA must be damaged so that some double-stranded breaks occur in the germline, but not enough to be lethal. This method of sterilization generally results in insects that still produce gametes, but their offspring fail to develop due to chromosomal abnormalities (aneuploidy). There is a trade-off between sterility and competitiveness with increasing radiation dose. The sterilized insects are released area-wide either from ground release stations or from planes fitted with specialized equipment that allows controlled release rates per linear mile flown. It is the area-wide approach that distinguishes SIT from farm-by-farm approaches such as insecticide sprays (Dowell et al., 2021).

The SIT was developed in the 1950s by Knipling, Bushland, and colleagues at the USDA to suppress the New World screwworm, *Cochliomyia hominivorax*, a pest of livestock that had ravaged the Southern United States for decades (Alexander, 2006; Klassen et al., 2021). As the screwworm SIT program has been reviewed elsewhere (Gutierrez et al., 2019; Mastrangelo, 2012; Scott et al., 2017; Vargas-Terán et al., 2021), we will only briefly summarize the important findings in this chapter. *C. hominivorax* is a species of blow fly (Calliphoridae) native to the Americas. The screwworm is an obligate parasite of mammals. Females lay eggs in wounds or moist areas of the host, often around the face. Larvae burrow into the flesh, feeding and growing for several days before dropping from the host to pupate into adult blow flies. They prefer to feed on large grazers such as cattle, sheep, and deer, but they have been known to infest humans (thus the scientific name 'hominivorax' or man-eater). The large, open wounds left by screwworm larvae can lead to infections and even the death of the host. The majority of animal infestations are not life-threatening, but weight loss and hide damage cause economic losses (Alexander, 2006).

The screwworm SIT eradication program developed methods for mass rearing, irradiation, and release of sterilized insects. Beginning in Florida in

1958, the campaign was expanded to other states, with eradication achieved in the United States by 1966. Subsequently, the SIT was used to eradicate screwworm from Mexico by 1991, and steadily down through Central America until Panama was declared screwworm-free in 2006 (Klassen et al., 2021). The total cost of this endeavor is estimated to surpass USD$1 billion, but savings to farmers are estimated to surpass that amount every year (Vargas-Terán et al., 2021). A barrier zone has been active on the Panama–Colombia border since 2004. A mass-rearing facility in Panama supplies sterilized insects that are continually released along the border to prevent the re-introduction of screwworm from South America (USDA, 2019).

The success of SIT for screwworm suppression paved the way for SIT-based control of many other agricultural and human pests, including pink bollworm (*Pectinophora gossypiella*), codling moth (*Cydia pomonella*), painted apple moth (*Teia anartoides*), false codling moth (*Thaumatotibia leucotreta*), cactus moth (*Cactoblastis cactorum*), onion maggot (*Delia antiqua*), Mediterranean fruit fly (*Ceratitis capitata*) and other Tephritid fruit flies, and tsetse flies (*Glossina* spp.) (Klassen et al., 2021). The SIT has been used to successfully manage populations of these pest species, prevent their spread into new territories, and even eliminate them from an area, as in the case of pink bollworm in the United States. Pink bollworm was fully eradicated in the United States by 2018, thanks to a combination approach involving SIT, mating-disrupting sex pheromones, and transgenic Bt cotton (Tabashnik et al., 2021; USDA, 2018). The SIT was the first system to deliver on genetic biocontrol's promise of effective suppression without insecticides. By relying only on the sterilized pest organism, the technique poses no threat to human health or to non-target species. However, the high release ratios and large production burden make SIT infeasible for controlling certain pests. Nevertheless, SIT was being used for insect pest management on six continents as of 2021 (Bakri et al., 2021).

One innovation that significantly increased the viability of SIT as a cost-effective strategy was the development of genetic sexing strains (GSS) that facilitate the mass separation of the sexes such that only sterile males are released. A GSS carries a pupal color and/or a temperature-sensitive lethal mutation and a translocation of the wild-type gene(s) to the Y chromosome. The sexes can be sorted en masse either with a pupal color sorting machine or by rearing embryos at an elevated temperature. Male-only releases are advantageous as sterile females may impede SIT suppression by competing with fertile females for sterile male mates. Indeed, field tests with the Mediterranean fruit fly, *C. capitata*, found that male-only releases are three to five times more effective than bisexual releases (Rendón et al., 2004). Although developing sexing strains represents a large initial investment, a functional GSS can drastically reduce labor costs and increase the efficiency of SIT. This approach has been quite successful for suppressing *C. capitata* in Croatia, Spain, Morocco, Israel, and South Africa (Klassen et al., 2021).

3 Sterile release systems: precision-guided sterile insect technique

The advent of CRISPR/Cas9-based gene editing allowed researchers to develop alternative forms of SIT that use transgenes, not radiation, to sterilize males before they are released. CRISPR/Cas9, which was originally derived from the immune systems of bacteria and archaea, has been adopted as a new biotechnological tool to precisely edit or insert any DNA fragment in the genome (Doudna and Charpentier, 2014; Jinek et al., 2012). The widely used genetic scissor, SpCas9, derived from *Streptococcus pyogenes*, is an RNA-guided endonuclease that induces DNA double-strand breaks (DSBs) at the targeted site (Sternberg et al., 2014). A single-guide RNA (sgRNA) precisely matches a specific sequence in the insect genome and targets the Cas9 endonuclease to make a cut at that site. The DSB can then be repaired through homologous recombination with a provided template, resulting in a specific sequence edit. Alternatively, if the goal is to disrupt the function of a gene, no repair template is provided.

Precision-guided SIT, or pgSIT, uses CRISPR to simultaneously sterilize males and kill or disable females (Kandul et al., 2019) (Fig. 2). To do this, two transgenic lines are created and maintained separately in the laboratory. One line is homozygous for a Cas9 construct. The other line is homozygous for sgRNA genes driven by a U6 gene promoter. The sgRNAs target Cas9 to cleave genes required for male fertility and female development/survival. To obtain sterilized males for a field release, the two transgenic lines are crossed, producing offspring that inherit both CRISPR elements (Cas9 and sgRNA). CRISPR activity leads to loss-of-function mutations, such that females die or are disabled, while males survive but are rendered sterile. The sterilized males can then be released into the field to suppress the wild population, similar to conventional SIT.

pgSIT has one potentially major advantage over conventional SIT: with pgSIT, it may be possible to release insects as eggs. Conventional SIT programs generally sterilize holometabolous insects at the pupal stage (Bakri et al., 2021). Consequently, these insects can only be released as pupae or adults. However, by the time an insect has reached the pupal stage, it will have consumed many times its own body weight in food in the mass-rearing facility. Providing enough food for the (typically) millions of larvae reared for SIT releases can be expensive, and dealing with the waste these larvae generate can pose health risks and logistical challenges (Parker et al., 2021). Eggs consume no food, generate no waste, take up less space than pupae, and can be produced in vast quantities. If

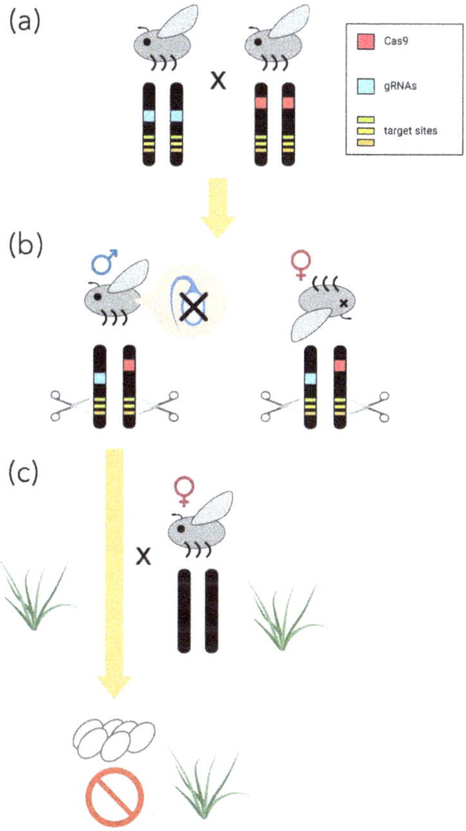

Figure 2 Precision-guided SIT (pgSIT). (a) Two transgenic lines are established and crossed to each other. One line is homozygous for a gene expressing Cas9. The other line is homozygous for a construct expressing sgRNAs that target genes required for male fertility and female viability or development. (b) The offspring inherit one copy of the Cas9 construct and one copy of the sgRNA construct. Clustered Regularly Interspaced Short Palindromic Repeats/CRISPR-associated protein 9 (CRISPR–Cas9) is active in the germline and somatic cells of these offspring. Male offspring are rendered sterile due to CRISPR-induced mutations in genes essential for fertility; in this example, males do not produce viable sperm. Female offspring die due to CRISPR-induced mutations in genes essential for female survival. (c) The surviving, but sterile, male offspring encounter and mate with wild females. No viable offspring are produced, leading to population decline.

pgSIT eggs hatch in the field, develop to adulthood, disperse, and compete for mates successfully – all assumptions that remain to be tested – then pgSIT may be more cost-effective than conventional SIT. pgSIT could therefore potentially be used to control a broader range of pests than conventional SIT, particularly insects with specialized larval diets. The other advantage of pgSIT is that it avoids the use of radiation, which can lower male fitness and hinder the success of SIT (Bakri et al., 2021; Crystal, 1979). However, the transgenes used for pgSIT can impose fitness costs of their own (Kandul et al., 2022), so it will be important for individual pgSIT lines to be thoroughly tested for fitness parameters before any field releases are conducted.

pgSIT is a fairly recent technology, having first been described in 2019 in the model organism *Drosophila melanogaster* (Kandul et al., 2019). The authors of this initial study used CRISPR to target a suite of genes involved in sex determination and spermatogenesis. For example, one pgSIT strain targeting *sex-lethal* and β*Tubulin 85D* killed all females and sterilized all males tested. To date, pgSIT systems have only been developed in two other species: the crop pest *Drosophila suzukii* (Kandul et al., 2022) and the disease vector *Aedes aegpyti* (Li et al., 2021b; Li et al., 2024). In all three species where pgSIT has been tested, very high rates of male sterilization and female lethality/disability have been reported; furthermore, small-scale population suppression trials have shown that pgSIT releases can successfully control or eliminate caged pest populations. While these laboratory experiments are encouraging, pgSIT remains to be tested in the field. A major challenge with implementing a pgSIT program is that sex-sorting systems will be needed to obtain virgin females and males for the cross between Cas9 and sgRNA strains (Bouyer and Vreysen, 2019). Sex-specific expression of fluorescent protein genes could provide a means for mass separation of sexes (Catteruccia et al., 2005; Liu et al., 2024; Weng, et al., 2023). The crossing requirement for pgSIT could also be advantageous, as it would minimize any fitness costs due to the inbreeding that occurs when breeding a transgenic line to homozygosity.

4 Sterile release systems: release of insects carrying a dominant lethal/female-specific release of insects carrying a dominant lethal techniques

As noted earlier, early practitioners of SIT knew that radiation could reduce male fitness (Bakri et al., 2021; Crystal, 1979). The production of fit, competitive males is essential for successful population suppression with SIT. Researchers, therefore, began to explore alternatives to conventional SIT that would use genetic factors, not radiation, to sterilize or kill pests. In the 1960s–1980s, various genetic control systems were theorized (recessive lethal strains in the boll weevil, *Anthonomus grandis*) (LaChance and Knipling, 1962) or implemented (field female-killing strains in the Australian sheep blowfly, *Lucilia cuprina*) (Foster et al., 1985). Modeling indicated that these systems could be as effective as conventional SIT, in some cases more effective, while avoiding the drawbacks of radiation (Foster et al., 1988). Indeed, the field female-killing system in *L. cuprina*, which used chromosomal translocations to render males semi-sterile and to introgress recessive female-lethal genes into the field population, was highly successful in early trials (Foster et al., 1993). However, a combination of economic factors and the breakdown of the translocation strain led to the eventual abandonment of this program (Scott, 2014).

By the turn of the century, advances in insect transgenesis and gene expression technology allowed researchers to design more stable, translocation-free genetic control systems. In 2000, two research groups independently reported conditional female-killing strains in *D. melanogaster* (Heinrich and Scott, 2000; Thomas et al., 2000). These strains were dubbed RIDL (release of insects carrying a dominant lethal) (Thomas et al., 2000), although they were technically fsRIDL strains (female-specific RIDL), a term which came into use later. For the sake of clarity, we refer to systems that kill both sexes as RIDL and to systems that kill only females as fsRIDL (Figs 3 and 4). The 'RIDL' (fsRIDL) strains took advantage of a tetracycline-dependent conditional gene expression system (Gossen and Bujard, 1992), originally described in mammalian cells but quickly adapted for use in insects. This system uses the tetracycline-controlled transactivator protein (tTA) to induce expression of a gene of interest, which is positioned downstream of tetO (tetracycline operator) sequences. The early RIDL strains used female-specific *yolk protein* gene promoters to drive the expression of tTA and placed a cytotoxic gene downstream of the tetO sequences. Because tTA was expressed only in females, the cytotoxic gene was also expressed only in females, leading to the complete elimination of female progeny in lines carrying both the tTA 'driver' construct and the tetO 'effector' construct (Heinrich and Scott, 2000; Thomas et al., 2000). Critically, these female-killing systems could be turned off by adding tetracycline to the diet, allowing researchers to rear both male and female carriers indefinitely in the laboratory on tetracycline. To control insect pest populations with RIDL (or fsRIDL), homozygous RIDL/fsRIDL insects would be released en masse into the field, where they would mate with wild insects and produce heterozygous RIDL/fsRIDL offspring. These offspring, not encountering tetracycline in the field, would die due to the dominant lethal effects of the transgenes (Figs. 3 and 4). Both systems are predicted to be as powerful (RIDL) or more powerful (fsRIDL) than conventional SIT (Schliekelman and Gould, 2000; Vella et al., 2021).

RIDL and fsRIDL offer several benefits over conventional radiation-based SIT. They avoid all regulatory and safety issues associated with radiation. In addition, the genetic components of a RIDL or fsRIDL strain developed in one species can often be ported successfully to other species (Ant et al., 2012; Fu et al., 2007; Jin et al., 2013). The ability to reuse genetic components can save researchers a great deal of time and effort when designing RIDL/fsRIDL strains, particularly in non-model species. RIDL/fsRIDL strains can also be easily designed to carry a fluorescent marker gene. These markers allow transgenic, released insects (and their marker-inheriting offspring) to be tracked in the field, providing valuable data on dispersal and mating success (Harris et al., 2011).

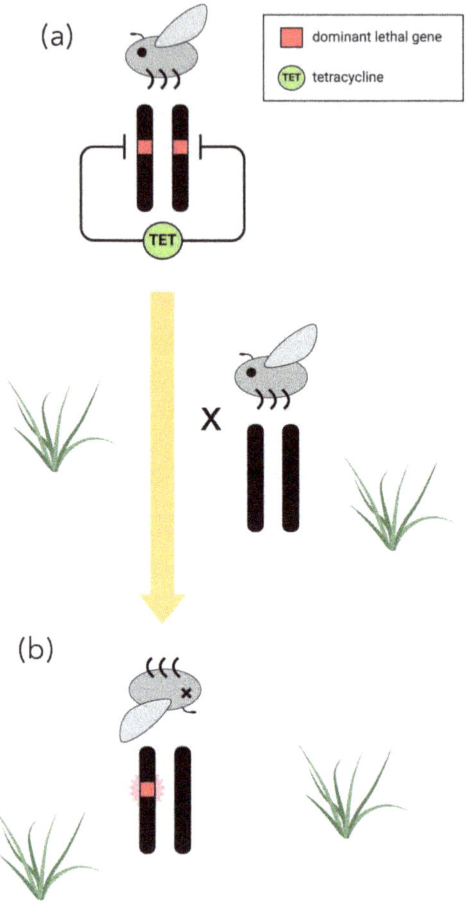

Figure 3 Release of insects carrying a dominant lethal (RIDL). (a) Transgenic insects homozygous for a dominant lethal gene are reared on tetracycline, which prevents the lethal gene from being expressed. To control the wild insect population, the transgenic insects are released into the field, where they mate with their wild counterparts. (b) The offspring inherit one copy of the dominant lethal gene. Because these offspring are not exposed to tetracycline in the field, the lethal gene is active, and the offspring die. RIDL insects need to be repeatedly released into the field to suppress or eliminate pest populations.

Despite these advantages, RIDL and fsRIDL share many of the same limitations as conventional SIT and have some unique drawbacks. Like SIT, RIDL/fsRIDL are self-limiting forms of population control that require repeated, mass releases of insects. This necessitates a large and ongoing investment in insect production infrastructure, which may only be possible in developed countries (or in partnership with developed countries). RIDL/fsRIDL insects are not irradiated and thus avoid any radiation-associated fitness costs, but the genes used for

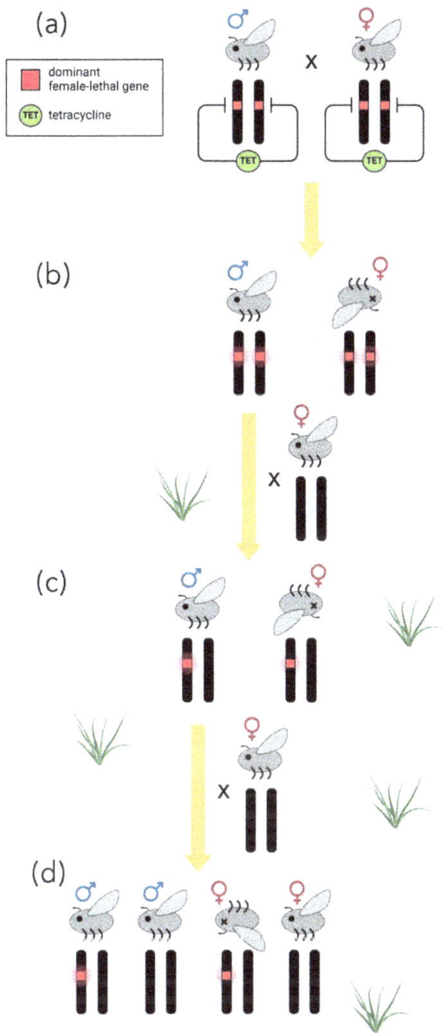

Figure 4 Female-specific release of insects carrying a dominant lethal (fsRIDL). (a) Insects homozygous for a dominant, tetracycline-repressible, female-lethal gene are bred in a mass-rearing facility on a diet containing tetracycline. (b) For the release generation, the fsRIDL strains are supplied with a diet without tetracycline. For female embryo lethal strains, tetracycline is also omitted from the maternal diet. This causes the females to die, leaving behind only males. The males are released into the field, where they mate with wild females. (c) The offspring of released fsRIDL males and wild females inherit one copy of the dominant female-lethal gene. Consequently, only males survive. These heterozygous fsRIDL males go on to mate with wild females. (d) The female-lethal gene introgresses into the wild population in subsequent generations, killing any females that inherit it. This introgression effect makes fsRIDL a more potent form of population suppression than conventional SIT or RIDL, particularly if more than one chromosome carries an fsRIDL transgene.

RIDL/fsRIDL may impose fitness costs of their own (Bargielowski et al., 2011; Irvin et al., 2004). Furthermore, fsRIDL and RIDL involve the release of transgenic insects, whereas SIT is not a transgenic technology. Some communities that accept the release of radiation-sterilized SIT insects may reject RIDL or fsRIDL programs, and the path to regulatory approval can be convoluted. A pertinent example is Oxitec's journey to obtain approval to conduct field trials with *A. aegypti* mosquitoes in the United States. Oxitec, a biotechnology company founded in 2002, has developed RIDL and fsRIDL strains in many medically and agriculturally important insect pests. Oxitec initially sought approval from the USDA in 2010 to conduct field trials with its RIDL *A. aegypti* mosquitoes (Waltz, 2015). It was not until 2021, however, that the first Oxitec mosquitoes were actually released in the United States. In the interim, Oxitec's application changed hands between three different federal agencies (Schairer et al., 2021). While this is an extreme example, it illustrates how transgenic technologies like RIDL and fsRIDL might not be able to be rapidly deployed against emerging insect threats to health or agriculture.

Another potential drawback of RIDL/fsRIDL is that these systems typically use tetracycline, a potent antibiotic, to prevent the expression of a lethal gene in the mass-rearing facility. Many insect pests are sensitive to antibiotics such as tetracycline and have been shown to suffer consequences ranging from slightly reduced body weight to complete lethality or sterility when reared on antibiotics (Costa et al., 1997; Kuriwada et al., 2010; Pais et al., 2008; Pietri and Liang, 2020; Schlein, 1977). A RIDL/fsRIDL program may nonetheless be able to mass-rear these sensitive species by optimizing the dosage and timing of tetracycline administration, although this would have to be investigated on a species-by-species basis.

Lastly, RIDL, like SIT, lacks an inherent sex-separating mechanism, limiting both its effectiveness and the number of species in which it can be used. When both males and females are released in the field, the released males may mate with the released females instead of wild females, which can dramatically impair population suppression (McInnis et al., 1994; Rendón et al., 2004). Furthermore, the females of some insect species, such as mosquitoes and *D. suzukii*, transmit diseases or damage crops, making bisexual releases problematic. In these species, male-only releases are the only option. fsRIDL was developed in conjunction with RIDL to solve this problem.

In fsRIDL, the dominant lethal gene is female-specific. That is, female carriers of the lethal gene die, while male carriers survive and pass on the gene to their offspring. This property grants fsRIDL several advantages over both RIDL and conventional SIT. An fsRIDL strain acts as a genetic sexing strain in a mass-rearing facility, allowing males to be easily separated from females. When some controlling factor (typically tetracycline) is removed from the diet of the insects in the mass-rearing facility, the females die, leaving behind only males, which are then released into the field. If the females die early enough in development, they will not consume food or take up space in the mass-rearing facility, freeing up these resources for additional male production. The male-only releases that can be achieved with fsRIDL are more efficient at suppressing pest populations than bisexual releases with RIDL or conventional SIT. Another key advantage of fsRIDL is that the female-lethal gene can introgress into the wild population through male carriers. This multigenerational female-killing effect makes fsRIDL a powerful system for population suppression while still being controllable and confinable. Modeling shows that fsRIDL strains with multiple female-killing alleles (on separate chromosomes) can be much more efficient than SIT at population suppression (Schliekelman and Gould, 2000), assuming the transgenes do not impose too high a fitness cost.

Over the past 25 years, RIDL and fsRIDL strains have been developed in at least a dozen insect species, including several major agricultural pests. Shortly after the initial RIDL/fsRIDL systems were described in *D. melanogaster*, scientists developed the first RIDL (Gong et al., 2005) and fsRIDL (Fu et al., 2007) strains in a crop pest, the medfly *C. capitata*. Other agriculturally important pests in which RIDL or fsRIDL strains have been developed include the pink bollworm (*P. gossypiella*) (Morrison et al., 2012), diamondback moth (*Plutella xylostella*) (Shelton et al., 2020), fall armyworm (*Spodoptera frugiperda*) (Reavey et al., 2022), olive fly (*Bactrocera oleae*) (Ant et al., 2012), the Carribean fruit fly (*Anastrepha suspensa*) (Schetelig and Handler, 2012), the Mexican fruit fly (*Anastrepha ludens*) (Schetelig et al., 2016), *L. cuprina* (Yan and Scott, 2020), *C. hominivorax* (Concha et al., 2016), and spotted-wing Drosophila (*D. suzukii*) (Li et al., 2021a). The strains developed in these species all use the tTA/tetO system to conditionally express genes that kill both sexes (RIDL) or females only (fsRIDL). Many of these strains have shown great promise in laboratory studies, which often report complete bisexual or female lethality (for RIDL and fsRIDL, respectively) with no loss of male fitness. Some of the aforementioned fsRIDL strains were created as genetic sexing strains to augment a conventional SIT program; these strains would kill females early in development, leaving behind males, which would then be irradiated and released in the field. Other fsRIDL (and RIDL) strains were designed to work independently, achieving population suppression without the use of radiation. However, field performance data is still lacking for these strains in most agricultural pests. Field releases of an fsRIDL strain of the diamondback moth (*P. xylostella*) were initiated in 2017, but these releases were intended to evaluate moth dispersal and survival, not population suppression (Shelton et al., 2020). Encouragingly, field releases of mosquitoes have shown that RIDL/fsRIDL strains can efficiently suppress wild populations (Carvalho et al., 2015; Spinner et al., 2022), and semi-field (greenhouse or large cage) trials have shown similar results in agricultural pests (Asadi et al., 2020; Harvey-Samuel et al., 2015; Leftwich et al., 2014). We therefore anticipate that

RIDL and fsRIDL strains will continue to be developed and eventually deployed to control both medically and agriculturally important insects.

5 Fertile-release systems: X-poisoning, X-shredder, Y-linked editor, and homing gene drive approaches

X-poisoning systems use CRISPR to mutate X-linked, haploinsufficient genes during spermatogenesis (Fig. 5). These X-linked genes, rendered non-functional due to the CRISPR-induced mutations, are passed on to female offspring via the father's X chromosome. These female offspring die, typically early in development, as with only one functional copy of the X-linked target gene the females cannot produce enough gene product to survive. In contrast, male offspring, which inherit a single functional copy of the X-linked target gene from their mother, survive because the expression of X-linked genes is naturally upregulated in the male sex by the X chromosome dosage compensation system.

Researchers have understood for decades that male-biasing systems like X-poisoning could be used to control insect pests (Hamilton, 1967). However, artificial X-poisoning systems have only been developed within the last 5 years. Two such systems have been created in the vinegar fly *D. melanogaster* (Fasulo et al., 2020; Lawler et al., 2024) and one in the malaria mosquito *Anopheles gambiae* (Haber et al., 2024). These X-poisoning systems used CRISPR/Cas9 to target haploinsufficient X-linked ribosomal protein genes (Fasulo et al., 2020; Haber et al., 2024) or the haploinsufficient X-linked wing muscle development gene *wupA* (Lawler et al., 2024). Cas9 was expressed in the male germline and located on an autosome, as were the sgRNA genes driven by a *U6* gene promoter. These proof-of-concept experiments were successful, with certain transgenic lines demonstrating significant (67–92%) biases toward male offspring. As of the time of writing, however, no X-poisoning systems have been developed in any agricultural pest species.

Like other genetic pest control strategies, one potential advantage of X-poisoning is that the genetic components developed in model organisms may translate more or less seamlessly to non-model pest species. For example, the X-linked haplolethal gene *wupA* shares >98% amino acid similarity between *D. melanogaster* and *D. suzukii* (Lawler et al., 2024), so *wupA* might be an excellent target for *D. suzukii* suppression using an X-poisoning system. However, the dynamics of X-poisoning may differ across species, even when the genetic components of the system are similar. This was recently demonstrated by Haber et al. (2024), who explicitly compared X-poisoning dynamics in *D. melanogaster* and *A. gambiae*. In *Anopheles*, unlike in *D. melanogaster*, the X-poisoning system behaved similarly to an X-shredding system (see Section 5.1 X-shredder), causing male mosquitoes to produce an overabundance of Y-carrying sperm.

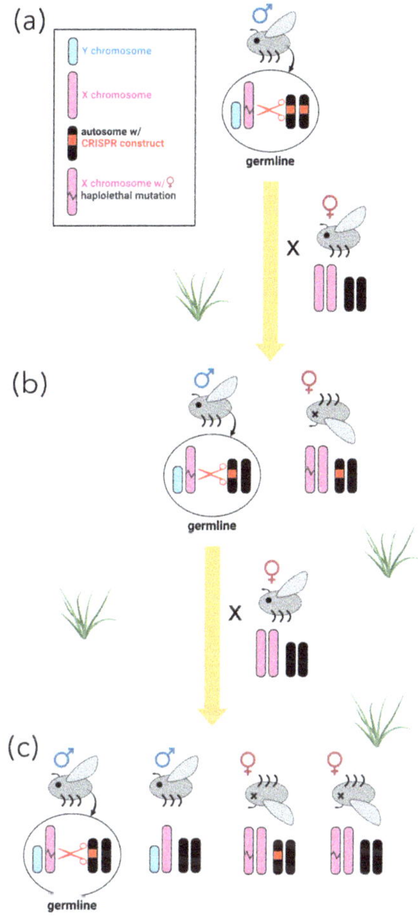

Figure 5 Autosomal X-poisoning. (a) Transgenic males carry autosomal CRISPR constructs that target an X-linked, haplolethal gene during spermatogenesis. These males are released into the field, where they mate with wild females. (b) The female offspring inherit a mutated X chromosome from their father, leading to their deaths due to loss of function of the haploinsufficient genes. The male offspring inherit a single copy of the CRISPR construct from their father. CRISPR is active during spermatogenesis in these males, causing them to produce sperm containing a mutated X chromosome. These males go on to mate with wild females. (c) The next generation of female offspring die due to inheriting a mutated X chromosome from their father. The autosomal CRISPR construct introgresses into the wild population through males in subsequent generations.

Importantly from the perspective of pest control, X-shredders have different population suppression dynamics than X-poisoners: X-shredders have gene drive activity when the CRISPR components are Y-linked, whereas X-poisoners do not (Burt and Deredec, 2018). This fact highlights the need to carefully evaluate X-shredding systems, especially those employing a Y-linked CRISPR construct.

5.1 X-shredder

The X-shredder is a system that, like X-poisoning, reduces the ratio of females to males in an individual's offspring (Galizi et al., 2014). The system was first developed in *A. gambiae* mosquitoes based on observations of natural meiotic drivers in insects (Wood and Newton, 1991). With these naturally occurring drivers, mosquito populations would experience sex ratio distortion in favor of males (Hickey and Craig, 1966). With more males than females in a population, population size is reduced in the next generation, and the application of an X-shredder over time could reduce pest populations effectively.

In the first synthetic X-shredder, the gene for a homing endonuclease, I-PpoI, was inserted into the genome of *A. gambiae*. The endonuclease 29 bp recognition sequence was only found within the repetitive ribosomal gene sequences on the X chromosome (Windbichler et al., 2008). The endonuclease gene was expressed during spermatogenesis from a beta-2-tubulin gene promoter active in testes. As a consequence, the X chromosome would be cut at each repeat of the recognition sequence, shredding it into several pieces (Windbichler et al., 2008). Because expression is confined to spermatogenesis only, the X chromosome is preserved in the male's cells during development and in all tissues except the developing sperm. When a male expresses the endonuclease in this way, the majority of his offspring receive the Y chromosome instead of the X, and males begin to outnumber females (Windbichler et al., 2008) (Fig. 6). The I-PpoI X-shredder was efficient enough to work in *A. gambiae*; however, its success depends on the presence of repetitive I-PpoI recognition sequences on the X chromosome only. In order to improve the X-shredder's efficacy in other species, the I-PpoI nuclease was exchanged for a Cas9 nuclease with sgRNAs chosen to target a different repetitive sequence found on the X chromosome. X-shredding using Cas9 was as effective as with I-PpoI, which indicated that it could be applied in other species that did not have the same repetitive ribosomal sequences on the X chromosome (Galizi et al., 2016). Cas9 can be targeted to many different sequences, and even multiple sequences at once. As long as the X chromosome of the target species has some repetitive sequence that is not found on other chromosomes, it can be shredded without reducing the fitness of the males to be released. Since their introduction in *A. gambiae*, X-shredder systems have been developed in other species, including *C. capitata* (Meccariello et al., 2021) and *D. melanogaster* (Fasulo et al., 2020), suggesting X-shredder systems could be developed for other crop pests.

Variations on the X-shredder system can impact the genetic load imposed on its carriers and its ability to spread and persist through a population. An autosomal X-shredder could be selected against in the population if there is expression in somatic cells that produces deleterious effects on the males that carry it. If an X-shredder is moved to the Y chromosome, either through

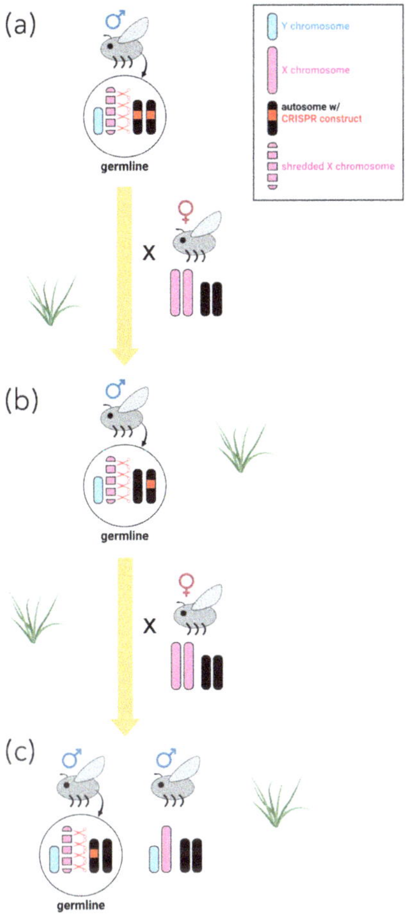

Figure 6 Autosomal X-shredder. (a) Transgenic males carry autosomal CRISPR constructs that target multiple, repetitive sequences on the X chromosome during spermatogenesis, 'shredding' the X chromosome. Sperm carrying shredded X chromosomes fertilize fewer (or no) eggs compared to Y-bearing sperm. The transgenic males are released into the field to mate with wild females. (b) Only (or predominantly) male offspring are produced, all of which carry one copy of the X-shredding CRISPR construct. CRISPR is active during spermatogenesis in these male offspring, causing them to produce sperm with shredded X chromosomes. (c) In the next generation, only (or predominantly) male offspring are produced, half of which carry one copy of the X-shredding CRISPR construct. The CRISPR construct continues to introgress into the wild population, enhancing population suppression through its male-biasing effect.

spontaneous relocation or targeted engineering, the system could show gene drive characteristics. If fewer than half of the offspring receive the shredded X chromosome, then the Y chromosome can bias its own inheritance, or drive, and become more common than before in the wild population (Alcalay et al.,

2021). A Y-linked X-shredder requires fewer releases to control a population and releases do not need to contain as many insects as in standard SIT (Schliekelman et al., 2005). However, a Y-linked X-shredder will only be successful if certain conditions are met. The Y chromosome must allow for expression during spermatogenesis, and locating the nuclease gene on the Y chromosome should not incur a large fitness cost in males (Alcalay et al., 2021). In order to fulfill these requirements, careful engineering is required, and the target species' genome must be well understood. For some species, in which the Y chromosome is not expressed at all during spermatogenesis, this approach may not be possible (Turner, 2007).

The main challenges when creating an X-shredder in a new pest species are finding shreddable target sites on the X chromosome and ensuring that the expression of the X-shredding nuclease is specifically confined to spermatogenesis. Further constraints are placed on the design when the nuclease is localized to the Y chromosome. However, an X-shredder can be a very effective population suppressor alone or in combination with other technologies (Simoni et al., 2020). Because X-shredders target many sites on the X chromosome for cutting, resistance is unlikely to develop among wild populations (Geci et al., 2022). An autosomal X-shredder alone poses less threat of spreading beyond its intended range. A Y-linked X-shredder is capable of quickly spreading through a target population. Both are capable of reducing the number of females in a wild population and bringing total numbers under control.

5.2 Y-linked editors

The Y-linked editor approach uses nucleases to make loss-of-function mutations in X-linked haploinsufficient genes, dominant-negative mutations in X-linked genes, or dominant mutations in genes necessary for female development and reproduction (Fig. 7) (Burt and Dederec, 2018; Geci et al., 2022). These female-essential genes do not necessarily need to be on the X chromosome; disruption of a female fertility gene on an autosome is just as functional in this system. If edits are targeted to genes necessary for female development and fertility during spermatogenesis, all female offspring of a particular Y-linked editor male will die or be unable to reproduce (Burt and Deredec, 2018). Modeling showed that Y-linked editors were more efficient for population suppression than fsRIDL or X-shredder systems (Burt and Dederec, 2018). Further, the Y-linked editor was self-limiting and less invasive in neighboring populations than a homing gene drive. Although the components for creating a Y-linked editor have been tested in *D. melanogaster* (Fasulo et al., 2020; Gamez et al., 2021), a Y-linked editor strain has not yet been assembled and tested.

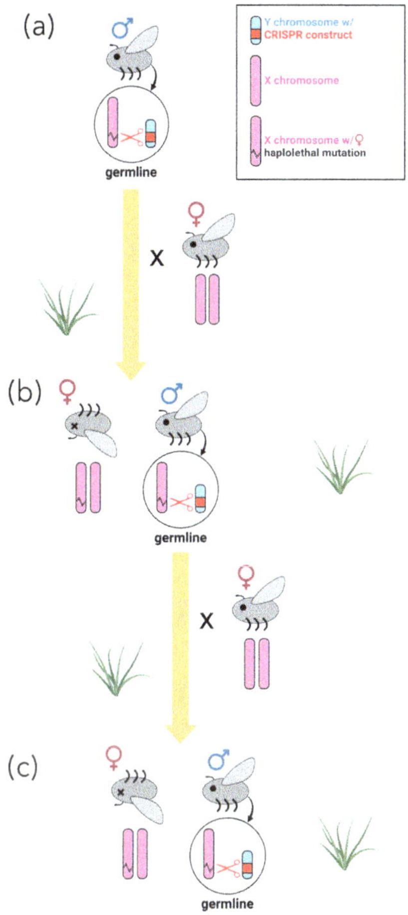

Figure 7 Y-linked editor). (a) Transgenic males carry a Y-linked CRISPR construct that targets a haploin-sufficient gene required for female development on the X chromosome during spermatogenesis. These males are released into the field, where they mate with wild females. (b) All female offspring die because they have inherited a mutated X chromosome from their father. All male offspring survive. The males inherit the Y-linked editor from their father, which targets the essential female gene on the X chromosome inherited from their mother. (c) Subsequent generations of male offspring inherit the Y-linked editor and survive, while female offspring die. By skewing the sex ratio toward males, Y-linked editors are predicted to efficiently suppress pest populations, especially if repeated releases of transgenic males are conducted.

The efficiency of the Y-linked editor can be increased by combining it with another sex distortion mechanism that favors inheritance of the Y chromosome (Willis and Burt, 2021). Effectiveness can also be modulated by changing the time point when females are removed from the population. If female offspring carrying the edit die after a density-dependent phase of the life cycle – a phase when members of the wild population must compete for resources – then a Y-linked editor can more quickly suppress the population (Burt and Deredec, 2018).

A major barrier to the creation of Y-linked editor strains is obtaining transgenic lines with the nuclease located on the Y chromosome in a location where the level of expression during spermatogenesis is sufficient for effective gene editing. The Y chromosome is typically largely heterochromatic with few expressed genes. During routine germline transformation experiments with transposon vectors that integrate relatively randomly, such as *piggyBac*, Y-chromosome lines are rarely obtained. Consequently, it may be necessary to target the nuclease to Y-chromosome locations (Gamez et al., 2021). Expression of the nuclease (e.g. Cas9 and sgRNAs) should not reduce male fitness but be sufficient for a high level of editing in the targeted genes.

5.3 Homing gene drives

Genetically engineered gene drive strains have been shown to effectively suppress insect populations in lab cages (Buchman et al., 2018; Hammond et al., 2021b; Terradas et al., 2021). However, none of the gene drive strains have yet gone to a field trial (Raban et al., 2023). Gene drives exhibit biased inheritance (super-Mendelian, >50%) and can quickly spread across a target population (Sinkins and Gould, 2006). In the context of agricultural insect pest control, gene drives have been developed and evaluated for population suppression through disrupting the genes essential for female fertility and development in *C. capitata* and *D. suzukii* (Meccariello et al., 2024; Yadav et al., 2023). A gene drive can be categorized as a 'high threshold drive' or 'low threshold drive' depending on its spread potential. A low threshold drive, also referred to as an efficient gene drive, can suppress the wild population upon fewer releases of the genetically edited insects (Legros et al., 2021). Hence, it could be a more efficient and cost-effective approach than conventional techniques like SIT and fsRIDL (Scott et al., 2018). A homing gene drive works by copying (homing) the genetic elements onto the homologous (opposite) chromosome in the germ cells (Burt and Crisanti, 2018). Thus, homing converts a germ cell from heterozygous to homozygous for the drive allele. The rate of conversion of a wild-type allele to the drive allele is called 'homing efficiency'. Efficient homing in the germ cells at each generation rapidly raises the drive allele's frequency in a population (Burt and Crisanti, 2018; Noble et al., 2018). In 2003, Austin Burt first proposed the concept of homing gene drive using homing endonuclease genes (HEGs) for insect population control (Burt, 2003). For a population suppression drive, an engineered HEG drive should target a gene essential for insect viability or fertility. A HEG-encoded endonuclease induces DNA DSBs at a predefined site, which undergo homology-directed repair (HDR) to copy the HEG-drive into the target gene. Following Burt's approach, *I-SceI* and *I-OnuI* HEG-based homing gene drives were developed and evaluated in *D. melanogaster* and *A. gambiae* (Chan et al., 2011; Chan et al., 2013b). These studies revealed that homing efficiency is spatio-temporally regulated and that efficient homing can be achieved using an appropriate regulatory element such as a germline promoter and 3′ UTR in the drive constructs (Chan et al., 2013a). Similarly, homing elements were also designed using TALENs (transcription activator-like effector nucleases) and ZFNs (zinc finger nucleases), which showed a higher level of transmission than natural homing endonuclease *I-SceI* (Simoni et al., 2014). However, the development of HEG or TALEN/ZFN-based homing gene drives was found technically infeasible as it is difficult to engineer the nuclease proteins to target pre-selected genes, and the use of repetitive DNA arrays in these systems can result in drive instability (Verkuijl et al., 2022).

The recent advances and technical simplicity of the genome editing tool CRISPR/Cas9 have revolutionized homing gene drive research (Bier, 2022; Raban et al., 2023). Like HEG systems, the CRISPR/Cas9-based homing gene drives (HGDs) also rely on DSB repair through HDR, which results in copying or 'homing' of the drive allele into the targeted gene in the germ cells. In 2015, Gantz and colleagues demonstrated the first CRISPR/Cas9-based homing in *D. melanogaster,* which generated *yellow* mutants with a high homing efficiency (97%) (Gantz and Bier, 2015). However, to date, the homing drive has been tested in very few agricultural pests (Meccariello et al., 2024; Xu et al., 2022; Yadav et al., 2023). The CRISPR/Cas9 homing construct typically consists of a Cas9 gene, sgRNA, and a selectable marker gene inserted between the homologous sequences (~1 kb) flanking the cut site of the chosen target (Hammond et al., 2016). The use of a marker gene such as GFP (green fluorescent protein) or RFP (red fluorescent protein) helps in sorting the drive insects and measuring the homing efficiency. Several additions to the drive construct have been shown to improve homing efficiency, such as limiting Cas9 expression to germ cells using germ-cell-specific promoters and UTRs, and adding two or more NLS (nuclear localization sequences) to enhance Cas9 activity (Hammond et al., 2021a; Shui et al., 2022; Yadav et al., 2023).

A fully autonomous homing gene drive has all the drive components mentioned above linked such that they are all inherited together (Fig. 8). In a split gene drive, the drive components such as Cas9 and sgRNA are present distantly at two separate loci inherited independently, and with only one component, usually the sgRNA, showing biased inheritance (Fig. 9) (Champer et al., 2019). A split homing gene drive is a 'self-limited' drive system which offers an efficient and environment-friendly genetic-based approach to control a target population. In contrast, a fully autonomous gene drive, being a 'low-threshold' and 'self-sustaining' drive, has potential to spread to a non-targeted population but is more cost-effective (Champer et al., 2019; Terradas et al.,

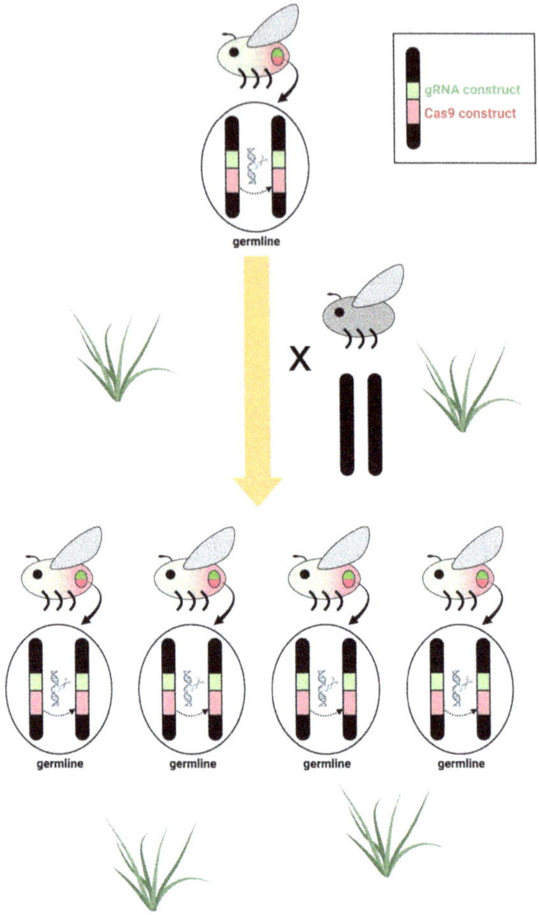

Figure 8 An autonomous CRISPR-based homing gene drive.Transgenic insects are designed to carry a sgRNA construct and Cas9 construct, which are physically linked on the same chromosome. These CRISPR elements are copied onto the homologous chromosome in the germline via the homology-directed repair (HDR) pathway. When the transgenic insects are released into the field and mate with their wild counterparts, all or most of the resulting offspring inherit one copy of the gRNA-Cas9 element, which is again copied over to the homologous chromosome in the germline. The sgRNA-Cas9 element thus biases its own inheritance and 'drives' itself throughout the population. By targeting genes that are essential for female development or fertility, a CRISPR-based gene drive can be used to manage or eliminate crop pests.

2021). Consequently, during development, it is recommended that the lab meets at least arthropod containment level 2 (ACL2) (Benedict et al., 2018). Interestingly, mathematical models show that a fully autonomous gene drive which leads to dominant sterility in females would be less invasive and cannot easily spread to a non-target population from the escapees (Papathanos et al.,

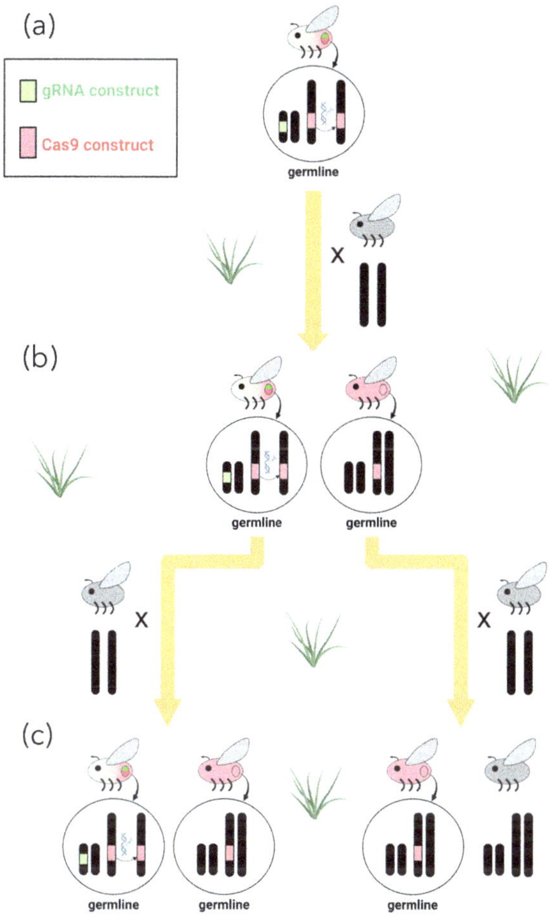

Figure 9 Split CRISPR-based homing gene drive. (a) Transgenic insects carry sgRNA and Cas9 constructs that are 'split' (not copied over together); in this example, the constructs are located on different chromosomes. CRISPR is active in the germline, causing one construct (in this case, the Cas9 construct) to copy itself onto the homologous chromosome during gametogenesis. The transgenic insects are released into the field, where they mate with wild conspecifics. (b) 50% of the offspring inherit a copy of the sgRNA construct, while 100% of the offspring inherit a copy of the Cas9 construct. The offspring that possess both the sgRNA and Cas9 constructs undergo CRISPR-mediated HDR during gametogenesis, copying the Cas9 construct onto the homologous chromosome. (c) The frequency of the sgRNA construct is diluted with each new generation, as this construct has no means of copying itself onto the homologous chromosome. As a result, split gene drives are less invasive and more confinable than autonomous gene drives.

2009; Yadav et al., 2023). The presence of drive resistance alleles can stop the spread of a gene drive across a population. In addition to naturally occurring allelic variants, resistance alleles can also be generated by the mutagenic Non-Homologous End Joining (NHEJ) pathway if used to repair the DSB at the target site. NHEJ repair modifies the original target site sequences, preventing further recognition and cleavage by the sgRNA/Cas9 complex. The resistance alleles can be either functional (r1) due to in-frame deletion/insertion mutations (still encoding a functional protein) or non-functional (r2) alleles as result of a frameshift mutation or a larger indel. The formation of 'r2' alleles is less likely to prevent drive allele propagation, while an 'r1' allele could be problematic for drive allele spread. If an 'r1' allele has low or no fitness cost, it can remain prevalent in the population over time and can inhibit the drive spread effectively (Champer et al., 2017). Indeed, it was the accumulation of 'r1' that impeded homing drive propagation in a caged population of *A. gambiae* (Hammond et al., 2017). Careful selection of a highly conserved target site or the use of multiplexed sgRNAs can minimize the chances of functional resistance allele formation (Kyrou et al., 2018; Yang et al., 2022).

In recent years, significant efforts have been made to develop homing gene drives for agricultural insect pests. The first split homing drive in an agricultural pest targeting the *doublesex* gene has been developed for *D. suzukii* population control (Yadav et al., 2023). Subsequently, homing gene drive targeting the *transformer* gene in *C. capitata* was established and evaluated for population suppression (Meccariello et al., 2024).

6 Case study: split homing gene drive in an agricultural pest (*Drosophila suzukii*)

Genetic engineering of insect pests remains difficult, although advances in gene editing tools (e.g. CRISPR/Cas9) and genome sequencing methods (e.g. high-throughput NGS) have made it easier (Gantz & Akbari, 2018). It becomes important to have full genetic sequences of an organism for targeted genetic manipulation. The genome of the crop pest *D. suzukii* was first published independently by two groups (Chiu et al., 2013; Ometto et al., 2013) in 2013, and a high-quality near-chromosome level genome assembly was made available in 2020 (Paris et al., 2020). The availability of a high-quality reference genome assembly facilitated the design of gene editing experiments in pest species. The first use of CRISPR/Cas9 in *D. suzukii* was to edit the *white* (*w*) and *sex-lethal* (*Sxl*) genes, paving the way for the development of CRISPR-based genetic control systems (Li and Scott, 2016). In this direction, CRISPR/Cas9-based split homing gene drive strains have been established for the genetic suppression of this pest (Yadav et al., 2023).

The sex determination genes *Sxl*, *transformer* (*tra*), and *doublesex* (*dsx*) are essential for female viability and development and,consequently, are potential targets for the development of suppression gene drives (Deredec et al., 2008). The master gene *Sxl* is X-linked, so drive would be limited to females. However, as *Sxl* is required in the female germline, it is not a suitable target for gene drive (Scott et al., 2017). The *tra* gene is poorly conserved in *Drosophila* (O'Neil and Belote, 1992) and consequently r1 alleles easily arise and inhibit drive (Carrami et al., 2018). A multi-sgRNA strategy could be used to prevent resistance alleles from accumulating, but *tra* would not appear to be an ideal target for a gene drive in *D. suzukii*. The *dsx* gene is transcribed in both sexes but undergoes alternative splicing to produce male- and female-specific transcripts. The established split homing drive disrupts the female-specific exon of the *dsx*, resulting in female sterility (Yadav et al., 2023). The homing cassette consisted of a DmU6:3 promoter-driven *dsx*-sgRNA and a ubiquitously expressed DsRed marker gene flanked on each side by ~1 kb of sequence matching the *dsx* gene (homology arms). The construct was designed to knock in at 52 bp downstream from the beginning of the *dsx* female exon. The *dsx* female exon is well conserved across *D. suzukii* populations. The homing cassette, along with a *dsx*-sgRNA/Cas9 complex, was microinjected into *D. suzukii* early embryos to obtain homing strains. From this construct, four independent homing strains named 3A3, 15A2, 30B1, and 1C2 were established. However, the homing cassette inserted into the female exon (CDS) resulted in dominant female sterility in most of the established lines, except one (1C2). Since the males were fertile, the transgenic strains are being maintained by crossing them back to wild-type virgin females each generation. In the dominant sterile lines, heterozygous females produce both the male-specific dsx^M and female-specific dsx^F transcripts. The encoded male-specific DSX^M protein could interfere with the DSX^F isoform from the wild-type allele and cause dominant female sterility. Interestingly, in the fertile strain (1C2), rather than produce the male transcript, an unusual splice variant was found with a repetition of exon 2. The encoded protein was likely not functional. Consequently, the 1C2 strain shows recessive female sterility when homozygous for the drive allele. It is not known why the rare splicing event occurs in the 1C2 strain, but several intronic SNPs unique to this strain were found.

To rescue dominant female sterility, the original homing cassette was redesigned using an optimal splice acceptor site to prevent the use of the downstream male-exon splice acceptor (SA) site in the females (Yadav et al., 2023). The revised homing cassette is the same as the first cassette but contains an optimal SA site from a muscle myosin heavy chain gene, an in-frame HA epitope tag, and a C-degron sequence. The purpose of using a C-degron sequence is to promote proteolysis of the encoded DSX-HA protein to minimize its interference with wild type DSX protein. The modified homing cassette was integrated into the same location in the female *dsx* exon, and four independent strains ($23AI1^{SA}$, $6CT1^{SA}$, $42BI1^{SA}$, and $49BI1^{SA}$) were established. All the new homing strains obtained show recessive female sterility when homozygous for the drive allele. An analysis of RNAs produced in these strains confirmed the use of the SA site, thus preventing the production of *dsx*^{M} transcripts in females. However, as the SA site was also used in the males, homozygotes developed as sterile intersexes. For efficient drive, it would be preferable if homozygous males were fertile. However, heterozygous males developed normally, were fertile, and were competitive with wild type males in mating competition assays.

Next, the germline-specific Cas9 expressing strains were established (Yadav et al., 2023). First, *Dmnos-Cas9-NLS* and *Dmvasa-Cas9-NLS* strains were engineered to express Cas9 with a single NLS in the germ cells under *D. melanogaster nanos* and *vasa* gene promoters, respectively. Second, *Dsnos-NLS-Cas9-NLS* strains were generated to express Cas9 with N- and C-terminal NLS under the native *D. suzukii nanos* gene promoter and 3′UTR elements. Transgenic lines were made using *piggyBac* vectors that integrate randomly in the *D. suzukii* genome. Cas9 activity was assessed by crossing the Cas9 strains with strains that express sgRNA for *white* or *yellow* genes. The offspring produced easily scorable white eye or yellow body color mutant phenotypes. The Cas9 was active in all strains; however, the lines that produced Cas9 with two NLS sequences showed a higher editing efficiency (Yadav et al., 2023). Next, homing efficiencies were measured for both the dominant and recessive female sterile strains using X-linked *Dmnos-Cas9-NLS* (#36), *Dsnos-NLS-Cas9-NLS* (#27AI1), and autosomal *Dmvasa-Cas9-NLS* (#4) lines. Virgin females of each Cas9 strain were crossed with hemizygous males of a homing strain, and their male (male-drive) and female (female-drive) offspring that carried both the Cas9 and homing cassette were collected. It is in the germline of these flies that gene drive occurs. To measure drive efficiency, male-drive and female-drive flies were each crossed to wild type and the number of offspring that showed red fluorescence were counted. With perfect homing drive, 100% of the offspring would have red fluorescence. Without drive, 50% would be fluorescent. The homing drive inheritance observed with *Dmnos-Cas9-NLS (#36)* and *Dmvasa-Cas9-NLS* (#4) ranged from 56% to 71%, but with *Dsnos-NLS-Cas9-NLS* (#27AI1) it was measured at 94–99%. The use of a germline-specific native promoter (*Dsnos*) and Cas9 with two NLS significantly increased the homing efficiency. Furthermore, mathematical modeling was performed to assess the population suppression potential of these drive lines in a caged population. The modeling results suggest that a lab cage population can be suppressed with repeated releases at low release ratios (1:4) of these split drive strains. It remains to be seen how well these drive strains will perform in lab cage population suppression experiments. The findings of this study will be useful for the development of homing drive systems in other agricultural insect pests.

7 From the lab to the field: risk assessment and cage and open-field trials

A phased approach has been recommended for the evaluation of a transgenic mosquito strain (World Health Organization, 2021; Benedict et al., 2008). In this approach, the initial assessment of a transgenic strain is performed in a controlled laboratory setting using lab cages. Should a strain pass all lab evaluation criteria, the next step would be to evaluate it in large cages in a field setting or a large contained environment. The final step would be evaluation in an open field test.

Since the development of fsRIDL strains began over 20 years ago, most strains that have completed the phased assessment process are fsRIDL strains. It is estimated that over 1 billion transgenic insects have been released in the field, mostly Oxitec's *A. aegypti* RIDL and fsRIDL strains in Brazil (Beech et al., 2022). Regarding insects of agricultural importance, Oxitec's medfly (*C. capitata*) and olive fly (*B. oleae*) fsRIDL strains were initially evaluated in a lab setting and subsequently in large cages (8 m^3) in a greenhouse (Leftwich et al., 2014; Ant et al., 2012). The cages contained a small tree to provide a more natural environment for the fruit flies. Although successful, to our knowledge the strains were never evaluated in an open field test. A diamondback moth (*P. xylostella*) fsRIDL strain that showed promise in lab evaluations (Jin et al., 2013) was subsequently assessed in two large cage (1–2 m^3) trials in greenhouses (Harvey-Samuel et al., 2015). Since it performed well in the large cage trials, the strain was subsequently evaluated in an open field trial (Shelton et al., 2020). The trial provided information on the longevity and dispersal of the transgenic insects in the field, which was valuable information for constructing more realistic models (Shelton et al., 2020).

At all phases of assessment, a risk assessment is undertaken. A thorough discussion of the risk assessment of transgenic insects is beyond the scope of this chapter. The topic is covered in several excellent reviews (Beech et al., 2022; Hayes and Quinlan, 2022; Schwindenhammer, 2020; Devos et al., 2022; International Atomic Energy Agency, 2002; EFSA Panel on Genetically Modified Organisms, 2013). Gene drive insects present additional risk assessment challenges in part due to the large spatial and temporal scales that may need to be considered (Hayes and Quinlan, 2022). The approval of the open field test of the diamondback moth fsRIDL strain provides an example of the factors that are considered during a risk assessment process. In determining the 'finding of no significant impact' or FONSI (USDA-APHIS, 2017), the regulatory authority (USDA-APHIS) considered potential environmental (e.g. wildlife and biodiversity) and human health impacts (USDA-APHIS, 2016). The transgenic strain was found to not encode any known allergen or toxin and was therefore unlikely to impact any wildlife or people that may consume the insects. Further, the insects would not overwinter, and it was considered unlikely that wildlife would acquire DNA from the transgenic insects through horizontal gene transfer. The regulatory agency responded to 286 individual written comments submitted during a 30-day public review and comment period. Stakeholder engagement plays an important role in public perception and acceptability of the technology (Thizy et al., 2022).

8 Conclusion and future trends

The insect pests of crops and livestock threaten food security and public health. About 20% of the global crops valued at ~$470 billion are lost annually to insect pests alone (Culliney, 2014). Efficient and cost-effective control strategies are needed to reduce the global burden of agricultural pests. Current pest control methods rely heavily on broad-spectrum insecticides that also kill vital pollinators and non-target species, thus endangering the entire ecosystem. Genetically enhanced biological systems offer environmentally safe and efficient pest control methods that could reduce the reliance on toxic pesticides. The successful use of the SIT in eradicating and limiting major agricultural pests has inspired the development of more genetically advanced systems. Gene-edited sterile male pgSIT and conditional dominant female-lethal fsRIDL strains could offer more efficient control than traditional SIT. SIT, pg-SIT, and fsRIDL are 'self-limiting' in nature and require repeated releases of an excess of transgenic males to suppress a pest population. The recently demonstrated self-limiting X-shredder and X-poisoning systems are cost-effective as well as highly efficient. Y-linked editor strains have the potential to confer particularly efficient population suppression but remain self-limiting, with little potential to disrupt neighboring populations. Modeling indicates that fully autonomous CRISPR/Cas9-based homing gene drives could provide the most efficient and economical approach for pest population suppression. More research is needed to understand the spread dynamics of gene drives to avoid any unforseen adverse ecological impacts. Self-limiting split-homing gene drive strains could be an attractive alternative for initial field testing. There remain major challenges in the successful development of strains and in obtaining regulatory approval for field testing. Regarding the latter, several fsRIDL strains have been approved for testing in open field trials, which is encouraging for the future implementation of other self-limiting genetic strains such as Y-linked editors. Some of the challenges remaining include developing mass sex-sorting techniques for implementing pgSIT, finding X-chromosome-specific repetitive sequences for X-shredder, and inserting transgenes on the Y chromosome for Y-linked editors. Further, for CRISPR-based systems, the well-known genetic diversity of insect populations could reduce effectiveness in the field (e.g. polymorphisms in the sgRNA target site). As with SIT, other genetic biocontrol strategies will fail if the fitness of the released males is poor. Thus, we anticipate programs will spend considerable effort to improve the fitness of transgenic strains. We are optimistic that these problems will be solved and limitations overcome to achieve effective agricultural pest control using genetic biocontrol approaches.

9 Where to look for further information

9.1 Reviews

- Dhole, S., et al., 2020. Gene drive dynamics in natural populations: the importance of density dependence, space, and sex. *Annual Review of Ecology, Evolution and Systematics* 51, 505–531. https://doi.org/10.1146/annurev-ecolsys-031120-101013
- Raban, R., et al., 2023. Manipulating the destiny of wild populations using CRISPR. *Annual Review of Genetics* 57, 361–390. https://doi.org/10.1146/annurev-genet-031623-105059

9.2 Books

- Dyck, V.A., et al. (Eds.), 2021. *Sterile Insect Technique: Principles And Practice In Area-Wide Integrated Pest Management*, 2nd ed. CRC Press, Boca Raton, FL. https://doi.org/10.1201/9781003035572
- Benedict, M.Q. and Scott, M.J., 2022. *Transgenic Insects: Techniques and Applications*, 2nd ed. CABI International, Wallingford, UK.

9.3 Seminal articles

- Knipling, E.F., 1955. Possibilities of insect control or eradication through the use of sexually sterile males1. *Journal of Economic Entomology* 48, 459–462. https://doi.org/10.1093/jee/48.4.459
- Heinrich, J.C. and Scott, M.J., 2000. A repressible female-specific lethal genetic system for making transgenic insect strains suitable for a sterile-release program. *Proceedings of the National Academy of Sciences* 97, 8229–8232. https://doi.org/10.1073/pnas.140142697
- Thomas, D.D., et al., 2000. Insect population control using a dominant, repressible, lethal genetic system. *Science* 287, 2474–2476. https://doi.org/10.1126/science.287.5462.2474
- Burt, A., 2003. Site-specific selfish genes as tools for the control and genetic engineering of natural populations. *Proceedings of the Royal Society of London. Series B: Biological Sciences* 270, 921–928. https://doi.org/10.1098/rspb.2002.2319

- Windbichler, N., et al., 2008. Targeting the X Chromosome during Spermatogenesis induces Y Chromosome transmission ratio distortion and early dominant embryo lethality in Anopheles gambiae. *PLoS Genetics* 4, e1000291. https://doi.org/10.1371/journal.pgen.1000291
- Esvelt, K.M., et al., 2014. Concerning RNA-guided gene drives for the alteration of wild populations. *eLife* 3, e03401. https://doi.org/10.7554/eLife.03401
- Burt, A. and Deredec, A., 2018. Self-limiting population genetic control with sex-linked genome editors. *Proceedings of the Royal Society B: Biological Sciences* 285, 20180776. https://doi.org/10.1098/rspb.2018.0776
- Kandul, N.P., et al., 2019. Transforming insect population control with precision guided sterile males with demonstration in flies. *Nature Communications* 10, 84. https://doi.org/10.1038/s41467-018-07964-7

9.4 Websites

- GeneConvene (https://fnih.org/our-programs/geneconvene-global-collaborative/)
- IAEA (International Atomic Energy Agency) on SIT and pest control (https://www.iaea.org/topics/insect-pest-control),

Biotech companies developing genetic biocontrol solutions:

- Oxitec, https://www.oxitec.com/
- Agragene, https://www.agragene.com/
- Biocentis, https://www.biocentis.com/
- Novoclade, https://www.novoclade.com/

9.5 Conferences

- Gordon Genetic Biocontrol Conference (https://www.grc.org/genetic-biocontrol-conference/)

Major International Projects:

- COPEG (screwworm) (https://www.copeg.org/en/)
- Moscamed (fruit fly) (https://moscamed-guatemala.org.gt/#/)
- While not directly working on crop pests, the international programs proposing to eliminate malaria through genetic control of the vectors are world-leading in the development and implementation of gene drive strains.

- Target Malaria (https://targetmalaria.org/)
- Transmission Zero (https://www.transmissionzero.org/)
- Stop Malaria (https://stopmalaria.org/)

10 Acknowledgments

Research in the Scott lab on screwworm, sheep blowfly, and spotted wing Drosophila has been supported by grants from USDA-NIFA Biotechnology Risk Assessment Research (BRAG) program grants 2016-33522-25625, 2020-33522-32317, and 2021-33522-35341 and agreements with USDA-APHIS (award numbers AP19IS000000C001, AP20IS000000C002, and AP21IS000000C003). S.E.H. was supported by a grant from the United States – Israel Binational Agricultural Research and Development Fund (Research Grant No. IS-5180-19). A.L.K. is supported by a NIFA AFRI Education and Workforce Development Predoctoral Fellowship (award number 2023-67011-40404). A.T. is supported by the National Institute of General Medical Sciences of the National Institutes of Health under Award Number T32GM133366.

11 References

Agragene, 2024. Homepage. (https://www.agragene.com/); accessed March 22, 2024.

Alcalay, Y., et al., 2021. The potential for a released autosomal X-shredder becoming a driving-Y chromosome and invasively suppressing wild populations of Malaria Mosquitoes. *Frontiers in Bioengineering and Biotechnology* 9. https://doi.org/10.3389/fbioe.2021.752253

Alexander, J.L., 2006. Screwworms. *Journal of the American Veterinary Medical Association* 228, 357–367. https://doi.org/10.2460/javma.228.3.357

Ant, T., et al., 2012. Control of the olive fruit fly using genetics-enhanced sterile insect technique. *BMC Biology* 10, 51. https://doi.org/10.1186/1741-7007-10-51

Asadi, R., et al., 2020. Preventative releases of self-limiting *Ceratitis capitata* provide pest suppression and protect fruit quality in outdoor netted cages. *International Journal of Pest Management* 66, 182–193. https://doi.org/10.1080/09670874.2019.1601293

Bakri, A., et al., 2021. Sterilizing insects with Ionizing radiation. In Dyck, V. A. et al. (Eds.), *Sterile Insect Technique*. CRC Press, Boca Raton, FL.

Bargielowski, I., et al., 2011. Cost of mating and insemination capacity of a genetically modified Mosquito *Aedes aegypti* OX513A compared to its wild type counterpart. *PLoS One* 6, e26086. https://doi.org/10.1371/journal.pone.0026086

Beech, C., et al., 2022. *Regulation of Transgenic Insects*. CABI Books, London, pp. 493–517. https://doi.org/10.1079/9781800621176.0025

Benedict, M., et al., 2008. Guidance for contained field trials of vector mosquitoes engineered to contain a gene drive system: recommendations of a scientific working group. *Vector-Borne and Zoonotic Diseases* 8, 127–166. https://doi.org/10.1089/vbz.2007.0273

Benedict, M.Q., et al., 2018. Recommendations for laboratory containment and management of gene drive systems in arthropods. *Vector-Borne and Zoonotic Diseases* 18, 2–13. https://doi.org/10.1089/vbz.2017.2121

Bier, E., 2022. Gene drives gaining speed. *Nature Reviews Genetics* 23, 5–22. https://doi.org/10.1038/s41576-021-00386-0

Biocentis, 2024. Science and Technology. (https://www.biocentis.com/science-and-technology/); accessed March 22, 2024.

Bouyer, J. and Vreysen, M.J.B., 2019. Concerns about the feasibility of using "precision guided sterile males" to control insects. *Nature Communications* 10, 3954. https://doi.org/10.1038/s41467-019-11616-9

Buchman, A., et al., 2018. Synthetically engineered *Medea* gene drive system in the worldwide crop pest *Drosophila suzukii*. *Proceedings of the National Academy of Sciences* 115, 4725–4730. https://doi.org/10.1073/pnas.1713139115

Burt, A., 2003. Site-specific selfish genes as tools for the control and genetic engineering of natural populations. *Proceedings of the Royal Society of London. Series B: Biological Sciences* 270, 921–928. https://doi.org/10.1098/rspb.2002.2319

Burt, A. and Crisanti, A., 2018. Gene drive: Evolved and synthetic. *ACS Chemical Biology* 13, 343–346. https://doi.org/10.1021/acschembio.7b01031

Burt, A. and Deredec, A., 2018. Self-limiting population genetic control with sex-linked genome editors. *Proceedings of the Royal Society B: Biological Sciences* 285, 20180776. https://doi.org/10.1098/rspb.2018.0776

Carrami, E.M., et al., 2018. Consequences of resistance evolution in a Cas9-based sex conversion-suppression gene drive for insect pest management. *Proceedings of the National Academy of Sciences* 115, 6189–6194. https://doi.org/10.1073/pnas.1713825115

Carvalho, D.O., et al., 2015. Suppression of a field population of *Aedes aegypti* in Brazil by sustained release of transgenic male Mosquitoes. *PLoS Neglected Tropical Diseases* 9, e0003864. https://doi.org/10.1371/journal.pntd.0003864

Catteruccia, F., et al., 2005. An anopheles transgenic sexing strain for vector control. *Nature Biotechnology* 23, 1414–1417. https://doi.org/10.1038/nbt1152

Champer, J., et al., 2019. Molecular safeguarding of CRISPR gene drive experiments. *eLife* 8, e41439. https://doi.org/10.7554/eLife.41439

Champer, J., et al., 2017. Novel CRISPR/Cas9 gene drive constructs reveal insights into mechanisms of resistance allele formation and drive efficiency in genetically diverse populations. *PLoS Genetics* 13, e1006796. https://doi.org/10.1371/journal.pgen.1006796

Chan, Y.-S., et al., 2011. Insect population control by homing endonuclease-based gene drive: an evaluation in *Drosophila melanogaster*. *Genetics* 188, 33–44. https://doi.org/10.1534/genetics.111.127506

Chan, Y.-S., et al., 2013a. Optimising homing endonuclease gene drive performance in a semi-refractory species: the *Drosophila melanogaster* Experience. *PLoS One* 8, e54130. https://doi.org/10.1371/journal.pone.0054130

Chan, Y.-S., et al., 2013b. The design and in Vivo evaluation of engineered I-OnuI-based enzymes for HEG gene drive. *PLoS One* 8, e74254. https://doi.org/10.1371/journal.pone.0074254

Chiu, J.C., et al., 2013. Genome of *Drosophila suzukii*, the Spotted Wing Drosophila. *G3 Genes|Genomes|Genetics* 3, 2257–2271. https://doi.org/10.1534/g3.113.008185

Concha, C., et al, 2016. A transgenic male-only strain of the New World screwworm for an improved control program using the sterile insect technique. *BMC Biology* 14, 72. https://doi.org/10.1186/s12915-016-0296-8

COPEG, 2024. Panama-United States Commission for the eradication and prevention of Screwworm. (https://www.copeg.org/en/); accessed March 22, 2024.

Costa, H.S., et al., 1997. Effects of antibacterial materials on *Bemisia argentifolii* (Homoptera: Aleyrodidae) oviposition, growth, survival, and sex ratio. *Journal of Economic Entomology* 90, 333–339. https://doi.org/10.1093/jee/90.2.333

Crystal, M.M., 1979. Sterilization of screwworm flies (Diptera: Calliphoridae) with gamma rays: restudy after two decades. *Journal of Medical Entomology* 15, 103–108. https://doi.org/10.1093/jmedent/15.2.103

Culliney, T.W., 2014. Crop losses to arthropods. In Pimentel, D., and Peshin, R. (Eds.), *Integrated Pest Management: Pesticide Problems*, Vol.3. Springer Netherlands, Dordrecht, pp. 201–225. https://doi.org/10.1007/978-94-007-7796-5_8

Deredec, A., et al., 2008. The population genetics of using homing endonuclease genes in vector and pest management. *Genetics* 179, 2013–2026. https://doi.org/10.1534/genetics.108.089037

Devos, Y., et al., 2022. Potential use of gene drive modified insects against disease vectors, agricultural pests and invasive species poses new challenges for risk assessment. *Critical Reviews in Biotechnology* 42, 254–270. https://doi.org/10.1080/07388551.2021.1933891

Dhole, S., et al., 2020. Gene drive dynamics in natural populations: the importance of density dependence, space, and sex. *Annual Review of Ecology, Evolution and Systematics* 51, 505–531. https://doi.org/10.1146/annurev-ecolsys-031120-101013

Doudna, J.A. and Charpentier, E., 2014. The new frontier of genome engineering with CRISPR-Cas9. *Science* 346, 1258096. https://doi.org/10.1126/science.1258096

Dowell, R.V., et al., 2021. Supply, emergence, and release of sterile insects. In Dyck, V. A. et al. (Eds.), *Sterile Insect Technique*. CRC Press, Boca Raton, FL.

Dyck, V.A., et al. (Eds.), 2021. *Sterile Insect Technique: Principles And Practice In Area-Wide Integrated Pest Management*, 2nd ed. CRC Press, Boca Raton, FL. https://doi.org/10.1201/9781003035572

EFSA Panel on Genetically Modified Organisms (GMO), 2013. Guidance on the environmental risk assessment of genetically modified animals. *EFSA Journal* 11, 3200. https://doi.org/10.2903/j.efsa.2013.3200

Esvelt, K.M., et al., 2014. Concerning RNA-guided gene drives for the alteration of wild populations. *eLife* 3, e03401. https://doi.org/10.7554/eLife.03401

Fasulo, B., et al., 2020. A fly model establishes distinct mechanisms for synthetic CRISPR/Cas9 sex distorters. *PLoS Genetics* 16, e1008647. https://doi.org/10.1371/journal.pgen.1008647

FNIH, 2024. GeneConvene Global Collaborative. (https://fnih.org/our-programs/geneconvene-global-collaborative/); (accessed March 22, 2024).

Foster, G.G., et al., 1985. Genetic analysis of field trials of sex-linked translocation strains for genetic control of the Australian sheep blowfly *Lucilia cuprina* (Wiedemann). *Australian Journal of Biological Sciences* 38, 275–293. https://doi.org/10.1071/bi9850275

Foster, G.G., et al., 1988. Computer simulation of genetic control. Comparison of sterile males and field-female killing systems. *Theoretical and Applied Genetics* 76, 870–879. https://doi.org/10.1007/BF00273675

Foster, G.G., et al., 1993. *Advances in Sheep Blowfly Genetic Control in Australia.* International Atomic Energy Agency (IAEA): IAEA.

Fu, G., et al., 2007. Female-specific insect lethality engineered using alternative splicing. *Nature Biotechnology* 25, 353–357. https://doi.org/10.1038/nbt1283

Galizi, R., et al., 2014. A synthetic sex ratio distortion system for the control of the human malaria mosquito. *Nature Communications* 5, 3977. https://doi.org/10.1038/ncomms4977

Galizi, R., et al., 2016. A CRISPR-Cas9 sex-ratio distortion system for genetic control. *Scientific Reports* 6, 31139. https://doi.org/10.1038/srep31139

Gamez, S., et al., 2021. Exploiting a Y chromosome-linked Cas9 for sex selection and gene drive. *Nature Communications* 12, 7202. https://doi.org/10.1038/s41467-021-27333-1

Gantz, V.M. and Akbari, O.S., 2018. Gene editing technologies and applications for insects. *Current Opinion in Insect Science, Vectors and Medical and Veterinary Entomology * Social Insects* 28, 66–72. https://doi.org/10.1016/j.cois.2018.05.006

Gantz, V.M. and Bier, E., 2015. The mutagenic chain reaction: A method for converting heterozygous to homozygous mutations. *Science* 348, 442–444. https://doi.org/10.1126/science.aaa5945

Geci, R., et al., 2022. Gene drive designs for efficient and localisable population suppression using Y-linked editors. *PLoS Genetics* 18, e1010550. https://doi.org/10.1371/journal.pgen.1010550

Gong, P., et al., 2005. A dominant lethal genetic system for autocidal control of the Mediterranean fruitfly. *Nature Biotechnology* 23, 453–456. https://doi.org/10.1038/nbt1071

Gossen, M. and Bujard, H., 1992. Tight control of gene expression in mammalian cells by tetracycline-responsive promoters. *Proceedings of the National Academy of Sciences of the United States of America* 89, 5547–5551. https://doi.org/10.1073/pnas.89.12.5547

Gutierrez, A.P., et al., 2019. Deconstructing the eradication of new world screwworm in North America: retrospective analysis and climate warming effects. *Medical and Veterinary Entomology* 33, 282–295. https://doi.org/10.1111/mve.12362

Haber, D.A., et al., 2024. Targeting mosquito X-chromosomes reveals complex transmission dynamics of sex ratio distorting gene drives. https://doi.org/10.1101/2023.08.20.554004

Hamilton, W.D., 1967. Extraordinary sex ratios. *Science* 156, 477–488.

Hammond, A., et al., 2016. A CRISPR-Cas9 gene drive system targeting female reproduction in the malaria mosquito vector *Anopheles gambiae*. *Nature Biotechnology* 34, 78–83. https://doi.org/10.1038/nbt.3439

Hammond, A.M., et al., 2017. The creation and selection of mutations resistant to a gene drive over multiple generations in the malaria mosquito. *PLoS Genetics* 13, e1007039. https://doi.org/10.1371/journal.pgen.1007039

Hammond, A., et al., 2021a. Regulating the expression of gene drives is key to increasing their invasive potential and the mitigation of resistance. *PLoS Genetics* 17, e1009321. https://doi.org/10.1371/journal.pgen.1009321

Hammond, A., et al., 2021b. Gene-drive suppression of mosquito populations in large cages as a bridge between lab and field. *Nature Communications* 12, 4589. https://doi.org/10.1038/s41467-021-24790-6

Harris, A.F., et al., 2011. Field performance of engineered male mosquitoes. *Nature Biotechnology* 29, 1034–1037. https://doi.org/10.1038/nbt.2019

Harvey-Samuel, T., et al., 2015. Pest control and resistance management through release of insects carrying a male-selecting transgene. *BMC Biology* 13, 49. https://doi.org/10.1186/s12915-015-0161-1

Hayes, K.R. and Quinlan, M.M., 2022. *Risk Analysis of Transgenic Insects*. Transgenic Insects, CABI Books, London pp. 552–578. https://doi.org/10.1079/9781800621176.0028

Heinrich, J.C. and Scott, M.J., 2000. A repressible female-specific lethal genetic system for making transgenic insect strains suitable for a sterile-release program. *Proceedings of the National Academy of Sciences* 97, 8229–8232. https://doi.org/10.1073/pnas.140142697

Hickey, W.A. and Craig, G.B., Jr., 1966. Genetic distortion of sex ratio in a mosquito, *Aedes Aegypti*. *Genetics* 53, 1177–1196. https://doi.org/10.1093/genetics/53.6.1177

International Atomic Energy Agency, 2002. Status and risk assessment of the use of transgenic arthropods in plant protection, IAEA-TECDOC-1483, IAEA, Vienna (2006). (https://www.iaea.org/publications/7204/status-and-risk-assessment-of-the-use-of-transgenic-arthropods-in-plant-protection); accessed May 2, 2024.

International Atomic Energy Agency, 2024. Insect Pest Control. (https://www.iaea.org/topics/insect-pest-control); accessed March 22, 2024.

Irvin, N., et al., 2004. Assessing fitness costs for transgenic *Aedes aegypti* expressing the GFP marker and transposase genes. *Proceedings of the National Academy of Sciences of the United States of America* 101, 891–896. https://doi.org/10.1073/pnas.0305511101

Jin, L., et al., 2013. Engineered female-specific lethality for control of pest lepidoptera. *ACS Synthetic Biology* 2, 160–166. https://doi.org/10.1021/sb300123m

Jinek, M., et al., 2012. A Programmable dual-RNA–guided DNA endonuclease in adaptive bacterial immunity. *Science* 337, 816–821. https://doi.org/10.1126/science.1225829

Kandul, N.P., et al., 2022. Precision guided sterile males suppress populations of an invasive crop pest. *GEN Biotechnology* 1, 372–385. https://doi.org/10.1089/genbio.2022.0019

Kandul, N.P., et al., 2019. Transforming insect population control with precision guided sterile males with demonstration in flies. *Nature Communications* 10, 84. https://doi.org/10.1038/s41467-018-07964-7

Klassen, W., et al., 2021. History of the sterile insect technique. In Dyck, V. A. et al. (Eds.), *Sterile Insect Technique*. CRC Press, Boca Raton, FL.

Knipling, E.F., 1960. The eradication of the screw-worm fly. *Scientific American* 203, 54–61.

Knipling, E.F., 1955. Possibilities of insect control or eradication through the use of sexually sterile males1. *Journal of Economic Entomology* 48, 459–462. https://doi.org/10.1093/jee/48.4.459

Kuriwada, T., et al., 2010. Biological role of nardonella endosymbiont in its weevil host. *PLoS One* 5, e13101. https://doi.org/10.1371/journal.pone.0013101

Kyrou, K., et al., 2018. A CRISPR–Cas9 gene drive targeting *doublesex* causes complete population suppression in caged *Anopheles gambiae* mosquitoes. *Nature Biotechnology* 36, 1062–1066. https://doi.org/10.1038/nbt.4245

LaChance, L.E. and Knipling, E.F., 1962. Control of insect populations through genetic manipulations. *Annals of the Entomological Society of America* 55, 515–520. https://doi.org/10.1093/aesa/55.5.515

Lawler, C.D., et al., 2024. The haplolethal gene *wupA* of Drosophila exhibits potential as a target for an X-poisoning gene drive. *G3 Genes|Genomes|Genetics* jkae025. https://doi.org/10.1093/g3journal/jkae025

Leftwich, P.T., et al., 2014. Genetic elimination of field-cage populations of mediterranean fruit flies. *Proceedings of the Royal Society B: Biological Sciences* 281, 20141372. https://doi.org/10.1098/rspb.2014.1372

Legros, M., et al., 2021. Gene drive strategies of pest control in agricultural systems: challenges and opportunities. *Evolutionary Applications* 14, 2162–2178. https://doi.org/10.1111/eva.13285

Li, F. and Scott, M.J., 2016. CRISPR/Cas9-mediated mutagenesis of the *white* and *Sex lethal* loci in the invasive pest, *Drosophila suzukii*. *Biochemical and Biophysical Research Communications* 469, 911–916. https://doi.org/10.1016/j.bbrc.2015.12.081

Li, F., et al., 2021a. A conditional female lethal system for genetic suppression of the global fruit crop pest *Drosophila suzukii*. *Pest Management Science* 77, 4915–4922. https://doi.org/10.1002/ps.6530

Li, M., et al., 2021b. Suppressing mosquito populations with precision guided sterile males. *Nature Communications* 12, 5374. https://doi.org/10.1038/s41467-021-25421-w

Li, M., et al., 2024. Targeting sex determination to suppress mosquito populations. *eLife* 12, RP90199. https://doi.org/10.7554/eLife.90199

Liu, J., et al., 2024. A fluorescent sex-sorting technique for insects with the demonstration in *Drosophila melanogaster*. *GEN Biotechnology* 3, 35–44. https://doi.org/10.1089/genbio.2023.0041

Mastrangelo, T. and Welch, J.B., 2012. An overview of the components of AW-IPM campaigns against the new world screwworm. *Insects* 3, 930–955. https://doi.org/10.3390/insects3040930

McInnis, D.O., et al., 1994. Population suppression and sterility rates induced by variable sex ratio, sterile insect releases of *Ceratitis capitata* (Diptera: Tephritidae) in Hawaii. *Annals of the Entomological Society of America* 87, 231–240. https://doi.org/10.1093/aesa/87.2.231

Meccariello, A., et al., 2024. Gene drive and genetic sex conversion in the global agricultural pest *Ceratitis capitata*. *Nature Communications* 15, 372. https://doi.org/10.1038/s41467-023-44399-1

Meccariello, A., et al., 2021. Engineered sex ratio distortion by X-shredding in the global agricultural pest *Ceratitis capitata*. *BMC Biology* 19, 78. https://doi.org/10.1186/s12915-021-01010-7

Morrison, N.I., et al., 2012. Engineered repressible lethality for controlling the pink bollworm, a lepidopteran pest of cotton. *PLoS One* 7, e50922. https://doi.org/10.1371/journal.pone.0050922

Noble, C., et al., 2018. Current CRISPR gene drive systems are likely to be highly invasive in wild populations. *eLife* 7, e33423. https://doi.org/10.7554/eLife.33423

Ometto, L., et al., 2013. Linking genomics and ecology to investigate the complex evolution of an invasive drosophila pest. *Genome Biology and Evolution* 5, 745–757. https://doi.org/10.1093/gbe/evt034

O'Neil, M.T. and Belote, J.M., 1992. Interspecific comparison of the *transformer* gene of Drosophila reveals an unusually high degree of evolutionary divergence. *Genetics* 131, 113–128. https://doi.org/10.1093/genetics/131.1.113

Oxitec, 2024. Homepage. (https://www.oxitec.com/); accessed March 22, 2024.

Pais, R., et al., 2008. The obligate mutualist *Wigglesworthia glossinidia* influences reproduction, digestion, and immunity processes of its host, the tsetse fly. *Applied and Environmental Microbiology* 74, 5965–5974. https://doi.org/10.1128/AEM.00741-08

Papathanos, P.A., et al., 2009. The vasa regulatory region mediates germline expression and maternal transmission of proteins in the malaria mosquito *Anopheles gambiae*: a versatile tool for genetic control strategies. *BMC Molecular Biology* 10, 65. https://doi.org/10.1186/1471-2199-10-65

Paris, M., et al., 2020. Near-chromosome level genome assembly of the fruit pest *Drosophila suzukii* using long-read sequencing. *Scientific Reports* 10, 11227. https://doi.org/10.1038/s41598-020-67373-z

Parker, A.G., et al., 2021. Mass-rearing for the sterile insect technique. In Dyck, V. A. et al. (Eds.), *Sterile Insect Technique*. CRC Press, Boca Raton, FL.

Pietri, J.E. and Liang, D., 2020. Insecticidal activity of doxycycline against the common bedbug. *Antimicrobial Agents and Chemotherapy* 64, 10.1128/aac.00005-20. https://doi.org/10.1128/aac.00005-20

Raban, R., et al., 2023. Manipulating the destiny of wild populations using CRISPR. *Annual Review of Genetics* 57, 361–390. https://doi.org/10.1146/annurev-genet-031623-105059

Reavey, C.E., et al., 2022. Self-limiting fall armyworm: a new approach in development for sustainable crop protection and resistance management. *BMC Biotechnology* 22, 5. https://doi.org/10.1186/s12896-022-00735-9

Rendón, P., et al., 2004. Medfly (Diptera:Tephritidae) genetic sexing: large-scale field comparison of males-only and bisexual sterile fly releases in Guatemala. *Journal of Economic Entomology* 97, 1547–1553. https://doi.org/10.1603/0022-0493-97.5.1547

Robinson, A.S., 2021. Genetic basis of the sterile insect technique. In Dyck, V. A. et al. (Eds.), *Sterile Insect Technique*. CRC Press, Boca Raton, FL.

Schairer, C.E., et al., 2021. Oxitec and MosquitoMate in the United States: lessons for the future of gene drive mosquito control. *Pathogens and Global Health* 115, 365–376. https://doi.org/10.1080/20477724.2021.1919378

Schetelig, M.F. and Handler, A.M., 2012. A transgenic embryonic sexing system for *Anastrepha suspensa* (Diptera: Tephritidae). *Insect Biochemistry and Molecular Biology* 42, 790–795. https://doi.org/10.1016/j.ibmb.2012.07.007

Schetelig, M.F., et al., 2016. Tetracycline-suppressible female lethality and sterility in the Mexican fruit fly, *Anastrepha ludens*. *Insect Molecular Biology* 25, 500–508. https://doi.org/10.1111/imb.12238

Schlein, Y., 1977. Lethal effect of tetracycline on tsetse flies following damage to bacterioid symbionts. *Experientia* 33, 450–451. https://doi.org/10.1007/BF01922204

Schliekelman, P., et al., 2005. Pest control by genetic manipulation of sex ratio. *Journal of Economic Entomology* 98, 18–34. https://doi.org/10.1093/jee/98.1.18

Schliekelman, P. and Gould, F., 2000. Pest control by the release of insects carrying a female-killing allele on multiple loci. *Journal of Economic Entomology* 93, 1566–1579. https://doi.org/10.1603/0022-0493-93.6.1566

Schwindenhammer, S., 2020. The rise, regulation and risks of genetically modified insect technology in global agriculture. *Science, Technology and Society* 25, 124–141. https://doi.org/10.1177/0971721819890042

Scott, M.J., 2014. Development and evaluation of male-only strains of the Australian sheep blowfly, *Lucilia cuprina*. *BMC Genetics* 15, S3. https://doi.org/10.1186/1471-2156-15-S2-S3

Scott, M.J., et al., 2017. Review of research advances in the screwworm eradication program over the past 25 years. *Entomologia Experimentalis et Applicata* 164, 226–236. https://doi.org/10.1111/eea.12607

Scott, M.J., et al., 2018. Agricultural production: assessment of the potential use of Cas9-mediated gene drive systems for agricultural pest control. *Journal of Responsible Innovation* 5, S98–S120. https://doi.org/10.1080/23299460.2017.1410343

Shelton, A.M., et al., 2020. First field release of a genetically engineered, self-limiting agricultural pest insect: evaluating its potential for future crop protection. *Frontiers in Bioengineering and Biotechnology* 7. https://doi.org/10.3389/fbioe.2019.00482

Shui, S., et al., 2022. Systematic investigation of the effects of multiple SV40 nuclear localization signal fusion on the genome editing activity of purified SpCas9. *Bioengineering* 9, 83. https://doi.org/10.3390/bioengineering9020083

Simoni, A., et al., 2020. A male-biased sex-distorter gene drive for the human malaria vector *Anopheles gambiae*. *Nature Biotechnology* 38, 1054–1060. https://doi.org/10.1038/s41587-020-0508-1

Simoni, A., et al., 2014. Development of synthetic selfish elements based on modular nucleases in *Drosophila melanogaster*. *Nucleic Acids Research* 42, 7461–7472. https://doi.org/10.1093/nar/gku387

Sinkins, S.P. and Gould, F., 2006. Gene drive systems for insect disease vectors. *Nature Reviews Genetics* 7, 427–435. https://doi.org/10.1038/nrg1870

Spinner, S.A.M., et al., 2022. New self-sexing *Aedes aegypti* strain eliminates barriers to scalable and sustainable vector control for governments and communities in dengue-prone environments. *Frontiers in Bioengineering and Biotechnology* 10. https://doi.org/10.3389/fbioe.2022.975786

Sternberg, S.H., et al., 2014. DNA interrogation by the CRISPR RNA-guided endonuclease Cas9. *Nature* 507, 62–67. https://doi.org/10.1038/nature13011

Tabashnik, B.E., et al., 2021. Transgenic cotton and sterile insect releases synergize eradication of pink bollworm a century after it invaded the United States. *Proceedings of the National Academy of Sciences* 118, e2019115118. https://doi.org/10.1073/pnas.2019115118

Target Malaria, 2024. Homepage. (https://targetmalaria.org/); accessed March 22, 2024.

Terradas, G., et al., 2021. Inherently confinable split-drive systems in Drosophila. *Nature Communications* 12, 1480. https://doi.org/10.1038/s41467-021-21771-7

Thizy, D., et al., 2022. *Public Acceptability and Stakeholder Engagement for Genetic Control Technologies*. Transgenic Insects, CABI Books, London, 474–492. https://doi.org/10.1079/9781800621176.0024

Thomas, D.D., et al., 2000. Insect population control using a dominant, repressible, lethal genetic system. *Science* 287, 2474–2476. https://doi.org/10.1126/science.287.5462.2474

Transmission Zero, 2024. Homepage. (https://www.transmissionzero.org); accessed March 22, 2024.
Turner, J.M.A., 2007. Meiotic sex chromosome inactivation. *Development* 134, 1823–1831. https://doi.org/10.1242/dev.000018
USDA-APHIS, 2016. Proposal to permit the field release of genetically engineered diamondback moth in New York. (https://www.regulations.gov/document/APHIS-2014-0056-0293); accessed May 2, 2024.
USDA-APHIS, 2017. Finding of no significant impact for the field release of genetically engineered diamondback moths. (https://www.regulations.gov/document/APHIS-2014-0056-0964); accessed May 2, 2024.
USDA, 2018. USDA announces pink bollworm eradication significantly saving cotton farmers in yearly control costs. (https://www.usda.gov/media/press-releases/2018/10/19/usda-announces-pink-bollworm-eradication-significantly-saving); accessed March 22, 2024.
USDA, 2019. USDA Celebrates the 25th Anniversary of the Screwworm Barrier Maintenance Program. (https://www.aphis.usda.gov/aphis/newsroom/news/sa_by_date/sa-2019/screwworm-program); accessed March 22, 2024.
Vargas-Terán, M., et al., 2021. Impact of screwworm eradication programmes using the sterile insect technique. In Dyck, V. A. et al. (Eds.), *Sterile Insect Technique*. CRC Press, Boca Raton, FL.
Vella, M.R., et al., 2021. Mathematical modeling of genetic pest management through female-specific lethality: Is one locus better than two? *Evolutionary Applications* 14, 1612–1622. https://doi.org/10.1111/eva.13228
Verkuijl, S.A.N., et al., 2022. The challenges in developing efficient and robust synthetic homing endonuclease gene drives. *Frontiers in Bioengineering and Biotechnology* 10. https://doi.org/10.3389/fbioe.2022.856981
Waltz, E., 2015. A face-lift for biotech rules begins. *Nature Biotechnology* 33, 1221–1222. https://doi.org/10.1038/nbt1215-1221
Weng, S.-C., et al., 2023. Efficient sex separation by exploiting differential alternative splicing of a dominant marker in *Aedes aegypti*. *PLoS Genetics* 19, e1011065. https://doi.org/10.1371/journal.pgen.1011065
Willis, K. and Burt, A., 2021. Double drives and private alleles for localised population genetic control. *PLoS Genetics* 17, e1009333. https://doi.org/10.1371/journal.pgen.1009333
Windbichler, N., et al., 2008. Targeting the X chromosome during spermatogenesis induces Y chromosome transmission ratio distortion and early dominant embryo lethality in *Anopheles gambiae*. *PLoS Genetics* 4, e1000291. https://doi.org/10.1371/journal.pgen.1000291
Wood, R.J. and Newton, M.E., 1991. Sex-ratio distortion caused by meiotic drive in Mosquitoes. *The American Naturalist* 137, 379–391.
World Health Organization, 2021. Guidance framework for testing genetically modified mosquitoes, second edition. (https://www.who.int/publications/i/item/9789240025233); accessed May 2, 2024.
Wyss, J.H., 2000. Screwworm eradication in the Americas. *Annals of the New York Academy of Sciences* 916, 186–193. https://doi.org/10.1111/j.1749-6632.2000.tb05289.x
Xu, X., et al., 2022. Toward a CRISPR-Cas9-based gene drive in the diamondback moth *Plutella xylostella*. *The CRISPR Journal* 5, 224–236. https://doi.org/10.1089/crispr.2021.0129

Yadav, A.K., et al., 2023. CRISPR/Cas9-based split homing gene drive targeting doublesex for population suppression of the global fruit pest *Drosophila suzukii*. *Proceedings of the National Academy of Sciences* 120, e2301525120. https://doi.org/10.1073/pnas.2301525120

Yan, Y. and Scott, M.J., 2020. Building a transgenic sexingstrain for genetic control of the Australian sheep blow fly *Lucilia cuprina* using two lethal effectors. *BMC Genetics* 21, 141. https://doi.org/10.1186/s12863-020-00947-y

Yang, E., et al., 2022. A homing suppression gene drive with multiplexed gRNAs maintains high drive conversion efficiency and avoids functional resistance alleles. *G3 Genes|Genomes|Genetics* 12, jkac081. https://doi.org/10.1093/g3journal/jkac081

Chapter 3

Developing plant-based insect biocontrol agents

Daniel Tapia Garcia, María Fé Andrés and Azucena Gonzalez Coloma, Instituto de Ciencias Agrarias – Consejo Superior de Investigaciones Científicas (CSIC), Spain

1 Introduction

Synthetic pesticides dominate the agricultural market due to their highly effective role in crop protection. However, this well-established practice results in some hazardous effects derived from the high persistence of residues in the environment, including agricultural soils (Demeter et al., 2021; Khursheed et al., 2022). Another major concern that is not always addressed when designing a pest management programme is the generation of resistance. The current approach to this problem involves limiting the dose and rate of pesticide application within the framework of integrated pest management. However, even though sublethal doses can prevent and mitigate insect damage, they may lead to the accumulation of resistance factors, resulting in a resistant population. Lower doses may delay single resistance gene evolution but may accelerate multifactorial polygenic resistance (Hawkins et al., 2019; Muniz-Junior et al., 2023).

http://dx.doi.org/10.19103/AS.2024.0149.05

Two reasons underlie the problem of synthetic pesticides: the use of formulations based on single (or few) chemical compounds and their use at rates that detrimentally impact ecosystems as a whole. As for the latter, changes in application rates and logistics are, to an extent, easy to integrate into control programmes as technological advances enable new and better ways of delivering chemicals in more precise and targeted ways. Despite novel synthetic pesticides posing less risk than their predecessors (Isman et al., 2011), the search for new unique compounds that act as control agents is a priority due to the emergence of resistance to traditional chemicals and governments legislating for safer agricultural practices. As a consequence, there is an urgent need to find novel substances and botanical extracts that are sources of new products for crop protection (Demeter et al., 2021; Zettler and Arthur, 2000).

Natural products derived or extracted from plant material are widely acknowledged as a green alternative to conventional pesticides, with a long list of studies reporting their effects on pest insects over the decades. Other than agriculture, botanical extracts are widely used in the cosmetic and food industries and regarded as safe (Koul et al., 2008), but certain natural products are harmful for humans such as certain alkaloids, e.g. atropine or nicotine (Isman et al., 2011). Hence, when discussing natural extracts as safer control agents, they must be grouped into one of the two new botanical insecticides available on the market, i.e. neem-based products or plant essential oils (Isman et al., 2011). In this chapter, we will focus on the latter.

2 Developing biocontrol techniques: the case of insect pests in black truffle orchards

The black truffle, *Tuber melanosporum* Vittad., like other species of the *Tuber* genus, is a hypogeous edible symbiotic fungus living in an ectomycorrhizal relationships with the roots of host trees such as oak, hazel, poplar, pine, among others. Its fruiting bodies, known as truffles, remain underground during the late autumn to winter and rely on volatile diffusion for fauna to locate and disperse them as a haploid spore. After germination, a new symbiotic relationship is formed when a compatible root is encountered, and mycelia will begin vegetative growth following a heterothallic behaviour, with separated sexes in separate hosts. Once hyphae of a complementary mating type are found, the process of forming the sexual fruiting body, an ascoma-bearing asci, begins and this allows the cycle to start over (Fischer et al., 2017; Lee et al., 2020). With this complex life cycle and dependency on appropriate biotic and abiotic conditions, traditional culturing practices (mainly forest collection) have been set aside in favour of optimized techniques.

In the wild, *T. melanosporum* is found in the natural woodlands of calcareous soils of Spain, France and northern Italy, along with other European countries as minor representatives (Fischer et al., 2017). In this context, Europe has seen a dramatic decline in truffle production over the last century, primarily as a result of habitat deterioration, overexploitation and climate change (Garcia-Barreda et al., 2018; Reyna, 2000).

Historically, black truffle was a non-timber forest product (NTFP) associated with oak trees (typically *Quercus ilex*), but it has now become a promising crop for cultivation. Truffle orchards are mostly found in the Celtiberian Range, a depopulated region of Spain including the provinces of Soria and Teruel (Carretero, 2017; Garcia-Barreda et al., 2018). This highly profitable crop has contributed to increasing the welfare and growth of these rural populations. Additionally, industrial networks revolving around truffle cultivation have been established in the surrounding territories, boosting the labour market (Palau, 2018).

Establishing a truffle monoculture presents the perfect scenario for mycophagus insects and mammals involved in dispersal to thrive, resulting in a population explosion. This is the case for animals such as wild boars (*Sus scrofa* L.), truffle flies (*Suillia tuberiperda* Rondani., *Suillia pallida* Fallen. *Helomyza tubervivora* R.-Desv.) and truffle beetles (*Leiodes cinnamomeus* Panzer). These organisms are responsible for reducing truffle production and quality (Andueza et al., 2015; Fischer et al., 2017; Martín-Santafé et al., 2014).

L. cinnamomeus is the most important truffle pest, with more than 20% of harvested truffle often being infested (Garcia-Barreda et al., 2020). The black truffle beetle is an obligatory feeder of *Tuber* species with a life cycle historically associated with truffles with a role in spore dispersion. This means its life cycle is completely linked to the presence of truffles and their stage of development (Martín-Santafé et al., 2014). Adults will mate in mid-September (they will remain in contact with truffles until May) and lay eggs in October. The first larvae (white yellowish) will then appear early in November and develop until March, damaging the fruiting bodies as a result of powerful jaws used to cavitate the truffle. Larvae of this species have no eyes, so the presence of volatile attractants is a likely means of finding a food source. After feeding, larvae will enter a diapause as early as late winter and spring to prepare for the summer, forming an earthy cocoon up until September, when pupation starts. One to one-and-a-half months allows the completion of a whole cycle, and several stages of *L. cinnamomeus* are known to coexist simultaneously in the field, exacerbating the difficulty in controlling this pest (Andueza et al., 2015; Kilian et al., 2022).

The mobility of these insects is aided by loose and moist soil and impeded by dry compacted soil (Andueza et al., 2015; Navarro-Llopis et al., 2021). Irrigation of truffle orchards creates an optimal environment for *L. cinnamomeus* to spread. The damage to truffles is two-fold, the beetle feeds on truffles and depreciates its market value due to aesthetic and organoleptic damage but also allows secondary infections from other fungi and truffle flies (Andueza et al., 2015; Garcia-Barreda et al., 2020; Martín-Santafé et al., 2014; Navarro-Llopis et al., 2021). The damage caused by this insect can currently reach 70% in terms of economic loss (Barriuso et al., 2012), necessitating the requirement for new environmentally safe pest control tools.

3 Biocontrol strategies for control of *Leiodes cinnamomeus* in black truffle orchards

The case of *L. cinnamomeus*, which was not considered a pest until recently, shows how control strategies can be designed (and adjusted) using cultural practices in combination with scientific research. On this occasion, the main topic of discussion is the application of essential oils. In this case study, updated information was gathered regarding present control strategies against *L. cinnamomeus*. The trends in novel pesticide research point to essential oils as suitable candidates to replace or enhance traditional chemical control. In this chapter, we will briefly address the chemistry of essential oils. We will also discuss their potential for the control of this specific pest. Additionally, several traditional and novel essential oil formulation methods are described.

3.1 Cultural techniques

There is currently no effective treatment for this insect pest, and the severity of damage is mitigated by means of cultural practices; superficial tillage is performed to prevent truffle development in the first centimetres of soil, where *L. cinnamomeus* presence is the highest (Martín-Santafé et al., 2014), as well as a thorough clearance of truffles in between the harvest seasons by starting collection earlier and finishing later (from late October to late March) to ensure that no fruiting bodies (and diapausant larvae) remain during the summer in the next season. Peat-based amendment was found to reduce, to a certain degree, the presence of *L. cinnamomeus* (Garcia-Barreda et al., 2020). This practice is similar to tilling, meaning that the superficial layer of soil not only is moved and aerated but also involves adding soil with a known spore concentration. As peat amendments rely on moisture to keep low thermal conductivity, lowering irrigation (hindering the survivability of *L. cinnamomeus*) is not an option.

3.2 Kairomone-based mass trapping

Dimethyl sulfide (DMS) is a compound detected in almost all hypogeous fungi that attract *L. cinnamomeus* (Pacioni et al., 1990; 1991). Recently, kairomone-based (DMS) mass trapping has been developed (Navarro-Llopis et al., 2021). Navarro-Llopis et al. assessed two trap networks/densities during two harvest seasons (40 traps/ha and 80 traps/ha from late September to early October until

April in the 2016–2017 and 2017–2018 seasons) in well-established orchards (17–25 years old) where the presence of *L. cinnamomeus* had been reported. Both trapping densities were able to collect male and female individuals. The total number of captures was higher in the 80 traps/ha plots, with a reduction in damage of around 60%, and fewer number of galleries/g truffle when compared with the 40 traps/ha plots. The authors reported that the cost of high-density trapping (€400/ha) was offset by the amount saved by damage reduction (€900/ha). However, only plantations in full production (around €200/ha/week) can support the cost of this high-density system. Trapping at 40 traps/ha density did not effectively prevent *L. cinnamomeus* damage (average reduction of 9%), nonetheless, a common accumulative effect of damage reduction at both densities was detected throughout the two harvest seasons.

The minimum area required for the treatment to be effective is highly dependent on insect biology and dispersal capacity, along with other important aspects of its ecology and ethology that remain uncertain. This coleopteran has a clear active flight capacity, with other *Leiodidae* covering distances from 40 m to 200 m. Given that orchards from different producers are usually adjacent to one another, it would be necessary to reach an agreement between all the producers involved to effectively reduce *L. cinnamomeus* to an acceptable threshold across the fields. Another point to consider is that truffles, as non-forest timber products, grow within a living and dynamic system, populated by a wide variety of taxa across the macroscopic and microscopic scales. In truffle orchards, DMS traps have been shown to attract arthropods from *Araneae, Pseudoscorpiones, Opiliones, Acarina, Isopoda, Collembola, Hemiptera, Hymenoptera, Lepidoptera, Diptera* and *Coleoptera*, with the latter three being the most frequent, and Coleoptera being the most frequent among the three, followed by Diptera (Pacioni et al., 1991). Thus, mass trapping against pests should consider the ecological impact of the different trap densities.

3.3 Entomopathogenic nematodes

The use of entomopathogenic nematodes to control this pest has been proposed (Julià et al., 2023). Nematodes of the *Steinernematidae* and *Heteorhabditidae* families were used due to their parasitic nature against insects (Argotti, 2019a,b; Lacey and Georgis, 2012; Shapiro-Ilan et al., 2017), with problems of controlling this pest related to environmental factors, such as low temperatures. *Steinernema carpocapsae* (Weiser), *Steinernema feltiae* (Filipjev) and *Heterorhabditis bacteriophora* (Poinar) juvenile individuals from commercial and wild origins were tested *in vitro* with parasite adults, larvae and diapausant larvae of *L. cinnamomeus* at 20°C, 10–15°C and 10°C at doses of 25 infective juveniles/cm^2. The results indicate that temperature affects nematode performance when infecting *L. cinnamomeus*, with *S. carpocapsae* as the most virulent and fastest to kill (100% mortality) against adults, at 20°C; against larvae, *S. feltiae* and *S. carpocapsae* performed similarly at 20°C, but the efficacy of the latter dropped at 10°C, leaving *S. feltiae* as the best choice to control the pest.

The authors indicated that the best application time would be September–October, where the temperature in the field is higher than 15°C for at least 6 h a day in order for the nematode application to be effective, with an additional treatment of *H. bacteriophora* during the summer (Julià et al., 2023).

4 Plant-based biocontrol agents: the use of essential oils

Volatiles have been mentioned several times in this chapter. Kairomone traps and olfactometry, though not being the panacea to solve this problem, have been regarded as a milestone in developing new techniques. If *L. cinnamomeus* adults are able to deviate from truffles and into kairomone traps, other volatile compounds present in the field have the potential to interfere with the perception of truffle volatiles from larvae and adults, protecting the crop.

Essential oils are botanical extracts, generally of terpenic nature, produced as a result of secondary metabolism and extracted via steam distillation. Historically, aromatic plants have been used as flavouring and cosmetic agents, and as protective agents from pests for stored products (Campolo et al., 2018; Demeter et al., 2021; Isman et al., 2011; Zettler and Arthur, 2000); essential oils are extracted mainly (but not exclusively) from such plants.

Essential oils contain a rich array of compounds, each with different toxicological properties that not only operate via multiple modes of action causing a variety of outcomes when applied to insects (contact or fumigant toxicity, knock-down effect, repellence, deterrence etc.) but also remain relatively non-toxic to mammals, particularly in the case of formulated, non-raw essential oil products (Isman et al., 2011). In addition, they have attracted considerable interest thanks to their high volatility and degradation sensitivity, leading to lower persistency in the field (Campolo et al., 2018; Isman et al., 2011); resulting in lower or zero residues.

Essential oils have been suggested as a novel control method, with similar or better activity than conventional pesticides not only *in vitro* but also in field experiments with other crops (Isman et al., 2011). The nature of these extracts make them great candidates for the control of pest insects. In the context of black truffle production in Spain, this is a minor but strategic crop, responsible for the economy of a now thriving depopulated region. Controlling emerging threats is crucial to ensure the welfare and prosperity of its people.

The acceptance of sustainable agriculture has promoted global research regarding the use of essential oils as insecticides. Insecticidal activity can be evaluated by means of absorption, passive ingestion or inhalation (e.g. fumigation), penetration of the cuticle (contact toxicity) or direct ingestion (de Lira et al., 2015; Nenaah, 2014) and their physiological effects can vary from repellence or growth inhibition to insecticidal or ovicidal activity (Garrido-Miranda et al., 2022).

As essential oils are composed of an array of molecules that belong to different terpene families, which are reasonably similar in structure, they can be expected to act somewhat similarly. However, in reality, these compounds mostly do not act independently but attack the insect's physiology via different mechanisms in a synergistic way, in contrast to a dramatic decline of biocidal activity when compounds are applied independently (Khursheed et al., 2022; Navarro-Rocha et al., 2020; Santana et al., 2014; Valcárcel et al., 2021).

Regarding the composition of essential oils, altering the growth conditions of the individual plant will alter the chemical composition of the plant due to physiological responses and, by extension, that of the extract (Isman and Grieneisen, 2014; Keefover-Ring et al., 2009; Murtagh and Smith, 1996). These differences persist to the extraction phase, even when harvesting different parts of the same plant and are responsible for variability in the results observed in the literature, because differences in growth conditions between different individuals of the same species are, usually, disregarded (Campolo et al., 2018). Cultivation and domestication of a species contributes to the traceability of the biomass material and production stability, leading to a higher chance of successful scaling into field experiments (Navarro-Rocha et al., 2020).

Chemotypes of aromatic plants are defined as the differences in chemical characterization within a population of morphologically indistinguishable individuals (Keefover-Ring et al., 2009); these are differences that align not with the category of subspecies, varieties or ecotypes but rather with the growth conditions of the individual plant (i.e. soil properties, watering regime, season of harvest and potential aggressions by herbivores, parasites, etc.) (Isman and Grieneisen, 2014; Keefover-Ring et al., 2009). Discussing chemotypes, rather than species, becomes essential when considering biological activity, as it refers to the chemical composition of the product rather than its evolutionary origin.

For these reasons, essential oils are promising candidates to act as agents of control; their low persistency in the field as well as low toxicity to mammals make them suitable for pest management. Nowadays, products based on rosemary, peppermint, cinnamon, clove, eucalyptus and thyme (among others) are already on the market, with some containing just one species as the active ingredient, while others contain mixes of several oils. These products should serve as a reference when developing essential oil-based pesticides (Isman, 2020). The risks regarding the use of such products involve the concentration needed to achieve the desired effect (sometimes relatively high when compared with synthetic counterparts) and the organoleptic impact on crops.

Both problems can be tackled with a proper formulation, application and post-application management regime.

An important disadvantage regarding essential oils in crop protection is the low extraction yield (unsuitable for large-scale control programmes); however, steady production is feasible, as demonstrated not only by the cosmetic industry but also by research on their potential use in crop protection against insect pests with oils extracted from pre-domesticated plant species (Demeter et al., 2021; Koul et al., 2008; Navarro-Rocha et al., 2020; Valcárcel et al., 2021).

5 Assessing the effects of essential oils on insect pests: the case of *Leiodes cinnamomeus*

To date, no bioassays with essential oils have been conducted against *L. cinnamomeus.* In addition, the seasonal nature of its occurrence and the difficulty of establishing a steady population of this insect in the laboratory given its diet currently undermines research in this field.

However, the effects of essential oils on other colepterans such as *Tribolium castaneum* or *Tribolium confusum* are well known. These species are pests of stored products but have historically served as models in terms of the ecology, ethology, genomics (and all the links in between) of beetles (Campbell et al., 2022; Pointer et al., 2021). Essential oils have shown activity against these models and other coleopterans (de Lira et al., 2015; Garrido-Miranda et al., 2022; Nenaah, 2014).

These extracts act via different modes of action with variable intensity related to the relative concentration of chemicals present in the mix (Khursheed et al., 2022). Using *Tribolium* models, essential oils have been reported to interfere with the metabolism and physiological functions of the beetle, decreasing the lipid, protein and glycogen reservoirs of larvae produced when larvae are produced with sublethal doses, thus reducing the total energy of adults, endangering their survival. In addition, varying degrees of histological and neurotoxic damage have been reported, with tissue necrosis and alteration of acetylcholinesterase enzyme function, as well as the enhanced activation of detoxifying systems, specifically their glutathione S-transferase activity, for some botanical species (Campolo et al., 2018).

Therefore, essential oils are a potential solution to reduce the damage by *L. cinnamomeus*, but one must keep in mind the volatility and complex composition of such extracts that may disappear from the field too soon or affect collateral targets. For this reason, an appropriate optimized formulation becomes necessary to efficiently tackle this problem.

Raw or ("unformulated") pesticides present problems such as low solubility in water, chemical instability or difficulties in transport or application in the field. Consequently, manufacturers add ingredients stated as "inert" (e.g. clay, solvents etc.) to serve as carriers. Usually, formulation involves a mixture of the pesticide that will control the target pest, a carrier (e.g. solvent or mineral clay), adjuvants (e.g. stickers, spreaders, etc.) and minor ingredients that improve its activity (e.g. stabilizers, safeners, dyes etc.) (Fishel, 2019; Kalyabina et al., 2021).

The non-polar nature of essential oils makes them difficult to formulate in liquid (dissolved in water). Emulsions or non-polar solvents in different formulas (e.g. aerosols, liquid baits, etc.), although effective in a wide variety of situations, would not be as effective when controlling underground insects, with the risk of affecting larger areas than needed. For these reasons, dry formulations are a potential candidate; they are manufactured as ready-to-use products or concentrates ready to mix with water and then spray (Fishel, 2019; Kalyabina et al., 2021).

On the ready-to-use side, there are dusts (used for seed treatments or open-air insects, but they can drift away during application and can hardly penetrate soil with a uniform distribution), granules (similar to dusts, but the particles are larger and designed for slow release; often under controlled application of moisture, they do not stick to surfaces, however, and may be hazardous to other species) and pellets (mainly used for fumigation, and they present the same pros and cons as the two previous cases). In addition, baits are commonly used (the active ingredient is placed along with food, they are able to control pests locally but may affect other species and serve as a food supply for the pest once the efficacy of the pesticide is lost) (Fishel, 2019; Kalyabina et al., 2021).

On the other hand, wettable powders are widely used, often mixed with water and sprayed on leaf surfaces. When the pesticide is not water soluble (as happens with essential oils), the powder formulation is not advisable. Water-dispersible granules break down in water and release the active ingredient and therefore could be used with essential oils (Fishel, 2019; Kalyabina et al., 2021).

Microencapsulated materials consist of liquid or dry particles covered in materials that will break down slowly and release the active ingredients. They are better than non-enclosed formulations when using biodegradable materials to encapsulate, and they present extended effectiveness due to their slow release, making them disappear later from the application site. This method can produce particles as small as pollen grains and, depending on the materials used and environmental conditions, one can expect different intervals for application (Fishel, 2019).

The encapsulation of essential oils is a thriving research field not only for pest control but also in industries such as cosmetics. Encapsulation will control the previously mentioned high volatility by physical or chemical interaction with a matrix that holds the substance for longer periods, enhancing their activity in the field (Baser and Buchbauer, 2015; Isman, 2020; Maes et al., 2019).

The typical pesticide release profile would be a rapid initial release followed by a steady prolonged release, this can be the consequence of simple diffusion, environmental intervention or matrix erosion. Different encapsulation techniques can render encapsulations in particles (matrix with the ingredients in dispersion), capsules (membrane surrounding the ingredients), complexes (spatial disposition of the ingredients into an open structure) or droplets (fine bubbles of ingredients dispersed in the solvent) on the micrometre to nanometre scale (Maes et al., 2019). The final choice depends on the required application.

In order to study product release, one can measure the targeted bioactivity (excluding the quantitative aspect of the release) or the quantitative release by static or dynamic headspace using gas chromatography–mass spectrometry analysis (excluding the quality of the encapsulation). Considering that essential oils are complex mixtures of compounds with different volatilities and knowing that the moment of release is as important as the chemical species found in the mix (certain compounds will act in synergy, if present, with others), a thorough analysis would involve recovering both sets of quantitative and qualitative data and nuancing the final comparison between different techniques of encapsulation (Maes et al., 2019).

A recent review, focusing on nanobiopesticides, discusses essential oils among other natural substances as potential active ingredients in phytosanitary treatments for modern agriculture (Machado et al., 2023). Nanoparticles (unbound or aggregate particles with a size range between 1 nm and 100 nm) are proposed as a promising alternative for newer phytosanitary strategies, and essential oils with different carriers were the most popular natural extract covered (with zein protein or montmorillonite as examples of how carriers can enhance the active ingredient's bioactivity). However, the final ecotoxicological activity of nanoparticles has not been explored as a whole (Machado et al., 2023).

Nanoproducts for insect control have already been tested against Curculionidae and other Coleoptera species with favourable results. These products are based on emulsions on the nanometric scale, polymeric biodegradable nanoparticles or biocompatible clay nanomaterials (via ionic adsorption) to retain essential oils and promote their gradual release (Garrido-Miranda et al., 2022).

There is still a long way to go for these products to be used in control programmes, with the main concerns being the development of safe and effective formulations and regulatory framework, the assessment of their stability and environmental persistence, as well as non-target toxicity (humans included), field assays that demonstrate the efficacy of these products and, finally, establishing a complete registration and standardization process for the market (Garrido-Miranda et al., 2022; Machado et al., 2023).

6 Conclusion

T. melanosporum cultivation has been consolidated as a strategic crop for depopulated areas in certain regions of Spain. It serves as a welfare engine that also supports the industrial cohesion of the region. However, the popularity of this crop has resulted in the proliferation of historically associated truffle pests, such as *L. cinnamomeus*. This coleopteran is responsible for large economical losses that are increasing annually and, to date, no effective control method has been developed despite the efforts made by farmers and the scientific community.

Essential oils have gained popularity in the field of novel green pesticides. These botanical extracts are neurotoxic against insects and their effectiveness to control pests is well studied, both *in vitro* and in the field, with often better results than traditional synthetic counterparts. A steady supply of essential oils is possible as demonstrated by the food and cosmetic industries for biological control; however, the process of plant domestication is also necessary to ensure a chemically stable plant population maintaining relatively constant concentrations of the active compounds in the oil.

Globally, there is effort to develop and design essential oil-based formulations to extend the effective life of the extract in the field and counter their volatility. Nanoformulations are one of the most promising results but an appropriate regulatory framework is needed to develop newer commercial products. With a proper formulation, essential oils may serve as a great solution against *L. cinnamomeus*, thanks to their low persistency, low ecotoxicity and potential to be integrated along with other control methods such as mass trapping or entomopathogenic nematodes.

7 Where to look for further information

For further information on truffle cultivation, one may consult the informational page of the Regional Government of Aragón or, for specific details on black truffles, refer to its dedicated webpage. Regarding the management of this crop, the Forest Resources Research Team at the Center for Research and Agro-Food Technology of Zaragoza (Centro de Investigación y Tecnología Agroalimentaria, CITA) possesses extensive expertise in studying forest fungal species. For inquiries related to pest biocontrol and the use of essential oils, the Biopesticides Research Group at the Institute of Agricultural Sciences (Instituto de Ciencias Agrarias, ICA) in Madrid serves as a key reference point. The case of *L. cinnamomeus* is particularly recent, and as such, the information presented at the time of writing this chapter represents the most up-to-date knowledge available.

8 References

Andueza, G., et al., 2015. Truffles´ Flies and Beatles - Suillia and Leiodes, in: *Truffle Farming Today, a Comprehensive World Guide* (pp. 287–299). Micofora. Micologia Forestal & Aplicada.

Argotti V., E.E., 2019a. Patogenicidad de nemátodos entomopatógenos del género Steinernema y Heterorhabditis sobre larvas de Tecia solanivora en Ecuador. *Bol. Téc. Ser. Zool.* 9(6).

Argotti V., E.E., 2019b. Caracterización ecológica de nemátodos entomopatógenos del género Heterorhabditis aislados de Tecia solanivora. *Bol. Téc. Ser. Zool.* 9(6).

Baser, K.H.C., and Buchbauer, G. (Eds.), 2015. *Handbook of Essential Oils: Science, Technology, and Applications*, Second Edition, 0 ed. CRC Press. https://doi.org/10.1201/b19393

Barriuso, Juan, Martín-Santafé, María, Sanchez, Sergio, & Palazón, Carlos. (2012). Plagas y enfermedades asociadas al cultivo de la trufa. En Santiago Reyna Domenech (Ed.), Truficultura: Fundamentos y técnicas (2a. ed. revisada y ampliada, pp. 275–302). Mundi-Prensa.

Campbell, J. F., Athanassiou, C. G., Hagstrum, D. W., & Zhu, K. Y. (2022). Tribolium castaneum: A Model Insect for Fundamental and Applied Research. Annual Review of Entomology, 67(1), 347–365. https://doi.org/10.1146/annurev-ento-080921-075157

Campolo, O., et al., 2018. Essential oils in stored product insect pest control. *J. Food Qual.* 2018, e6906105. https://doi.org/10.1155/2018/6906105

Carretero, N., 2017. *La Laponia española*. El País.

de Lira, C.S., et al., 2015. Evaluation of the toxicity of essential oil from Alpinia purpurata inflorescences to Sitophilus zeamais (maize weevil). *Crop. Prot.* 71, 95–100. https://doi.org/10.1016/j.cropro.2015.02.004

Demeter, S., et al., 2021. Insecticidal activity of 25 essential oils on the stored product pest, *Sitophilus granarius*. *Foods* 10, 200. https://doi.org/10.3390/foods10020200

Fischer, C., et al., 2017. *Best Practices for Cultivation of Truffles*. CTFC, Forest Sciences Centre of Catalonia, Solsona, Spain; Yaşama Dair Vakıf, Antalaya, Turkey.

Fishel, F.M., 2019. PI231/PI231: Pesticide Formulations [WWW Document]. IFAS Ext. Univ. Fla. URL https://edis.ifas.ufl.edu/publication/PI231 (accessed 12.12.23).

Garcia-Barreda, S., et al., 2018. Black truffle harvesting in spanish forests: trends, current policies and practices, and implications on its sustainability. *Environ. Manage.* 61, 535–544. https://doi.org/10.1007/s00267-017-0973-6

Garcia-Barreda, S., et al., 2020. Edaphic and temporal patterns of tuber melanosporum fruitbody traits and effect of localised peat-based amendment. *Sci. Rep.* 10, 4422. https://doi.org/10.1038/s41598-020-61274-x

Garrido-Miranda, K.A., et al., 2022. Essential oils and their formulations for the control of curculionidae pests. *Front. Agron.* 4, 876687. https://doi.org/10.3389/fagro.2022.876687

Hawkins, N.J., et al., 2019. The evolutionary origins of pesticide resistance. *Biol. Rev.* 94, 135–155. https://doi.org/10.1111/brv.12440

Hoyer, P.B., Keating, A.F. 2014. Xenobiotic effects in the ovary: temporary versus permanent infertility. Drug Metabolism and Toxicology (Expert Opinion). 10(4):511–23.

Isman, M.B., 2020. Bioinsecticides based on plant essential oils: a short overview. *Z. Für Naturforschung C* 75, 179–182. https://doi.org/10.1515/znc-2020-0038

Isman, M.B. and Grieneisen, M.L., 2014. Botanical research: many publications, limited useful data. *Trends Plant Sci*. 19, 140–145. https://doi.org/10.1016/j.tplants.2013.11.005

Isman, M.B., et al., 2011. Commercial opportunities for pesticides based on plant essential oils in agriculture, industry and consumer products. *Phytochem. Rev*. 10, 197–204. https://doi.org/10.1007/s11101-010-9170-4

Julià, I., et al., 2023. The development of the truffle beetle Leiodes cinnamomeus at low temperature, a determining factor for the susceptibility of adults and larvae to entomopathogenic nematodes. *Biol. Control* 180, 105197. https://doi.org/10.1016/j.biocontrol.2023.105197

Kalyabina, V.P., et al., 2021. Pesticides: formulants, distribution pathways and effects on human health – a review. *Toxicol. Rep*. 8, 1179. https://doi.org/10.1016/j.toxrep.2021.06.004

Keefover-Ring, K., et al., 2009. Beyond six scents: defining a seventh Thymus vulgaris chemotype new to southern France by ethanol extraction. *Flavour Fragr. J*. 24, 117–122. https://doi.org/10.1002/ffj.1921

Khursheed, A., et al., 2022. Plant based natural products as potential ecofriendly and safer biopesticides: a comprehensive overview of their advantages over conventional pesticides, limitations and regulatory aspects. *Microb. Pathog*. 173, 105854. https://doi.org/10.1016/j.micpath.2022.105854

Kilian, A., et al., 2022. Larval Morphological Adaptations of Leiodes cinnamomea (Panzer, 1793) (Coleoptera: Leiodidae: Leiodinae)–Obligatory Feeder of Tuber Species. *Insects* 13, 249. https://doi.org/10.3390/insects13030249

Koul, O., et al., 2008. Essential oils as green pesticides: potential and constraints. *Biopest. Int*. 4, 63–84.

Lacey, L.A. and Georgis, R., 2012. Entomopathogenic nematodes for control of insect pests above and below ground with comments on commercial production. *J. Nematol*. 44, 218–225.

Lee, H., et al., 2020. Potentials of truffles in nutritional and medicinal applications: a review. *Fungal Biol. Biotechnol*. 7, 9. https://doi.org/10.1186/s40694-020-00097-x

Machado, S., et al., 2023. Nanobiopesticides: are they the future of phytosanitary treatments in modern agriculture? *Sci. Total Environ*. 896, 166401. https://doi.org/10.1016/j.scitotenv.2023.166401

Maes, C., et al., 2019. Encapsulation of essential oils for the development of biosourced pesticides with controlled release: a review. *Molecules* 24, 2539. https://doi.org/10.3390/molecules24142539

Martín-Santafé, M., et al., 2014. Phytosanitary problems detected in black truffle cultivation: a review. *For. Syst*. 23, 307–316. https://doi.org/10.5424/fs/2014232-04900

Mayorga E.J., et al., 2020. Biology of heat stress; the nexus between intestinal hyperpermeability and reproduction in swine. Theriogenology. 154:73–83.

Muniz-Junior, G., et al., 2023. Are lower pesticide doses better? An evolutionary perspective on integrated pest management. *Ecol. Model*. 482, 110408. https://doi.org/10.1016/j.ecolmodel.2023.110408

Murtagh, G.J. and Smith, G.R., 1996. Month of harvest and yield components of tea tree. II. Oil concentration, composition, and yield. *Aust. J. Agric. Res*. 47, 817–827. https://doi.org/10.1071/ar9960817

Navarro-Llopis, V., et al., 2021. Control of *Leiodes cinnamomeus* (Coleoptera: Leiodidae) in Cultivated Black Truffle Orchards by Kairomone-based mass trapping. *J. Econ. Entomol.* 114, 801–810. https://doi.org/10.1093/jee/toaa317

Navarro-Rocha, J., et al., 2020. Composition and biocidal properties of essential oil from pre-domesticated Spanish Satureja Montana. *Ind. Crops Prod.* 145, 111958. https://doi.org/10.1016/j.indcrop.2019.111958

Nenaah, G.E., 2014. Chemical composition, toxicity and growth inhibitory activities of essential oils of three Achillea species and their nano-emulsions against Tribolium castaneum (Herbst). *Ind. Crops Prod.* 53, 252–260. https://doi.org/10.1016/j.indcrop.2013.12.042

Pacioni, G., et al., 1990. Odor composition of the Tuber melanosporum complex. *Mycol. Res.* 94, 201–204. https://doi.org/10.1016/S0953-7562(09)80614-X

Pacioni, G., et al., 1991. Insect attraction by tuber: a chemical explanation. *Mycol. Res.* 95, 1359–1363. https://doi.org/10.1016/S0953-7562(09)80385-7

Palau, N., 2018. La trufa crea empleo y combate el éxodo rural [WWW Document]. La Vanguardia. https://www.lavanguardia.com/natural/tu-huella/20181005/452177710143/sector-trufa-teruel-empleo-combate-despoblacion-motor-economico-cultivo-rural.html (accessed 10.11.23).

Pointer, M. D., Gage, M. J. G., & Spurgin, L. G. (2021). Tribolium beetles as a model system in evolution and ecology. Heredity, 126(6), Article 6. https://doi.org/10.1038/s41437-021-00420-1

Reyna, S., 2000. *Trufa, Truficultura y Selvicultura Trufera*, 1st ed. Ediciones Mundi-Prensa.

Santana, O., et al., 2014. Valorization of essential oils from moroccan aromatic plants. *Nat. Prod. Commun.* 9, 1934578X1400900. https://doi.org/10.1177/1934578X1400900812

Shapiro-Ilan, D., et al., 2017. Chapter 6 - Basic and Applied Research: Entomopathogenic Nematodes, in: Lacey, L.A. (Ed.), *Microbial Control of Insect and Mite Pests*. Academic Press, pp. 91–105. https://doi.org/10.1016/B978-0-12-803527-6.00006-8

Valcárcel, F., et al., 2021. Acaricidal and Insect antifeedant effects of essential oils from selected aromatic plants and their main components. *Front. Agron.* 3, 662802. https://doi.org/10.3389/fagro.2021.662802

Zettler, J.L. and Arthur, F.H., 2000. Chemical control of stored product insects with fumigants and residual treatments. *Crop Prot., XIVth International Plant Protection Congress* 19, 577–582. https://doi.org/10.1016/S0261-2194(00)00075-2

Chapter 4

Developments in peptide-based biocontrol agents to manage insect pests

Yousef M. Abul-Haija, Lewis J. Archibald, Alexander J. Finney, Hugh A. Nicholson, Julian A. T. Dow, and Shireen-Anne Davies, SOLASTA Bio Ltd., UK

1 Introduction

Insects represent an abundant and dominant life form on Earth, successfully colonising the widest range of habitats despite an ever-expanding human population and changing climatic conditions.[1] Given the incalculable sum of insects in existence, with habitats across all agricultural economies, insects represent a major threat to our food security. Harmful insects, which occupy an important subsection of all insect species, present markets with enormous social, health and economic pressures by harbouring and transmitting plant diseases and causing direct crop damage. Ultimately, this translates to an annual loss of 40% of global crop production, according to the World Economic Forum[2] highlighting the necessary role that expanding the repertoire of regulated Plant Protection Products (PPPs) will play in securing food for the future. Food security is becoming increasingly pressurised by existing and emerging pest and disease pressures and compounded by the increasing difficulty in gaining access to regulated insecticides, given the mounting legislation in leading national and supranational market economies. There are

http://dx.doi.org/10.19103/AS.2024.0149.06

currently 65 insecticides (low-risk active substances) approved in the European Union, of which 14 will expire in the next 10 years.[3] Greener regulatory policies will drive the need to develop environmentally friendly alternatives as part of IPM approaches, avoiding unwanted mortality in beneficial pollinating insect species and reducing residues within the soil and groundwater. There is also increasing pressure from consumers, who are ever more aware of the processes involved in food production and who increasingly consider conventional chemical intervention in crop protection to be unwelcome. A further major current challenge to the insecticide marketplace is the dynamic nature whereby pest species adapt to become resistant to both chemical agents in widespread use and the changing climate.

Warming air temperatures have resulted in the widening habitat of some non-native insect pest species, presenting a novel threat to native field crops in a variety of temperate regions whose populations depend on major crop types. These changes have also enhanced insect survival, growth and increased reproduction rates.[4, 5] Accordingly, there is a global dependence on insecticides. However, the efficacy of currently deployed chemical interventions is negated in several contexts; for example, more than 500 arthropod species worldwide (of which >50% are pest species) are known to have developed resistance to insecticides in the last half century.[6, 7]

There is thus increasing pressure to utilise new avenues to discover insect control agents that exist in a space outside of the historical insecticide targets and to develop biological insecticides that possess new modes of action to complement and ultimately replace chemical approaches to insect management. This will prioritise cutting-edge scientific advances in targeted approaches, which could, for example, hijack target pest insect integrative physiological pathways to invoke resultant behavioural changes and mortality. Utilising emerging technologies to our advantage will afford the industry a more targeted approach than ever before, with methodologies that possess the ability to design insect control agents that specifically target pests while avoiding non-pest species. As neuropeptides regulate critical life processes, new biological insecticides that utilise these relevant pathways may exhibit far-reaching modes of action by overstimulating a range of different biological responses.[8, 9] For further information on insect neuropeptides, readers can consult the references section towards the end of this chapter.

Using neuropeptides to disrupt physiological processes in insects has been considered a viable basis for the development of alternative strategies for insect pest management[10–12] and continues to be an area of promising research interest.[13] Developing neuropeptide-like biologics, such as synthetic derivatives, may offer a novel avenue to increase our repertoire of approaches that will meet the increasing need for greener and more targeted pest management products. These new products would have inherently specific and well-understood characteristics due to the integral properties of such signaling modalities with desired over-stimulation of signaling pathways and selective reduction of fitness resulting in selective mortality. Ideally, these new products would be applied to crops using similar application methods to conventional insect control agents, largely limiting the need to change farming practices.

2 Key issues in using neuropeptides as bioinsecticides

New bioinsecticides based on insect neuropeptides will provide growers with novel, ecologically benign insect management strategies that not only satisfy the requirements of legislative and regulatory bodies, but moreover provide safe alternatives to control an increasingly problematic insect population to ensure our food security for the future.

However, one significant drawback of using native neuropeptides as insect control agents is their vulnerability to rapid degradation by natural peptidase enzymes within insects.[8, 14, 15] In contrast, peptide mimetics can be engineered to resist degradation by these enzymes, ensuring their longevity and effectiveness. Additionally, peptide mimetics may overcome challenges related to penetrating the insect cuticle, ensuring that the active compounds reach their target sites.

Furthermore, the customizability of peptide mimetics might allow for the selective targeting of specific insect pest species, sparing beneficial insects and minimising the impact on non-target organisms. This precision is crucial for IPM programmes, which aim to reduce the use of broad-spectrum chemical insecticides.[16] By doing so, peptide mimetics contribute to the reduction of chemical residues in the environment and help mitigate the development of insecticide-resistant pests.

The application of peptide mimetics within IPM programs represents a significant advancement in sustainable agriculture. IPM involves an integrated approach to pest control, combining various strategies such as biological control, cultural practices and chemical interventions. Peptide mimetics, with their high selectivity and reduced environmental footprint, align perfectly with the principles of IPM.

3 Neuropeptide profiling and families

The knowledge base around insect neuropeptides is ever-increasing due to continued technological advancements and blue-sky research efforts, for example, through improvements in 'omics-based technologies and thus an increase in available data with relevant, accurate information. Genomics allows for the identification of peptide-encoding genes, while transcriptomics allows for a more specific understanding of when and where these genes are expressed. Peptidomics can provide information on the mature peptides produced by a specific cell or tissue. An often-deployed combination of techniques, transcriptomics and peptidomics are used to identify the expression of genes within a tissue while also defining the peptides active within the same sample. Recent examples of this were demonstrated in Pine weevil,[17] cockroaches and the Greater Wax Moth,[18, 19] with another recent example provided by the Chinese Oak Silk Moth,[20] albeit using a transcriptomics-only approach.[17]

With these peptides produced by cells within the nervous system of insects, targeted analysis of the CNS (central nervous system) of these organisms can allow for improved outcomes following computational analyses. One example of such methodology is the analysis of processed neuropeptides by direct tissue profiling, for example, by Matrix Assisted Laser Desorption/Ionization-time of flight mass spectrometry (MALDI-TOF MS) and extract analysis by, for example, Liquid Chromatography (LC)-Orbitrap MS system.[21] Further improvements in mass spectrometry imaging have allowed for peptide localisation within insect brains to be uncovered at high resolution.[22]

More systematic analyses of neuropeptides are now possible within specific model organisms used to study the basic underlying biology of these systems.[23] Both the fruit fly (*Drosophila melanogaster*) and the roundworm (*Caenorhabditis elegans*) have large databases associated with 'omics-based data, allowing for large systematic studies.

Table 1 Published insect neuropeptide families for peptide analogue development

Neuropeptide family	Function	References
Adipokinetic hormone	Lipid metabolism	[13, 26]
Allatostatins	Growth regulation	[27–29]
Capability CAP2b/CAPA	Diuresis, desiccation and cold tolerance	[30–33]
Diapause hormone (DH)	Development	[34]
Diuretic hormone 44 (DH44)	Diuresis, feeding	[35]
Kinin	Diuresis, feeding, environmental stress tolerance	[11, 13, 24, 31, 32, 35–38]
Myosuppresin	Feeding	[39]
Natalisin	Reproductive behaviours	[40]
Pyrokinin (PK)	Multifunctional	[31, 41, 42]
Proctolin	Locomotion, digestive processes	[25]
Pheromone biosynthesis-activating neuropeptide (PBAN)	Regulation of pheromone biosynthesis	[43–49]
Sulfakinin	Digestive processes	[50–52]
Tachykinin-like peptide	Gut myostimulatory activity	[47]

DH, diapause hormone; DH44, diuretic hormone 44; PBAN, pheromone biosynthesis-activating neuropeptide; PK, pyrokinin.

Neuropeptides direct many life-critical processes within insects and higher organisms. Research has identified many families of neuropeptides that are important to insect survival with various vital functions, for example, insect development, feeding or water homeostasis. Table 1 above summarises neuropeptide families, their functions and cites references of studies attempting to develop analogues aimed towards biopesticide design. Much of the recent focus of these studies has been to show the development of bee-safe analogues that still show desired effects on a chosen pest species.[24, 25]

4 Design of peptide bioinsecticides

Peptides, as a class of compounds occupying the chemical space between small molecules and high-molecular-weight biologics, possess attractive properties in the field of green bioinsecticides. Namely, they often display high target-selectivity, low/no off-target toxicity, minimal environmental residue and a reduced risk of evolved resistance.[53]

However, the use of peptides as bioinsecticides is restricted by their inherent instability to proteolytic degradation *in vivo* and limited ability to cross the biological barriers presented by the insect gut and cuticle. For peptides to function as viable insecticidal agents, these challenges must be overcome by rational design of peptide amino acid sequence, chemical modification of peptide structure and suitable formulation for field application.[54] Additional factors that must be addressed when considering the application of peptides as bioinsecticides are their solubility in aqueous solution and suitability for use in the field. Some of these key factors in the design of peptide-based bioinsecticides are shown in Fig. 1.

Improvement of peptide bioinsecticide properties can be achieved through generational, chemically modified analogue design. This strategy involves careful consideration of how incremental chemical modifications made to the peptide structure affect the properties of the resulting modified peptide analogue. Often, a balance is struck between improving the desired properties of a particular peptide sequence by chemical modification while maintaining activity against the desired target. Some examples of chemical modifications that can be made to peptides to increase their suitability for use as bioinsecticides are given in Fig. 2.

Application of these chemical modifications in isolation or combination has led to success in several reported studies of peptide bioinsecticide design. Shi et al. demonstrated the efficacy of an *N*-terminally-conjugated CAPA-peptide analogue against the peach-potato aphid *Myzus persicae* when applied in combination with another peptide analogue modified by non-canonical amino acid substitution.[33] The authors were also able to show the selectivity of this combinatorial treatment for inducing mortality in aphids over various beneficial

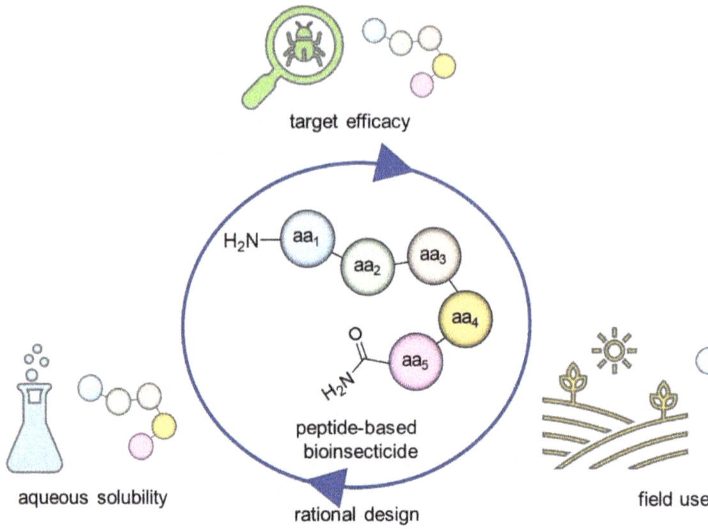

Figure 1 Key aspects of peptide-based bioinsecticide design. Rational candidate design is a continuous process focusing on successive and balanced optimisation of a combination of these factors.

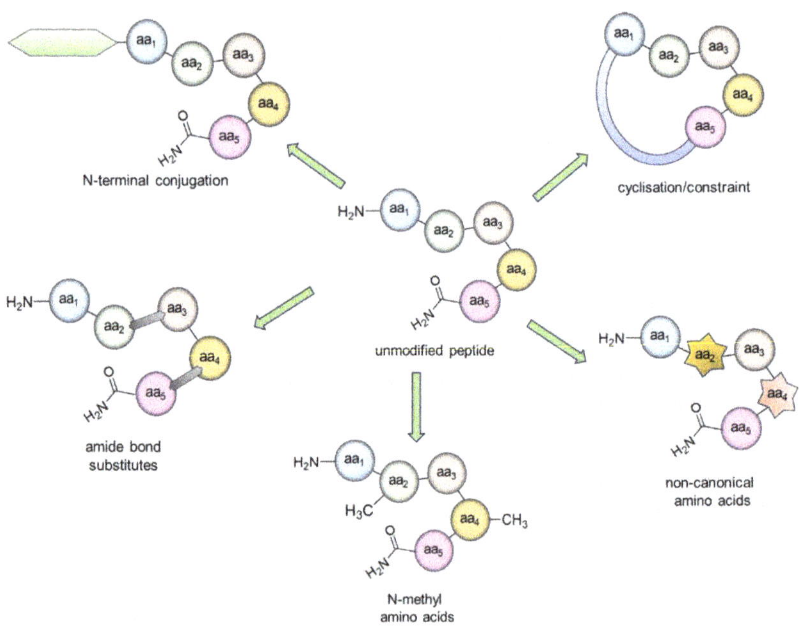

Figure 2 Examples of chemical modifications that can be made to peptides to enhance their properties for use as bioinsecticides.[55]

insect species such as the buff-tailed bumblebee *Bombus terrestris* and the two-spot ladybird *Adalia bipunctata*. This strategy of combining *N*-terminal conjugation with non-canonical amino acid substitution also led to the discovery of several kinin-peptide analogues, reported by Zhang et al., to display efficacy against the soybean aphid *Aphis glycines*.[56] In a different approach, Abdulganiyyu et al. used *N-C*-terminal cyclisation to conformationally constrain an AKH-peptide identified in the migratory locust *Locusta migratoria*.[57] They observed that the resulting cyclic peptide exhibited selective *in vitro* activity for the AKH (adipokinetic hormone) receptor of the desert locust *Schistocerca gregaria* over that of the honeybee *Apis mellifera*.

SOLASTA Bio has recently demonstrated the efficacy of chemically modified peptide analogues as bioinsecticides.[58] Modified peptides were assessed for their efficacy against the important agricultural pest *Drosophila suzukii* when administered *via* feeding, with an example of efficacy data shown in Fig. 3.[58] Also, several iterative generations of peptide designs were tested *in vivo* against *D. suzukii* larvae in a feeding assay. The results show increasing efficacy from first generation to third generation peptide treatments, see Fig. 4.

These chemical-modification strategies are now beginning to be complemented by *in silico* pre-screening of candidates, expediting compound discovery.[37, 59] While significant progress has been made in using chemical modifications to improve the efficacy and stability of peptide-based bioinsecticides, proven strategies for the enhancement of peptide absorption through the insect gut or enhancing cuticle penetration are relatively scarce. It has been observed that some larger peptides (>1 kDa) may be absorbed by the insect gut *via* endocytosis (in which the cells lining the gut absorb

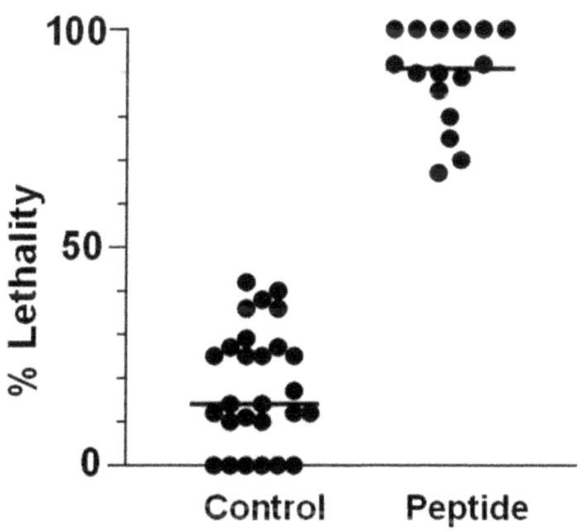

Figure 3 Peptide efficacies against *Drosophila suzukii* larvae, *via* feeding. *In vivo* efficacies data shown after 72-h treatment with the peptide. Median values are indicated. Statistical analysis (Welch *t*-test) indicates statistically significant mean efficacy compared to the control. Source: Reproduced from ref. [58].

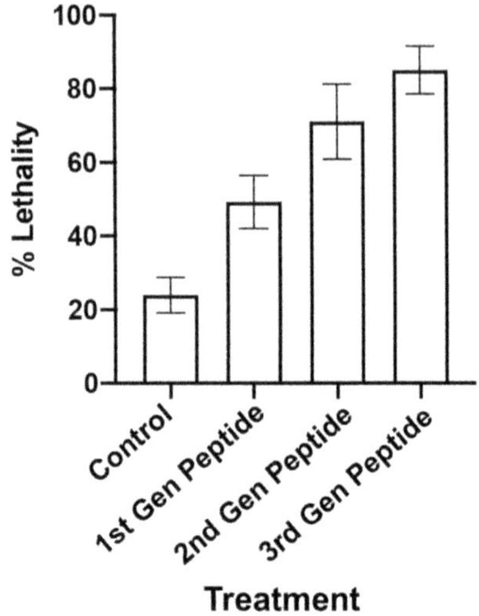

Figure 4 Generational peptide efficacies against *Drosophila suzukii* larvae, via feeding. *In vivo* efficacies data shown after 72-h treatment with the peptide. Source: Reproduced from ref. [58].

compounds by engulfing them in the cell membrane and forming an intracellular vesicle), while some smaller peptides (≤1 kDa) have been shown to pass the gut membrane *via* paracellular diffusion (by which a compound is able to passively diffuse through the septate junctions between gut cells).[60] The development of demonstrable chemical modification strategies, in combination with appropriate formulations,[61] that can improve penetration of a peptide bioinsecticide through the insect gut or cuticle will be vital to the enhancement of their efficacy in the field. Some formulation efforts towards the improved cuticle penetration of other classes of biomolecules such as double-stranded (dsRNA) have already been made.[62]

5 Scale up, manufacture and improving sustainability of peptide bioinsecticides

In order to be considered viable commercial bioinsecticides, peptides must be amenable to facile, cost-effective and environmentally sustainable synthesis in the bulk quantities required for field application (typically requiring a minimum of multiple grams per hectare, although this varies depending on the rate of application and concentration of the dose). The most common means of peptide production to this effect are recombinant bacterial expression (biological production) and chemical synthesis. Recently, *in planta* expression vectors have become of interest in the production of a special class of peptides known as cyclotides; however, this technology has not yet been widely adopted.[63, 64]

Of the two most common approaches, bacterial production is best suited to the manufacture of proteins composed wholly of canonical amino acids – with the introduction of non-canonical residues to peptide sequences produced in this manner involving careful engineering of the expression vectors used. For peptides that have been chemically modified in such a way as to improve their stability, solubility or field suitability, chemical synthesis remains the production method of choice. This is due to the advantage of the atom-by-atom molecular control over the final peptide structure afforded to the synthetic chemist.

However, challenges in producing peptides efficiently at scale remain at the frontier of peptide synthesis research. Recent advances have been made in continuous flow synthesis manifolds and hybrid synthetic approaches employing a combination of solid-phase peptide synthesis (SPPS) and liquid-phase peptide synthesis (LPPS). These have been extensively reviewed elsewhere.[65] Most of the efforts towards the improvement of peptide production at the kilogram scale have focused on pharmaceutically relevant peptides. For example, Frederick et al. demonstrated that the kilogram-scale production of the peptide Tirzepatide, used in chronic weight management, may be achieved using the hybrid strategy in good overall yield for a synthetic peptide (46%) and in large quantity (8.7 kg).[66] Industrial peptide manufacturing companies now offer custom synthesis up to the tonne-scale with cost decreasing per kilogram as the scale is increased.

The major drawback of producing peptides at scale *via* chemical synthesis is the relatively high environmental impact when compared to biological production methods. This is due to the amount of waste produced in the synthesis and downstream purification processes and the safety considerations of the reagents and solvents involved. However, in recent years, significant progress has been made toward addressing these issues.[67] In the majority, these efforts have focused on solvent/base substitution and waste reduction.

Traditionally, the solvent/base system used for peptide synthesis has been a combination of dimethylformamide (DMF) and piperidine. This has proven problematic in bulk synthesis due to the high toxicity and flammability of DMF, making it a poor solvent with regard to environmental sustainability. However, modern peptide synthesis techniques now favour the alternative solvent *N*-butyl pyrrolidone (NBP) in combination with the alternative base pyrrolidine.[68, 69] In addition, the efficiency of automated peptide synthesis in terms of reducing the amount of solvent waste produced has also been improved.[70]

While these advancements have improved the environmental sustainability of bulk peptide synthesis, the advent of fully aqueous peptide synthesis (a technology still in its relative infancy) and its widespread use in the industry remains the ultimate goal of green chemistry efforts in peptide synthesis.[71]

6 Conclusion and future trends

New mode-of-action PPPs will be vital in securing food for the future and will encompass novel active agents outside the repertoire of conventional chemistry. Mounting legislative pressure from governmental bodies and the general population puts the bioinsecticide market at the forefront of this new area. Such new PPPs might be based on insect neuropeptides and act on several distinct pathways in pest insect species. Based on rational design, these new mode-of-action PPPs could target pest insects while leaving beneficial and pollinator insect species unharmed. Understanding the biology of neuropeptides and profiling their abilities to affect insect behaviour and function is key to the development of better neuropeptide-based bioinsecticides. Increasing data on a few important pest species, while utilising basic physiological understanding from model organisms, has allowed for assay development for rapid, iterative neuropeptide mimetic design. The rational design of new peptide bioinsecticide candidates with improved stability has been well-explored, and several proven strategies for tackling this issue through chemical modification have been demonstrated. However, the issues of facilitating the delivery of peptides into the insect haemolymph and their formulation for application in the field remain challenges at the frontier of bioinsecticide design. Significant advances have been made in reducing the environmental impact of kilogram-scale peptide production, and these efforts will continue to be vital to their viability as small-molecule pesticide alternatives.

Further development of peptide bioinsecticides is reliant on the ability to understand how these molecules affect basic insect function and understand how chemical biology approaches can affect in-field efficacy as bioinsecticides. Data uncovered by 'omics-based experiments on insect species can inform testing platforms to aid the iterative design of improved peptide mimetics. Increased data on more economically important insect species (pests and beneficial) will aid in the design of specificity; this requires further funding of 'omics-based projects to collect regionally and internationally relevant data. With these data available, systemic experiments to identify extensive neuropeptide modes of action will allow for better insecticidal ratios in crop fields and within IPM scenarios. Multiple neuropeptide families are severely understudied, some of which might represent ideal targets for mimetic development based on the insects they are found in and the chemical properties of the specific neuropeptide.

Future efforts in the design of peptide bioinsecticide candidates should focus on the demonstrable facilitation of delivery to target insects. This is likely to be achieved through strategic chemical modification or appropriate formulation. With the discovery of candidates and formulations able to overcome these barriers, the remaining challenge will be the reduction of cost and environmental impact of kilogram-scale peptide production. These same challenges, and the pressure to overcome them, also exist for the application of peptides in the pharmaceutical industry. It is likely, then, that the agrochemical industry will continue to benefit from advancements made in this area – with both industries learning from the discoveries of the other.

7 Acknowledgements

SOLASTA Bio Ltd gratefully acknowledges the support of its investors.

8 Where to look for further information

- Website: www.solastabio.com
- Book: *Insect Neuropeptides: Chemistry, Biology, and Action*. Julius J. Menn, Edward P. Masler (Eds.), Thomas J. Kelly, September 30, 2021, by American Chemical Society.

9 References

1. Harvey JA, Tougeron K, Gols R, Heinen R, Abarca M, Abram PK, et al. Scientists' warning on climate change and insects. *Ecological Monographs*. 2023;93(1):Article e1553. https://doi.org/10.1002/ecm.1553
2. Bhalla N. *40% of global crop production is lost to pests. And it's getting worse*. World Economic Forum, Foundation TR; 2021.
3. Comission E. Active substances, safeners and synergists 2023. Available from: https://ec.europa.eu/food/plant/pesticides/eu-pesticides-database/start/screen/active-substances
4. Skendžić S, Zovko M, Pajač Živković I, Lešić V, Lemić D. Effect of climate change on introduced and native agricultural invasive insect pests in Europe. *Insects*. 2021;12(11):985.
5. Skendžić S, Zovko M, Živković IP, Lešić V, Lemić D. The Impact of climate change on agricultural insect pests. *Insects*. 2021;12(5):440.
6. Sparks TC, Nauen R. IRAC: mode of action classification and insecticide resistance management. *Pestic Biochem Physiol*. 2015;121:122–128.
7. Whalon M, Mota-Sanchez D, Hollingworth RM. *Global pesticide resistance in arthropods*. Oxford: CAB International; 2008.
8. Yeoh JGC, Pandit AA, Zandawala M, Nassel DR, Davies SA, Dow JAT. DINeR: database for insect neuropeptide research. *Insect Biochem Mol Biol*. 2017;86:9–19.
9. Altstein M, Nässel DR. Neuropeptide signaling in insects. *Adv Exp Med Biol*. 2010;692:155–165.
10. Altstein M. Novel insect control agents based on neuropeptide antagonists: the PK/PBAN family as a case study. *J Mol Neurosci*. 2004;22(1–2):147–157.

11. Nachman RJ, Pietrantonio PV. Interaction of mimetic analogs of insect kinin neuropeptides with arthropod receptors. *Adv Exp Med Biol*. 2010;692:27–48.
12. Van Hiel MB, Van Loy T, Poels J, Vandersmissen HP, Verlinden H, Badisco L, Vanden Broeck J. Neuropeptide receptors as possible targets for development of insect pest control agents. *Adv Exp Med Biol*. 2010;692:211–226.
13. Shi Y, Li H, Nachman RJ, Liu TX, Smagghe G. Insecticidal efficacy and risk assessment of different neuropeptide analog combinations against the peach-potato aphid following topical exposure. *Pest Manag Sci*. 2023;79(1):226–233.
14. Lamange NSN, Hayes TK, Strey A, Isaac RE. Hydrolysis of insect neuropeptides by an angiotensin-converting enzyme from the housefly, *Musca domestica. Peptides*. 1997;18(1):47–52.
15. Isaac RE, Bland ND, Shirras AD. Neuropeptidases and the metabolic inactivation of insect neuropeptides. *Gen Comp Endocrinol*. 2009;162(1):8–17.
16. Pecenka JR, Ingwell LL, Foster RE, Krupke CH, Kaplan I. IPM reduces insecticide applications by 95% while maintaining or enhancing crop yields through wild pollinator conservation. *Proc Natl Acad Sci U S A*. 2021;118(44). https://doi.org/10.1073/pnas.2108429118
17. Pandit AA, Ragionieri L, Marley R, Yeoh JGC, Inward DJG, Davies SA, et al. Coordinated RNA-Seq and peptidomics identify neuropeptides and G-protein coupled receptors (GPCRs) in the large pine weevil *Hylobius abietis*, a major forestry pest. *Insect Biochem Mol Biol*. 2018;101:94–107.
18. Jiang S, Marco HG, Scheich N, He S, Wang Z, Gade G, McMahon DP. Comparative analysis of adipokinetic hormones and their receptors in Blattodea reveals novel patterns of gene evolution. *Insect Mol Biol*. 2023.
19. Luo LL, Lin Y, Linghu JH, Gong W, Luo YH, Liu M, et al. Genomics, transcriptomics, and peptidomics of the greater wax moth *Galleria mellonella* neuropeptides and their expression in response to lead stress. *Insect Sci*. 2023. https://doi.org/10.1111/1744-7917.13264
20. Wang G, Wang J, Nie L. Transcriptome sequencing of the central nervous system to identify the neuropeptides and neuropeptide receptors of *Antheraea pernyi*. *Int J Biol Macromol*. 2023;244:125411.
21. Liessem S, Ragionieri L, Neupert S, Büschges A, Predel R. Transcriptomic and neuropeptidomic analysis of the stick insect, *Carausius morosus*. *J Proteome Res*. 2018;17(6):2192–204.
22. Ly A, Ragionieri L, Liessem S, Becker M, Deininger SO, Neupert S, Predel R. Enhanced coverage of insect neuropeptides in tissue sections by an optimized mass-spectrometry-imaging protocol. *Anal Chem*. 2019;91(3):1980–1988.
23. Beets I, Zels S, Vandewyer E, Demeulemeester J, Caers J, Baytemur E, et al. System-wide mapping of peptide-GPCR interactions in *C. elegans*. *Cell Rep*. 2023;42(9):113058.
24. Shi Y, Pandit A, Nachman RJ, Christiaens O, Davies SA, Dow JAT, Smagghe G. Transcriptome analysis of neuropeptides in the beneficial insect lacewing (*Chrysoperla carnea*) identifies kinins as a selective pesticide target: a biostable kinin analogue with activity against the peach potato aphid *Myzus persicae*. *J Pest Sci*. 2022;96(1):253–264.
25. Jindal V, Li D, Rault LC, Fatehi S, Singh R, Mating M, et al. Bee-safe peptidomimetic acaricides achieved by comparative genomics. *Sci Rep*. 2022;12(1):17263.
26. Marco HG, Simek P, Gade G. Unique members of the adipokinetic hormone family in butterflies and moths (Insecta, Lepidoptera). *Front Physiol*. 2020;11:614552.

27. Xie Y, Zhang L, Wu XQ, Zhang CL, Yang XL, Tobe SS. Probing the active conformation of FGLamide allatostatin analogs with N-terminal modifications using NMR spectroscopy and molecular modeling. *Peptides*. 2015;68:214–218.
28. Xie Y, Kai ZP, Tobe SS, Deng XL, Ling Y, Wu XQ, et al. Design, synthesis and biological activity of peptidomimetic analogs of insect allatostatins. *Peptides*. 2011;32(3):581–586.
29. Wang M, Li X, Chen M, Wu X, Mi Y, Kai Z, Yang X. 3D-QSAR based optimization of insect neuropeptide allatostatin analogs. *Bioorg Med Chem Lett*. 2019;29(7):890–895.
30. Nachman RJ, Wang XJ, Etzkorn FA, Kaczmarek K, Zabrocki J, Lopez J, Coast GM. Evaluation of insect CAP2b analogs with either an (E)-alkene, trans- or a (Z)-alkene, cis-Pro isostere identifies the Pro orientation for antidiuretic activity in the stink bug. *Peptides*. 2013;41:101–106.
31. Alford L, Marley R, Dornan A, Dow JAT, Nachman RJ, Davies SA. Desiccation, thermal stress and associated mortality in Drosophila fruit flies induced by neuropeptide analogue treatment. *J Pest Sci*. 2019;92(3):1123–1137.
32. Alford L, Marley R, Dornan A, Pierre JS, Dow JA, Nachman RJ, Davies SA. Assessment of neuropeptide binding sites and the impact of biostable kinin and CAP2b analogue treatment on aphid (*Myzus persicae* and *Macrosiphum rosae*) stress tolerance. *Pest Manag Sci*. 2019;75(6):1750–1759.
33. Shi Y, Nachman RJ, Gui SH, Piot N, Kaczmarek K, Zabrocki J, et al. Efficacy and biosafety assessment of neuropeptide CAPA analogues against the peach-potato aphid (*Myzus persicae*). *Insect Sci*. 2022;29(2):521–530.
34. Zhang Q, Nachman RJ, Denlinger DL. Diapause hormone in the Helicoverpa/Heliothis complex: a review of gene expression, peptide structure and activity, analog and antagonist development, and the receptor. *Peptides*. 2015;72:196–201.
35. Cannell E, Dornan AJ, Halberg KA, Terhzaz S, Dow JAT, Davies SA. The corticotropin-releasing factor-like diuretic hormone 44 (DH44) and kinin neuropeptides modulate desiccation and starvation tolerance in *Drosophila melanogaster*. *Peptides*. 2016;80:96–107.
36. Zhang C, Qu Y, Wu X, Song D, Ling Y, Yang X. Design, synthesis and aphicidal activity of N-terminal modified insect kinin analogs. *Peptides*. 2015;68:233–238.
37. Zhou YL, Li XL, Zhang YM, Shi Y, Li HH, Zhang Z, et al. A novel bee-friendly peptidomimetic insecticide: synthesis, aphicidal activity and 3D-QSAR study of insect kinin analogs at Phe(2) modification. *Pest Manag Sci*. 2022;78(7):2952–2963.
38. Zhou Y, Zhang Y, Zhang Y, Zhao Y, Xu W, Ye D, et al. Insect kinin mimics act as potential control agents for aphids: structural modifications of Trp(4). *J Pept Sci*. 2023;29(1):e3444.
39. Down RE, Matthews HJ, Audsley N. Oral activity of FMRFamide-related peptides on the pea aphid *Acyrthosiphon pisum* (Hemiptera: Aphididae) and degradation by enzymes from the aphid gut. *Regul Pept*. 2011;171(1–3):11–18.
40. Gui SH, Pei YX, Xu L, Wang WP, Jiang HB, Nachman RJ, et al. Function of the natalisin receptor in mating of the oriental fruit fly, *Bactrocera dorsalis* (Hendel) and testing of peptidomimetics. *PLoS One*. 2018;13(2):e0193058.
41. Nachman RJ, Hamshou M, Kaczmarek K, Zabrocki J, Smagghe G. Biostable and PEG polymer-conjugated insect pyrokinin analogs demonstrate antifeedant activity and induce high mortality in the pea aphid *Acyrthosiphon pisum* (Hemiptera: Aphidae). *Peptides*. 2012;34(1):266–273.

42. Xiong C, Wulff JP, Nachman RJ, Pietrantonio PV. Myotropic activities of tick pyrokinin neuropeptides and analog in feeding tissues of hard ticks (Ixodidae). *Front Physiol.* 2021;12:826399.
43. Hariton A, Ben-Aziz O, Davidovitch M, Nachman RJ, Altstein M. Bioavailability of insect neuropeptides: the PK/PBAN family as a case study. *Peptides*. 2009;30(6):1034–1041.
44. Hariton A, Ben-Aziz O, Davidovitch M, Zubrzak P, Nachman RJ, Altstein M. Bioavailability of beta-amino acid and C-terminally derived PK/PBAN analogs. *Peptides*. 2009;30(12):2174–2181.
45. Nachman RJ, Teal PE, Aziz OB, Davidovitch M, Zubrzak P, Altstein M. An amphiphilic, PK/PBAN analog is a selective pheromonotropic antagonist that penetrates the cuticle of a heliothine insect. *Peptides*. 2009;30(3):616–621.
46. Nachman RJ, Wang XJ, Etzkorn FA, Aziz OB, Davidovitch M, Kaczmarek K, et al. Evaluation of a PK/PBAN analog with an (E)-alkene, trans-Pro isostere identifies the Pro orientation for activity in four diverse PK/PBAN bioassays. *Peptides*. 2009;30(7):1254–1259.
47. Nachman RJ, Mahdian K, Nassel DR, Isaac RE, Pryor N, Smagghe G. Biostable multi-Aib analogs of tachykinin-related peptides demonstrate potent oral aphicidal activity in the pea aphid *Acyrthosiphon pisum* (Hemiptera: Aphidae). *Peptides*. 2011;32(3):587–594.
48. Altstein M, Ben-Aziz O, Zeltser I, Bhargava K, Davidovitch M, Strey A, et al. Inhibition of PK/PBAN-mediated functions in insects: discovery of selective and non-selective inhibitors. *Peptides*. 2007;28(3):574–584.
49. Hariton A, Ben-Aziz O, Davidovitch M, Altstein M. Bioavailability of backbone cyclic PK/PBAN neuropeptide antagonists–inhibition of sex pheromone biosynthesis elicited by the natural mechanism in *Heliothis peltigera* females. *FEBS J.* 2010;277(4):1035–1044.
50. Leander M, Heimonen J, Brocke T, Rasmussen M, Bass C, Palmer G, et al. The 5-amino acid N-terminal extension of non-sulfated drosulfakinin II is a unique target to generate novel agonists. *Peptides*. 2016;83:49–56.
51. Yu N, Benzi V, Zotti MJ, Staljanssens D, Kaczmarek K, Zabrocki J, et al. Analogs of sulfakinin-related peptides demonstrate reduction in food intake in the red flour beetle, *Tribolium castaneum*, while putative antagonists increase consumption. *Peptides*. 2013;41:107–112.
52. Yu N, Smagghe G. Characterization of sulfakinin receptor 2 and its role in food intake in the red flour beetle, *Tribolium castaneum*. *Peptides*. 2014;53:232–237.
53. Zhang Y-M, Ye D-X, Liu Y, Zhang X-Y, Zhou Y-L, Zhang L, Yang X-L. Peptides, new tools for plant protection in eco-agriculture. *Advanced Agrochem*. 2023;2(1):58–78.
54. Herzig V, de Araujo AD, Greenwood KP, Chin YK, Windley MJ, Chong Y, et al. Evaluation of chemical strategies for improving the stability and oral toxicity of insecticidal peptides. *Biomedicines*. 2018;6(3). https://doi.org/10.3390/biomedicines6030090
55. Muttenthaler M, King GF, Adams DJ, Alewood PF. Trends in peptide drug discovery. *Nat Rev Drug Discov*. 2021;20(4):309–25.
56. Zhang C, Li X, Song D, Ling Y, Zhou Y, Yang X. Synthesis, aphicidal activity and conformation of novel insect kinin analogues as potential eco-friendly insecticides. *Pest Manag Sci*. 2020;76(10):3432–3439.
57. Abdulganiyyu IA, Kaczmarek K, Zabrocki J, Nachman RJ, Marchal E, Schellens S, et al. Conformational analysis of a cyclic AKH neuropeptide analog that elicits

selective activity on locust versus honeybee receptor. *Insect Biochem Mol Biol.* 2020;125:103362.
58. Davies SA, Liskamp RJM, Dow JAT, inventors. *Insect neuropeptide analogues*. United Kingdom: WO2021245429A1.
59. Jackson GE, Gäde G, Marco HG. In silico screening for pesticide candidates against the desert locust *Schistocerca gregaria. Life (Basel).* 2022;12(3):387.
60. Denecke S, Swevers L, Douris V, Vontas J. How do oral insecticidal compounds cross the insect midgut epithelium? *Insect Biochem Mol Biol.* 2018;103:22–35.
61. Pinto TV, Silva CA, Siquenique S, Learmonth DA. Micro- and nanocarriers for encapsulation of biological plant protection agents: a systematic literature review. *ACS Agric Sci Technol.* 2022;2(5):838–857.
62. Zheng Y, Hu Y, Yan S, Zhou H, Song D, Yin M, Shen J. A polymer/detergent formulation improves dsRNA penetration through the body wall and RNAi-induced mortality in the soybean aphid *Aphis glycines. Pest Manag Sci.* 2019;75(7):1993–1999.
63. de Veer SJ, Kan MW, Craik DJ. Cyclotides: from structure to function. *Chem Rev.* 2019;119(24):12375–12421.
64. Jackson MA, Xie J, Nguyen LTT, Wang X, Yap K, Harvey PJ, et al. Plant-based production of an orally active cyclotide for the treatment of multiple sclerosis. *Transgenic Res.* 2023;32(1–2):121–133.
65. Ferrazzano L, Catani M, Cavazzini A, Martelli G, Corbisiero D, Cantelmi P, et al. Sustainability in peptide chemistry: current synthesis and purification technologies and future challenges. *Green Chem.* 2022;24(3):975–1020.
66. Frederick MO, Boyse RA, Braden TM, Calvin JR, Campbell BM, Changi SM, et al. Kilogram-scale GMP manufacture of tirzepatide using a hybrid SPPS/LPPS approach with continuous manufacturing. *Org Process Res Dev.* 2021;25(7):1628–1636.
67. Martin V, Egelund PHG, Johansson H, Thordal Le Quement S, Wojcik F, Sejer Pedersen D. Greening the synthesis of peptide therapeutics: an industrial perspective. *RSC Adv.* 2020;10(69):42457–42492.
68. Lopez J, Pletscher S, Aemissegger A, Bucher C, Gallou F. N-Butylpyrrolidinone as alternative solvent for solid-phase peptide synthesis. *Org Process Res Dev.* 2018;22(4):494–503.
69. Egelund PHG, Jadhav S, Martin V, Johansson Castro H, Richner F, Le Quement ST, et al. Fmoc-removal with pyrrolidine expands the available solvent space in green solid-phase peptide synthesis. *ACS Sustain Chem Eng.* 2021;9(42):14202–14215.
70. Collins JM, Porter KA, Singh SK, Vanier GS. High-efficiency solid phase peptide synthesis (HE-SPPS). *Org Lett.* 2014;16(3):940–943.
71. Jaradat DsMM, Al Musaimi O, Albericio F. Advances in solid-phase peptide synthesis in aqueous media (ASPPS). *Green Chem.* 2022;24(17):6360–6372.

Chapter 5

Using gene silencing (RNA interference) techniques to produce safe insecticidal compounds

Salvatore Arpaia, ENEA (Italian National Agency for New Technologies, Energy and Sustainable Economic Development), Italy

1 Introduction

In 2006, the Nobel Prize in Medicine was awarded to American scientists Craig Mello and Andrew Fire for the discovery of RNA interference (RNAi), a natural silencing mechanism leading to the temporary disruption of gene expression at the transcriptional or post-transcriptional level. This mechanism operates through the production of a micro-RNA (miRNA) molecule which specifically recognizes the messenger RNA (mRNA) of a target gene, preventing the formation of the corresponding protein. The initial discovery of RNAi occurred in the nematode *Caenorhabditis elegans*; subsequently, it was confirmed that this mechanism is widely utilized across eukaryotic organisms. The process is used as a natural defense system capable of identifying 'non-self' genetic material (environmental RNA) and can target invading nucleic acids, such as those from viruses.

http://dx.doi.org/10.19103/AS.2024.0149.07

The discovery that external double-stranded RNA molecules (dsRNA) could also trigger RNAi and interfere with gene expression in plants and animals has attracted attention from the academia and biotech industries. In the last decade, the activities in this field have increased and genetically modified (GM) crop plants resistant to pests and diseases or with altered chemical characteristics are commercially available in some areas (e.g. Americas, China, and Australia). Unlike, in other areas, the use of GM plants in the field is restricted due to institutional decisions, therefore alternative routes of delivery of dsRNA that do not involve the genetic transformation of plants have been explored.

2 RNA mechanisms in insects

2.1 Natural micro-RNA and double-stranded RNA

Native miRNA molecules are small endogenous single-stranded noncoding RNA molecules found in the transcriptome of plants, animals, and some DNA viruses. They are polymers encoded by eukaryotic nuclear DNA, about 20–22 nucleotides long and mainly active in the regulation of gene expression at the transcriptional and post-transcriptional levels. In eukaryotic cells, dsRNA longer than 30 bp are also produced and are key activators of the innate immune response against viral infections. Applications of RNAi have widely been developed by synthesizing in the laboratory dsRNA molecules designed to target specific mRNA in the recipient organisms.

2.2 Metabolic pathway for double-stranded RNA

The metabolic pathway leading to a successful gene silencing in eukaryotic cells requires the succession of several steps, whose responses are mediated by specific enzymatic complexes and proteins. When the cell detects the presence of dsRNA molecule, it activates what is known as the RNAi machinery to initiate the interference mechanism. The dsRNA sequence undergoes cleavage into shorter fragments, typically ranging from 19 to 21 base pairs, facilitated by an enzyme named Dicer. The resultant short dsRNA fragment, termed short interfering RNA (siRNA), binds with an enzyme complex known as the RNA-induced silencing complex (RISC). An unfolding of the dsRNA, possibly assisted by a helicase, ensues. Subsequently, only the antisense RNA strand remains in association with RISC, while the sense strand undergoes degradation. Activated RISC possesses the capability to scan numerous mRNAs within the cytosol until it identifies one that complements the antisense RNA fragment linked to the complex. When there is a near-perfect pairing between the siRNA and mRNA, a component within the RISC, referred to as the argonaute protein (Ago), cleaves the mRNA. This action results in the degradation of the mRNA fragments by the cell's intrinsic RNAses, leaving one fragment without a 5' cap and the other devoid of a 3' polyA tail. The different steps are schematically described in Table 1.

2.3 Intercellular transportation and systemic interference

In order to obtain a strong silencing response, it is necessary that a good number of siRNA copies are present in the body of the target species, so as to provoke a silencing response in a sufficiently large number of cells. In some particularly susceptible species, strong systemic RNAi responses can be obtained even with a single administration of a small amount of siRNA. In the case of interference obtained using GM plants expressing dsRNA, silencing is fueled by repeated administration of dsRNA through the trophic activity of the target species to plants.

In *C. elegans*, a metabolic pathway has been identified that causes an amplification of the silencing signal within the body. Briefly, this mechanism is based on RNA-dependent polymerases (RdRPs) that are able to produce secondary siRNAs from the target mRNA or from primary siRNAs (Zhang and Ruvkun, 2012). These secondary siRNAs differ from primary siRNAs in that they contain di- or triphosphate groups at the 5' end, whereas primary siRNAs possess a monophosphate group. In the nematode *C. elegans*, these secondary siRNAs constitute the vast majority of siRNAs found within the body and therefore play an important role in the immune response. The production of these siRNAs is under the control of several genes, leading to

Table 1 Metabolic steps necessary to induce gene silencing by double-stranded RNA

Step	Description
1. dsRNA detection	When a cell identifies a double-stranded RNA (dsRNA) molecule, the RNAi machinery is activated.
2. Cleavage by Dicer	Dicer enzyme facilitates the cleavage of dsRNA into shorter fragments, typically 19–22 base pairs.
3. Formation of siRNA	The resulting short dsRNA fragments become short interfering RNA (siRNA).
4. Binding with RISC	siRNA binds with the RNA-induced silencing complex (RISC).
5. Unfolding of dsRNA	The dsRNA unfolds, possibly aided by a helicase.
6. Sense strand degradation	The sense strand of RNA undergoes degradation, leaving the antisense RNA strand associated with RISC.
7. Scanning by activated RISC	RISC scans multiple mRNAs within the cytosol until it finds a complementary match to the antisense RNA fragment.
8. mRNA cleavage by Ago	With near-perfect pairing, the argonaute protein within RISC cleaves the mRNA, leading to its degradation.
9. mRNA fragment degradation	The cleaved mRNA fragments are broken down by the cell's RNAses, resulting in fragments with altered end structures.

the production of highly efficient siRNAs (Slicer activity) that degrade target mRNAs much more efficiently than primary siRNAs. Thus, the higher efficiency of RNAi is not only due to the higher amount of siRNA present but also due to a more efficient denaturation of the target mRNA. In arthropods, the presence of genes homologous to the RdRPs of *C. elegans* is known for some arachnids, such as the ticks *Rhipicephalus microplus* and *Ixodes scapularis* and the red spider *Tetranychus urticae* (Nganso et al., 2020). In the evaluation of the effects of maize GM for resistance to *Diabrotica virgifera virgifera*, RNAi was used to show the knockdown of target transcripts in beetle body tissues (e.g. the fat body) distant from the dsRNA entry site. In a few cases, the interfering effect was also verified in eggs laid by the treated female (in Coleoptera, Blattodea, Orthoptera, and Hemiptera), providing further evidence of the systemic effect of RNAi (Cooper et al., 2019).

3 Activity spectrum of RNA interference in insects

The effect of gene silencing via RNAi has been studied, at least under laboratory conditions on more than 200 insect species The most extensive work initially involved the model species *Drosophila melanogaster*, for which the entire genome was available (Christiaens et al., 2018). Subsequently, the expression of many genes of interest was studied in some species of important economic value such as *Apis mellifera* and *Bombyx mori*, some agriculturally harmful species such as *Tribolium castaneum* and *Helicoverpa armigera*, or dipterans of health interest such as *Aedes aegyptii* and *Anopheles gambiae* (Riabinina et al., 2022). Therefore, the technique has become a routine in many laboratories to elucidate mechanisms involved in gene expression/silencing; consequently, more studies were conducted in species of economic interest, leading to the identification of possible target genes to induce RNAi-mediated gene silencing.

Available results indicate that the efficiency of silencing varies greatly among different species and according to the mode of administration. In general, different susceptibilities can be observed among different orders of insects (Fig. 1). Although a certain scale of susceptibility can be identified among the different orders, within the same order great differences can be found among species, even taxonomically very close. Coleoptera represent the order with the highest sensitivity to dsRNA treatment even at very low doses; however, about half of the species assayed were found to be insensitive to silencing, highlighting the substantial variation within this order.

Exploration of different modes of administration revealed noteworthy differences in inducing silencing in different taxa. While beetles are commonly found to be sensitive to dsRNA treatment by ingestion, several lepidopteran species show high refractoriness to treatment, whether by ingestion or

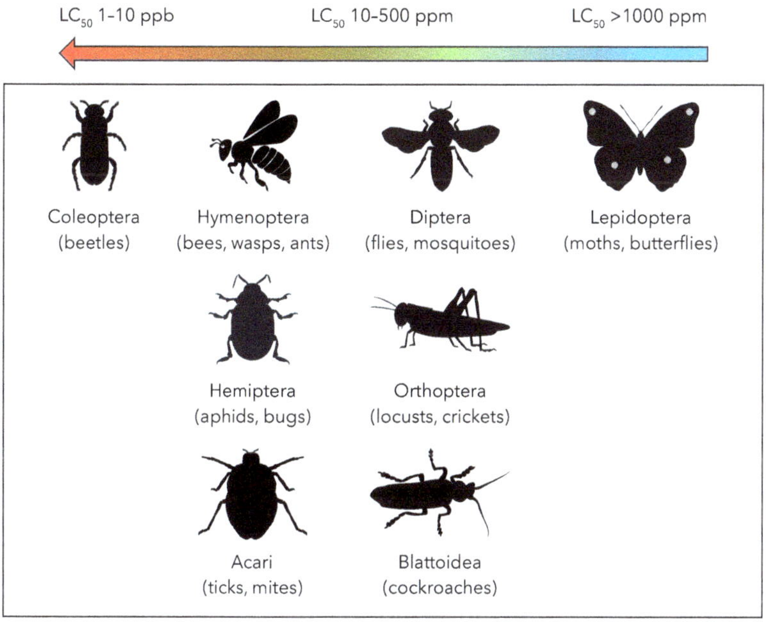

Figure 1 Variability of RNA interference efficiency among different arthropods.

injection. In the case of *Blattella germanica*, RNAi was found to be effective for many genes and at different life stages, demonstrated by both injection and feeding methods (Cruz et al., 2006). In the mosquito *Aedes aegypti*, dsRNA injection proved highly effective, on the contrary dsRNA orally administered yielded successful silencing in only a few cases (Lopez et al., 2019). For the desert locust *Schistocerca gregaria*, a significant knockdown in the expression of target genes was observed through dsRNA injection, but the response remained insensitive to feeding with a dsRNA-supplemented diet (Wynant et al., 2014).

Interestingly, in some instances, even genes expressed in specific tissues proved challenging to silence by RNAi, as observed in the case of ovaries in *S. gregaria* (Wynant et al., 2012). Moreover, it is noteworthy that certain genes appear refractory to silencing by RNAi, regardless of the other features discussed so far. The complexity of responses across different species and tissues underscores the nuanced and variable nature of RNAi efficacy in diverse organisms.

4 Physiological mechanisms preventing gene silencing in insects

As previously mentioned, the efficacy of RNAi exhibits significant variability in insects, particularly when dsRNA is administered orally. Several physiological factors contribute to the non-silencing action, including the stability of RNAs within the organism and their uptake at the cellular level from the intestinal lumen. Research has extensively explored these factors across various species. However, other potential influences on RNAi efficacy, such as the impact of viral infections or the response of the immune system, remain less understood and warrant further investigation.

4.1 Degradation by RNAses

The primary challenge dsRNA molecules face is a possible degradation within the gut lumen or hemolymph, where specific enzymes like RNAses play a pivotal role. In species like *Nezara viridula,* an important amount of dsRNA could already be degraded during feeding, due to the extra-oral digestion by saliva that is characteristic for several insect species (Mitchell et al., 2018). While the full spectrum of RNAses' functions in insects remains to be completely deciphered, they are recognized for several critical roles. For instance, they regulate molting during preimaginal development and are crucial in the cellular response to dsRNA, especially during events like viral infections in insects, as observed in *B. mori* (Liu et al., 2012). Experimental evidence has confirmed the active involvement of these enzymes in degrading dsRNA molecules post-ingestion across various species, including dipteran, hemipteran, and lepidopteran insects. The pH conditions within the insect gut significantly influence this degradation process. Notably, higher pH values within the digestive tract or hemolymph contribute to more efficient degradation of dsRNA molecules. In Lepidoptera, where the gut tends to have an alkaline pH, the degradation of RNA molecules within the digestive system is particularly pronounced (Singh et al., 2017).

4.2 Uptake efficiency

Research into the cellular uptake of dsRNA in *Leptinotarsa decemlineata*, a species highly sensitive to RNAi, revealed a notably efficient uptake mechanism facilitated by clathrin-mediated endocytosis (Cappelle et al., 2016). However, contrasting observations emerge when examining certain dipteran species, suggesting that the cellular uptake of dsRNA, particularly from the mesentery, is notably inefficient in some cases. Furthermore, among Hymenoptera and Orthoptera species, a variable efficiency in dsRNA uptake across different tissues has been observed, highlighting the variable nature of RNAi mechanisms across diverse insect taxa.

4.3 Differences in the genome

While RNAi serves as a conserved mechanism across all eukaryotes, notable genomic differences exist among taxa in regulating the production of key proteins involved in the functioning of the RNAi machinery, such as Dicer and Ago. For instance, insects possess a repertoire of five Ago-like proteins, a more limited count compared to other animals and plants. Additionally, significant variability is evident in the number of genes participating in siRNA amplification across the diverse species examined.

However, establishing a direct correlation between these genetic disparities and susceptibility to RNAi remains elusive at present. Within invertebrates, the RNAi mechanism emerges as a pivotal component of the immune response against viral infections. Numerous studies have demonstrated that the injection of nonspecific dsRNA can activate this immune response (Karpala et al., 2005). This activation is often evidenced by heightened expression levels of essential genes like *Dicer-2* and *Ago-2*, which play central roles in facilitating RNAi functionality.

The combination of these factors may explain the results of numerous studies during which species such as *B. mori*, *Drosophila suzukii*, and *N. viridula* showed a very sensitive RNAi response in the microinjection assays but exhibited a lower sensitivity in the oral feeding bioassays (Liu et al., 2012; Taning et al., 2016; Sharma et al., 2021).

5 Double-stranded RNA-based insecticides

Recent technical progress in the production and formulations of dsRNA to improve its efficacy, stability as well as cost-effectiveness has led to the first-ever registration of a commercial insecticide exploiting the mechanism of RNAi. The new foliar-applied insecticide active ingredient Ledprona (*L. decemlineata*-specific recombinant double-stranded interfering Oligonucleotide GS2) developed by Greenlight Biosciences has received the approval for a 3-years period of commercialization by the US environmental protection agency (EPA) on December 22, 2023. Ledprona is intended to control *L. decemlineata*, a ravaging coleopteran pest of potatoes; products formulated from Ledprona will be used as foliar applications, in which the insects will be killed upon feeding on treated plant material. Ledprona exploits RNAi machinery of *L. decemlineata* to specifically downregulate a protein that is essential for the organism's survival, by targeting the mRNA sequence encoding the *proteasome subunit beta type-5 (PSMB5)* gene of the target pest. The nucleic acids contained in the commercial product have not been chemically modified and are thus indistinguishable from those that are naturally occurring. The authorization by the EPA was published after the assessment of a dossier including data from more than 100 field trials over 4 years to develop a solution that proved to be effective at just 9.9 g/ha of product.

RNAi stands out as a potentially very specific method for plant pest control due to its underlying mechanism, which relies on the base pairing between exogenous dsRNA – typically in active forms consisting of segments 19–22 nucleotides long – and complementary endogenous mRNA. The concept of target specificity in RNAi is of paramount importance, especially given that many of the target genes identified are highly conserved across various eukaryotic organisms. For instance, ATPase is a ubiquitous enzyme found across all known organisms. However, the mere presence of a conserved gene does not guarantee that the designed dsRNA will induce the desired effects in any organism.

A groundbreaking study by Whyard et al. (2009) exemplified this specificity by utilizing dsRNAs tailored for V-ATPase from four distinct insect species spanning different orders: *D. melanogaster* (fruit fly), *Tribolium castaneum* (flour beetle), *Acyrthosiphon pisum* (aphid), and *Manduca sexta* (tobacco hornworm). The findings were instructive: when individuals from these species were exposed to diets enriched with species-specific dsRNA, pronounced mortality rates were observed. Importantly, this lethal effect was only evident when the dsRNA sequences were precisely complementary to the target gene of each respective species. Conversely, dsRNAs targeting the homologous gene sequences of the other species failed to elicit adverse effects, underscoring the requisite specificity of dsRNA-based interventions.

Generally, a complementarity stretch of 21 consecutive nucleotides is deemed adequate to initiate interference, thereby hindering mRNA translation. Interestingly, even sequences with less-than-perfect complementarity can, under certain circumstances, trigger this interference process. From studies conducted on mammals, it is established that a specific number of mismatches can be accommodated without compromising the silencing effect (He et al., 2020). However, the extent to which this phenomenon is prevalent among arthropods remains a topic of ongoing investigation and is not conclusively understood currently.

6 Target genes for RNA interference applications

From an applied perspective focused on insect pest management, the efficacy of RNAi has been assessed against various target genes. Ideally, target genes for RNAi applications are those critical for insect development or survival, lacking functional redundancy to mitigate potential secondary effects, and ideally specific to the intended target species. Experimental endeavors have explored genes integral to energy metabolism, the formation of specific cellular components, hormonal regulation, chitin synthesis, and digestive processes. Notably, genes associated with immune defense mechanisms, detoxification pathways, and broader resistance to insecticidal agents have garnered significant attention. Below are some illustrative examples highlighting the targeted genes and their relevance in the context of RNAi-based pest control strategies.

Energy metabolism is pivotal for the survival, proliferation, and regular physiological functions of cells. The silencing of genes associated with energy metabolism, like vacuolar ATPase (V-ATPase) and arginine kinase, has demonstrated the capability to induce significant mortality in various insect species. V-ATPase, an ATP-powered proton pump responsible for generating energy gradients across membranes, stands as one of the paramount enzymes in eukaryotic cells. When targeted by dsRNA, its silencing has resulted in noteworthy mortality across a range of insect taxa, including Coleoptera, Hemiptera, Lepidoptera, Diptera, Orthoptera, and even mites (Christiaens et al., 2018).

Similarly, arginine kinase, which plays a crucial role in maintaining energy homeostasis, has been targeted with specific dsRNA treatments. Juvenile stages of phytophagous Lepidoptera species, such as *Tuta absoluta*, *Plutella xylostella*, and *H. armigera*, exhibited mortality rates nearing 20% upon treatment (Ai et al., 2018; Bento et al., 2020; Fu et al., 2020).

Furthermore, other genes implicated in energy metabolism, like ADP/ATP translocase and NADH hydrogenase, have been pinpointed as potential targets. Experimental studies have revealed both lethal and sublethal effects on economically significant phytophagous insects, underscoring the potential of RNAi-based strategies in pest management (Lu et al., 2023).

The application of RNAi has been instrumental in successfully silencing genes responsible for crucial cellular components, including cytoskeleton proteins and ribosomal proteins. Silencing cytoskeleton proteins like actin and tubulin has demonstrated profound effects on various insect species. Specifically, the targeted silencing has resulted in both lethal and sublethal outcomes across a diverse range of insects, encompassing beetles, dipterans, blattoids, hemipterans, thrips, and lepidopterans (e.g. Ulrich et al., 2015; Thakur et al., 2016; Yu et al., 2017). Furthermore, when dsRNA targeting ribosomal proteins was ingested by larvae of specific species such as *D. virgifera virgifera* and *B. tabaci*, it led to larval mortality (Knorr et al., 2021). In the case of *Aethina tumida*, treatment with ribosomal protein dsRNA during the larval stage induced malformations in the resulting adult beetles (Bonina and Arpaia, 2023). These findings underscore the versatility and potential of RNAi-based approaches in targeting essential cellular components for effective pest management.

Hormones play a crucial role in insects' life cycle during the postembryonic development controlled by two key hormones, 20-hydroxyecdysone (20E) and juvenile hormone (JH), and in reproduction stages of insects. Knowledge about the hormonal system has already been used in the development of specific classes of synthetic insecticides, such as growth regulators. Even with the use of dsRNAs, it is possible to interfere with all stages of insect development, both juvenile and adult stages. Several genes are involved in insect hormone metabolism, and among them silencing of ecdysone receptor (*EcR*), *HR3* a transcription factor that regulates molting, *JHBP* juvenile hormone-binding protein, and JHR juvenile hormone esterase have been found to be effective in causing mortalities of even more than 50% of treated individuals (Stafford, 2015).

Chitin serves as a fundamental component of the insect cuticle and plays a crucial role in forming the peritrophic membrane, which lines the midgut of insects and serves vital functions in nutrition and defense against microbial pathogens. Given its significance, chitin formation has been a recognized target in insecticide development for over two decades. Inhibitors targeting enzymes involved in chitin synthesis have been developed and marketed as commercial products. RNAi has provided a promising avenue to target genes associated with chitin synthesis effectively. For instance, by silencing genes responsible for chitin synthase, significant mortality rates resulting from cuticle formation failures have been observed in various insect species. This outcome has been achieved through both topical applications and transgenic plant-mediated delivery methods, affecting species like *H. armigera, Aphys glycines, Liriomyza trifolii*, and *D. virgifera virgifera* (e.g. Lu et al., 2023).

Conversely, targeting chitin deacetylases (CDAs), which are enzymes involved in chitin degradation during molting processes, can also yield impactful results. Silencing the *SfCAD1* gene, for instance, has led to growth retardation and increased mortality in larvae of the species *Spodoptera frugiperda* (Lu et al., 2023). These findings highlight the potential of RNAi in disrupting essential biological processes like chitin synthesis and molting, thereby offering innovative strategies for insect pest control.

The exploration of genes encoding digestive enzymes as potential targets for controlling herbivore insects through the oral administration of dsRNA has yielded intriguing results. Specifically, the silencing of *SSJ* (smooth septate junction proteins) genes demonstrated remarkable efficacy, leading to 100% mortality in *D. virgifera virgifera* upon successful silencing (Hu et al., 2020).

However, the outcomes were more nuanced when targeting other digestive enzymes. For instance, the silencing of the alpha-amylase gene resulted in only partial mortality when tested against larvae of *H. armigera* (Vatanparast et al., 2017). This differential response underscores the variability in the effectiveness of RNAi-based strategies across different target genes and insect species.

The insect immune system plays a pivotal role as the primary defense mechanism against a variety of infections. Hemocytin, a lectin found in hemocytes, contributes significantly to both humoral and cellular immune responses, particularly in processes like encapsulation. Through RNAi-mediated silencing of hemocytin, studies have observed increased mortality rates in individuals of *Myzus persicae*, highlighting its critical role in insect defense mechanisms (Zhang et al., 2022).

Insects have evolved sophisticated detoxification systems to shield themselves from harmful xenobiotics. A recurring challenge arises from repeated exposure to specific active ingredients, leading to the development of resistance in targeted species. Notably, this resistance often correlates with elevated levels of detoxification enzymes, including cytochrome P450 monooxygenases, glutathione-S-transferases, and carboxykinases. In this context, RNAi technology offers a promising avenue for combatting insecticide resistance. By targeting key genes responsible for producing detoxification enzymes, such as CYP321A7 and CYP6AE43 in *S. frugipreda*, CYP6AE14 in *H. armigera*, and CYP6CM1 in *B. tabaci*, researchers have observed increased larval susceptibility to xenobiotics (Chen et al., 2023) offering a novel application of dsRNA-based products, as an innovative strategy to manage insect populations that have developed resistance to conventional active ingredients.

7 Biosafety issues and assessment

7.1 Genomic analyses

A key first step in ensuring the selectivity of a dsRNA molecule is to design the sequence in a way that avoids apriori interference toward other genes with similar sequences in the target organism and in nontarget organisms (NTOs). Similarity predictions with genomes of potentially dsRNA-exposed organisms help design molecules that effectively bind to a target with a low probability of off-target effects. Bioinformatics tools are available that can help to first identify potential off-target binding sites in the genome of the recipient species, through algorithms that can support optimization of dsRNA and siRNA design. These include both online applications and stand-alone and open-source software. However, due to the limited availability of genome sequences for many NTOs, bioinformatics can only represent an initial screening that will need to be further confirmed with more specific toxicity assays. In addition, the lack of clearly defined rules on sequence homology required for effective silencing limits the utility of bioinformatics analysis in environmental risk assessment.

7.2 Environmental persistence

Existing studies with nucleic acids indicate that for an RNA molecule in a soil or water matrix, the half-life can range from a few hours to a couple of days. Environmental factors such as ultraviolet radiation, runoff, and high temperatures can have a significant impact on the persistence and stability of the dsRNA molecule. Therefore, the possible exposure of organisms to dsRNA in field conditions is expected to be extremely limited. Once in contact with different matrices, an RNA molecule is subject to multiple abiotic, enzymatic or microbial degradation processes as well as adsorption at the solid–water interface. Abiotic degradation pathways include denaturation of single-stranded dsRNA and acid- or base-catalyzed hydrolysis of the ribose–phosphodiester bonds that make up the basic structure of RNA. Microorganisms in receiving environments can accelerate the degradation of dsRNA molecules either through the production of extracellular hydrolases or through direct uptake of the molecules. In soil microcosms, adsorption of 32P-labeled dsRNA was found to be rapid and extensive, particularly in finer-textured soils (Parker et al., 2019). Adsorption sites may be limited particularly in the presence of competing adsorbents, including other nucleic acids, phosphate, and organic acids present along with dsRNA molecules in different environmental matrices.

In the case of GM crops, the amount of RNA released into the environment has not yet been directly quantified. Using the known release rates for Cry proteins from plant tissues to receiving environments as a model, dsRNA molecules released into agricultural soil can be expected to be at levels of micrograms per hectare, which is 3–4 orders of magnitude lower than the release rates of Cry proteins. However, it is worth noting that in the case of dsRNA-based insecticide products, the use of co-formulants is expected to promote its penetration within plant and animal tissues. This action prolongs its efficacy under field conditions and increases its environmental persistence compared to the naked RNA molecule.

7.3 Risks for non-target organisms

One of the most attractive aspects of this technology is its presumably low hazard for most species due to the high selectivity determined by the choice of target gene. Due to its specific mode of action, dsRNAs are expected to ensure compatibility with the activity of natural enemies, pollinators, and other NTOs (Christiaens et al., 2022). While several studies have confirmed the specificity of action of these molecules, undesirable effects on NTOs have been observed in some cases. Two general mechanisms for these adverse effects have been proposed:

1. The first involves adverse effects due to sequence homology between the dsRNA and one or more genes expressed in the NTO.
2. The second concerns non-sequence-related mechanisms that could arise when exposure to nonspecific dsRNA disrupts certain metabolic functions of the exposed species. Such imbalances might include saturating the RNAi machinery, resulting in decreased abilities of the organism to combat viral infections or immunostimulation, potentially leading to a loss of fitness.

Experimental evidence has demonstrated saturation of the RNAi machinery in mammals but as of now, no such cases have been confirmed regarding insects (Rodrigues and Petrick, 2020).

In contrast, dsRNA-induced immunostimulation has been well documented in insects such as *A. mellifera, Bombus terrestris* through activation of different immune pathways (RNAi, JAK/STAT, antimicrobial peptide production (McMenamin et al., 2018). However, the extent to which this may cause an adverse effect on these insects is not known.

Taxonomic proximity between species is known to increase the likelihood of silencing effects caused by the same dsRNA molecule across different species. However, the magnitude of these effects can vary, even among species within the same family. For instance, studies on different coccinellid species treated with dsRNA against the ATPase of *D. virgifera virgifera* showed partial silencing and antimetabolic effects on nontarget species, although the outcomes varied among the tested taxa (Haller et al., 2019). In contrast, the lepidopteran *Ostrinia nubilalis* demonstrates highly inefficient responses to RNAi, regardless of the method of administration. On the other hand, its congeneric species, *Ostrinia furnacalis*, responds to dsRNA through methods like dipping and injection into embryos (Cooper et al., 2021). Consequently, while considering biosafety, it is essential to acknowledge the increased probability of nontarget effects among closely related species. Nevertheless, these effects must be experimentally assessed on the chosen indicator species.

7.4 Insect resistance to double-stranded RNA

The genetic plasticity of insects and their rapid development cycles can lead to resistance against active ingredients, as already observed in many cases with synthetic insecticides. This inherent ability of insects to develop resistance suggests that similar challenges might arise with dsRNA molecules, potentially reducing the technology's effectiveness over time. Experimental evidence supports this concern. For instance, a laboratory colony of *D. virgifera virgifera* developed resistance to dsRNA targeting *DvSnf7* after several generations of feeding on transgenic maize expressing *DvSnf7* dsRNA (Khajuria et al., 2018). The larvae from this resistant colony exhibited approximately 130-fold lower susceptibility compared to a control population. Alarmingly, this resistance was not limited to dsRNA *DvSnf7*; when fed different dsRNAs, the larvae still demonstrated resistance. The underlying mechanism of this resistance was linked to the failure of resistant individuals to uptake dsRNA into the cytoplasm. Consequently, if resistance develops against a dsRNA due to cellular uptake failure, there is a risk of cross-resistance to other dsRNAs. A similar pattern of resistance was observed in a colony of *L. decemlineata* under experimental conditions (Mishra et al., 2021).

7.5 Safety for humans and farm animals

Safety aspects regarding mammals possibly exposed to dsRNA-based pesticides have been extensively reviewed by a recent OECD document (OECD, 2023). In synthesis, there are several aspects that indicate that these products are unlikely to induce toxic effects on humans or higher animals. Indeed, humans and other vertebrates possess significant physiological and biochemical barriers that restrict the uptake and distribution of exogenous RNAs through oral and dermal routes. These barriers encompass various protective mechanisms:

- Nucleases are present in the saliva.
- Denaturation and depurination effects due to the acidic pH of the stomach.
- Nucleases are also found in the digestive tract.
- Pancreatic secretions containing bile salts and degradative enzymes.
- Cellular membrane barriers.
- The polysaccharide coating that protects the intestinal epithelium.
- Intracellular degradation processes occurring in endosomes and lysosomes.

The prevalence of these barriers likely stems from the pervasive presence of RNAs in the environment where the organism has been exposed.

Both long and short dsRNAs, bearing sequence identity to genes/exons in humans and other organisms, have a long history of safe consumption by humans and other mammals. Nucleic acids, whether long or short, are naturally present in foods and feeds derived from plants and animals. Consequently, humans and animals regularly consume these nucleic acids without adverse effects. Moreover, the expected application rate of these products in the field is low and residues on food will be limited, as dsRNA-based foliar insecticides are expected to undergo rapid degradation due to microbes in the environment once applied.

8 Regulatory aspects

Plant protection products are regulated differently under legislative frameworks worldwide. As mentioned before, the United States has authorized the first sprayable product based on this technology. The US EPA has treated the dsRNA-based insecticide as a biochemical pesticide. In contrast to conventional synthetic pesticides, biochemical pesticides are defined as naturally occurring compounds which are inherently less toxic (Dietz-Pfeilstetter et al., 2021).

In the European Union, a separate category for biochemical insecticides is not foreseen in the legislative framework; therefore, the authorization of these products should be currently conducted as for traditional pesticides. Similarly, in Australia a statement by the Office of the Gene Technology Regulator established that techniques involving the external application of RNAi to an organism do not constitute gene technology. By the effect of this amendment, RNAi-based products topically applied are included in the category of agricultural chemical products which are assessed and regulated by the Australian Pesticides and Veterinary Medicines Authority. In Table 2, the main differences between the United States and the European Union relevant regulatory frameworks are reported.

9 Conclusion: advantages, limitations, and opportunities

The use of insecticides in pest management has surged since the 1960s, paralleling the expansion of the chemical industry. Over subsequent decades, this growth led to the synthesis of novel classes of molecules, capitalizing on emerging insights into insect physiology to explore fresh mechanisms of action and administration modes. However, as time progressed, growing concerns arose regarding the adverse impacts on human health and the environment. This highlighted an urgent imperative for the development of insecticidal agents that are both effective and environmentally benign.

The contemporary emphasis on fostering sustainable yet productive agricultural practices, aimed at addressing the nutritional requirements of an expanding global population, underscores the need for innovative solutions. The goal is to equip farmers with efficacious tools that safeguard crops against both longstanding and emerging biotic challenges, ensuring food security while minimizing ecological footprint. In this evolving landscape, the quest continues for molecules that strike an optimal balance between agricultural productivity and environmental stewardship, enabling growers to navigate the complexities of modern farming responsibly and effectively.

Advancements in molecular biology have deepened our understanding of insect physiology, paving the way for innovative, environmentally compatible technologies in pest management. RNAi stands out as a promising technique, having already demonstrated success in developing plant defense approaches, including GM plants or field-distributed products.

RNAi offers several advantages, predominantly its high selective against target insects due to the sequence-specific mode of action. Additionally, the environmental impact of dsRNA products is mitigated by their low persistence in field conditions. This characteristic minimizes long-term ecological implications, aligning with sustainable agricultural practices. Furthermore, these molecules exhibit a high degree of safety toward mammals due to the presence of histological and physiological barriers that restrict cellular uptake.

Table 2 Regulatory aspects for the approval of dsRNA-based pesticides in the United States and European Union

Aspect/Parameter	USA	European Union
Classification	Considered as biochemical pesticides	No specific category; treated under traditional synthetic pesticide regulations.
Regulatory framework	Federal Insecticide, Fungicide, and Rodenticide Act (FIFRA) and the Federal Food, Drug, and Cosmetic Act (FFDCA)	Regulation (EC) 1107/2009 governs the authorization of plant protection products (PPPs).
Registration process	Requires US EPA registration under FIFRA and FFDCA. Technical grade material of the pesticide is tested along with product formulation.	Two-step process: (1) Active compound evaluation by EFSA, (2) PPP assessment by a member state (MS).
Authorization criteria	Must declare no unreasonable adverse effects on humans or the environment	Active compound undergoes EU-wide risk evaluation by EFSA and EU Commission approval; MS retains authority over PPP authorizations.
Maximum residue levels	FFDCA allows setting of maximum residue levels for pesticides	Not explicitly mentioned in the context of dsRNA-based products; governed by individual member states.
Zonal approach	Not applicable	Introduced in Europe: Authorization by a single zonal rapporteur member state is sufficient for the entire zone (Northern, Central, Southern Europe); individual MS can assert national specificities.
Flexibility in risk management	Limited flexibility as per FIFRA and FFDCA	Individual member states can determine specific risk management measures based on national ecological or agricultural specificities.
Amendment	Not mentioned	Article 77 of Regulation No. 1107/2019 allows for possible amendments by the Commission, but none produced so far for dsRNA-based products.

Importantly, the unique mechanism of action associated with RNAi remains unaffected by resistance mechanisms to other chemical pesticides already present in target arthropod populations, offering a potential solution to combat resistance issues.

The adoption of this novel biotechnology, like any other pest management strategy, is better exploited if approached within the framework of integrated pest management (IPM). Historical insights from pest management practices emphasize that no single solution exists for effectively controlling pests. These organisms possess an evolutionary capacity that enables them to manifest mutated and adapted phenotypes within relatively few generations, necessitating continuous updates to control strategies.

While RNAi presents promising advantages, it is not without limitations. Foremost among these is the variable sensitivity observed across herbivorous insect species. Additionally, like other insecticidal agents, the potential for these insects to develop resistance mechanisms to RNAi interventions underscores the importance of employing diverse and adaptive strategies within an IPM approach. This holistic perspective ensures resilience against evolving pest populations while maximizing the efficacy and sustainability of control measures.

The practical application of RNAi in the field faces challenges related to the effective absorption of the product when applied via foliar spray. The external characteristics of many leaves, such as the presence of wax layers or hydrophobic cuticles, create barriers that hinder the efficient uptake of dsRNA. Additionally, factors like the presence of trichomes and the density and positioning of stomata further complicate the penetration of dsRNA into plant tissues.

To enhance the environmental stability and improve penetration capabilities through plant tissues, commercial formulations may incorporate various adjuvants. These could include chelators, liposomes, nanocarriers, peptides, cationic polymers, and other innovative delivery systems designed to optimize the efficacy and environmental compatibility of RNAi-based insecticides. Given these challenges, the environmental risk assessment of this emerging class of insecticides must be meticulously tailored, taking into account the specific formulation of the product.

By addressing these formulation considerations, researchers will be able to discover how to maximize the potential benefits of RNAi while minimizing unintended ecological impacts.

10 Where to look for further information

On the web site of the 'iPlanta' Cost action (https://iplanta.univpm.it/) presentations from a number of webinars on different aspects of RNAI are available; at the following link https://iplanta.univpm.it/node/45 there are podcasts and videos illustrating RNAi technology, biosafety and regulation. Procedures for the approval of biopesticides in the United States are available at https://www.epa.gov/pesticides/biopesticides and the US EPA opinion on the approval of the dsRNA-based insecticide Ledprona is available at https://www.epa.gov/pesticides/epa-registers-novel-pesticide-technology-potato-crops Regulation concerning approval of pesticides in the European Union can be accessed at https://food.ec.europa.eu/plants/pesticides/approval-active-substances-safeners-and-synergists_en

11 References

Ai, X., et al. (2018). Developmental control of *Helicoverpa armigera* by ingestion of bacteria expressing dsRNA targeting an arginine kinase gene. *Biocontrol Sci. Technol.*, 28(3), 253–267.

Bento, F. M., et al. (2020). Gene silencing by RNAi via oral delivery of dsRNA by bacteria in the South American tomato pinworm, *Tuta absoluta*. *Pest Manag. Sci.*, 76(1), 287–295.

Bonina, V. and Arpaia, S. (2023). The use of RNA interference for the management of arthropod pests in livestock farms. *Med. Vet. Entomol.*, 37(4), 631–646.

Cappelle, K., et al. (2016). The involvement of clathrin-mediated endocytosis and two Sid-1-like transmembrane proteins in double-stranded RNA uptake in the Colorado potato beetle midgut. *Insect Mol. Biol.*, 25(3), 315–323.

Chen, H.L., et al. (2023). Resistance monitoring and mechanism in the fall armyworm *Spodoptera frugiperda* (Lepidoptera: Noctuidae) for chlorantraniliprole from Sichuan Province, China. *Front. Physiol.*, 14, 1180655.

Christiaens, O., et al. (2018). Literature review of baseline information on RNAi to support the environmental risk assessment of RNAi-based GM plants. *EFSA Supporting Publications*, *15*(5). https://doi.org/10.2903/sp.efsa.2018.en-1424.

Christiaens, O., et al. (2022). Implementation of RNAi-based arthropod pest control: environmental risks, potential for resistance and regulatory considerations. *J. Pest Sci.*, 95(1), 1–15.

Cooper, A. M., et al. (2019). Molecular mechanisms influencing efficiency of RNA interference in insects. *Pest Manag. Sci.*, 75(1), 18–28.

Cooper, A. M., et al. (2021). Comparison of strategies for enhancing RNA interference efficiency in *Ostrinia nubilalis*. *Pest Manag. Sci.*, 77(2), 635–645.

Cruz, J., et al. (2006) Functions of the ecdysone receptor isoform-A in the hemimetabolous insect *Blattella germanica* revealed by systemic RNAi in vivo. *Dev. Biol.*, 297(1), 158–171.

Dietz-Pfeilstetter, A., et al. (2021). Considerations and regulatory approaches in the USA and in the EU for dsRNA-based externally applied pesticides for plant protection. *Front. Plant Sci.* 12, 974.

Fu, S., et al. (2020). Silencing arginine kinase/integrin β1 subunit by transgenic plant expressing dsRNA inhibits the development and survival of *Plutella xylostella*. *Pest Manag. Sci.*, 76(5), 1761–1771.

Haller, S., et al. (2019). Responses of two ladybird beetle species (Coleoptera: Coccinellidae) to dietary RNAi. *Pest manag. Sci.*, 75(10), 2652–2662.

He, W., et al. (2020). Low mismatch rate between double-stranded RNA and target mRNA does not affect RNA interference efficiency in Colorado potato beetle. *Insects,* 11(7), 449.

Hu, X., et al. (2020). Characterization of DvSSJ1 transcripts targeting the smooth septate junction (SSJ) of western corn rootworm (*Diabrotica virgifera virgifera*). *Sci. Rep.*, 10(1), 11139.

Karpala, A. J., et al. (2005). Immune responses to dsRNA: implications for gene silencing technologies. *Immunol. Cell Biol.*, 83(3), 211–216.

Khajuria, C., et al. (2018). Development and characterization of the first dsRNA-resistant insect population from western corn rootworm, *Diabrotica virgifera virgifera* LeConte. *PLoS One*, 13(5), e0197059.

Knorr, E., et al. (2021). Knockdown of genes involved in transcription and splicing reveals novel RNAi targets for pest control. *Front. Agron.*, 3, 715823.
Liu, J., et al. (2012). *Bombyx mori* DNA/RNA non-specific nuclease: expression of isoforms in insect culture cells, subcellular localization and functional assays. *J. Insect Physiol.*, 58, 1166–1176.
Lopez, S. B. G., et al. (2019) RNAi-based bioinsecticide for *Aedes* mosquito control. *Sci. Rep.*, 9, 4038.
Lu, Y., et al. (2023). The dsRNA Delivery, Targeting and Application in Pest Control. *Agronomy*, 13(3), 714.
McMenamin, A. J., et al. (2018). Honey bee and bumble bee antiviral defense. *Viruses*, 10(8), 395.
Mishra, S., et al. (2021). Selection for high levels of resistance to double-stranded RNA (dsRNA) in Colorado potato beetle (*Leptinotarsa decemlineata* Say) using non-transgenic foliar delivery. *Sci. Rep.*, 11(1), 6523.
Mitchell, P. L., et al. (2018). Probing behavior of *Nezara viridula* on soybean: characterization and comparison of electrical penetration graph (EPG) waveforms on vegetative and reproductive plant structures. *J. Agric. Urban Entomol.*, 34, 19–43.
Nganso, B. T., et al. (2020). A genome-wide screening for RNAi pathway proteins in Acari. *BMC Genom.*, 21, 1–16.
OECD (2023). Considerations for the Human Health Risk Assessment of Externally Applied dsRNA Based Pesticides Series on Pesticides No. 110.
Parker, K. M., et al. (2019). Environmental fate of RNA interference pesticides: adsorption and degradation of double-stranded RNA molecules in agricultural soils. *Environ. Sci. Technol.*, 53(6), 3027–3036.
Riabinina, O., et al. (2022) Genetic toolbox approaches in mosquitoes. *Cold Spring Harb. Protoc.*, 2022(11), pdbtop107691.
Rodrigues, T. B. and Petrick, J. S. (2020). Safety considerations for humans and other vertebrates regarding agricultural uses of externally applied RNA molecules. *Front. Plant Sci.*, 11, 407.
Sharma, R., et al. (2021). RNAi-mediated mortality in southern green stinkbug *Nezara viridula* by oral delivery of dsRNA. *Pest Manag. Sci.*, 77(1), 77–84.
Singh, I.K., et al. (2017). Comparative analysis of double-stranded RNA degradation and processing in insects. *Sci. Rep.*, 7, 17059.
Stafford, J. (2015). Juvenile Hormone esterase is a conserved regulator of starvation-induced behavior (Doctoral dissertation, University of British Columbia).
Taning, C. N. T., et al. (2016). Oral RNAi to control *Drosophila* suzukii: laboratory testing against larval and adult stages. *J. Pest Sci.* 89, 803–814.
Thakur, N., et al. (2016). RNAi–Implications in entomological research and pest control. *RNA Interference*, 6, 341–70.
Ulrich, J., et al. (2015). Large scale RNAi screen in *Tribolium* reveals novel target genes for pest control and the proteasome as prime target. *BMC Genom.*, 16, 1–9.
Vatanparast, M., et al. (2017). RNA interference-mediated knockdown of some genes involved in digestion and development of *Helicoverpa armigera*. *Bull. Entomol. Res.*, 107(6), 777–790.
Whyard, S., et al. (2009). Ingested double-stranded RNAs can act as species-specific insecticides. *Insect Biochem. Molec. Biol.*, 39(11), 824–832.

Wynant, N., et al. (2012). Tissue-dependence and sensitivity of the systemic RNA interference response in the desert locust, *Schistocerca gregaria*. *Insect Biochem.*, 42(12), 911–917.

Wynant, N., et al. (2014). Identification, functional characterization and phylogenetic analysis of double stranded RNA degrading enzymes present in the gut of the desert locust, *Schistocerca gregaria*. *Insect Biochem.*, 46, 1–8.

Yu, X., et al. (2017). Double-stranded RNA delivery through soaking mediates silencing of the muscle protein 20 and increases mortality to the Asian citrus psyllid, *Diaphorina citri*. *Pest Manag. Sci.*, 73(9), 1846–1853.

Zhang, C. and Ruvkun, G. (2012). New insights into siRNA amplification and RNAi. *RNA biology*, 9(8), 1045–1049.

Zhang, J., et al. (2022). RNA-interference-mediated aphid control in crop plants: a review. *Agriculture*, 12(12), 2108.

Chapter 6

Exploiting plant phenotypic plasticity in promoting crop resistance to insect pests

Michael J. Stout, Louisiana State University Agricultural Center, USA; Lina Bernaola, Texas A&M University AgriLife Research Center, USA; James M. Villegas, Louisiana State University Agricultural Center, USA; and Emily C. Kraus, University of California - Agriculture and Natural Resources, USA

1 Introduction

Plant resistance to herbivorous arthropods results from the expression of biochemical, morphological, and physiological traits that reduce the impact of herbivores on plant fitness (or, in the context of agriculture, crop yield or quality) (Kant et al. 2015). The inherent resistance of plants to the arthropods that feed on them is of interest to ecologists because plant resistance is an important factor in structuring terrestrial communities – that is, the patterns of

http://dx.doi.org/10.19103/AS.2024.0149.14

expression of resistance-related traits by plants help determine the abundance and distribution of herbivores, which in turn affects other organisms associated with herbivores (Glassmire et al. 2016; Visakorpi et al. 2019; Wimp et al. 2005). Plant resistance is also of interest to economic entomologists and agronomists because the activities of arthropod herbivores often constrain agricultural productivity, and the use of resistant crop varieties can reduce these impacts (Smith 2021). Consequently, two parallel bodies of research and theory have developed to understand plant resistance. The study of plant resistance by applied entomologists and agronomists is guided by a framework developed over 70 years ago by Painter (1951) in his monograph, *Insect Resistance in Crop Plants* (Stout et al. 2024). The ecological and evolutionary framework for the study of plant resistance was developed at roughly the same time (Fraenkel 1959; Ehlich and Raven 1964, Hairston et al. 1960), and the study of the ecology and evolution of plant resistance has since emerged as one of the most vigorous and fertile subdisciplines in ecology (Stout et al. 2024). However, insights from the ecological literature have not been consistently related to the applied literature and vice versa (Kogan 1986; Stout et al. 2024).

The last few decades have witnessed a progressive unfolding in the ecological literature on the extent and importance of phenotypic plasticity in plant resistance against arthropods (Wang et al. 2023). Plant phenotypic plasticity in general refers to the capacity of plants to alter their phenotypes in response to changes in their biotic or abiotic environments (Agrawal 2001; Gratani 2014); with respect specifically to plant resistance against herbivores, plant phenotypic plasticity refers to the ability of plants to modify their resistance phenotype in response to changes in environmental conditions. Both ecologists and applied entomologists have long noted that the resistance of plants to insects is influenced by environmental factors such as light intensity, water availability, and soil fertility (Painter 1951). However, current interest among ecologists in plant phenotypic plasticity was spurred by the discovery, first made 50 years ago, that plants respond to feeding by herbivorous arthropods by increasing their allocation to resistance-related traits (Karban and Baldwin 1997). Since that time, the study of herbivore-induced resistance has become a very active area of ecology (Karban 2020). Herbivore-induced resistance has been found in virtually all plants in which the phenomenon has been investigated, and a wide array of resistance-related morphological, biochemical, and physiological traits have been implicated (Stout 2023). Tremendous progress has been made in understanding many aspects of induced resistance, including perception of herbivore attack, early signaling following attack, phytohormonal mediation of responses, involvement of higher trophic levels in induced resistance, and coordinated changes in plant physiology that allow plants to tolerate herbivore injury. However, our understanding of the role and importance of phenotypic plasticity in *crop* resistance to pests in managed ecosystems remains inchoate;

very few of the insights from the ecological literature have been applied to the management of pest arthropods (Lu et al. 2016; Pappas et al. 2017; Karban 2020; Poelman et al. 2023).

In this chapter, we argue that more effective use of plant resistance in future pest management programs will be partly a matter of understanding and learning to manage phenotypic plasticity in crop plants (Fig. 1). Resistance is not always expressed optimally in crop plants under managed conditions, for various reasons; expression of crop plant resistance to herbivores is influenced by environmental factors, including herbivory but also soil and climatic factors; farmers, because they exert some control over the environment in which their crops grow, can manipulate (to some degree) expression of resistance. Thus, phenotypic plasticity can be exploited to improve pest management. We will explore how an understanding of phenotypic plasticity in crops might inform pest management practices such as scouting or economic thresholds, and we will suggest how phenotypic plasticity might be manipulated to manage pests. The objective of the chapter is not to provide a comprehensive review of the phenomenon of phenotypic plasticity in plant resistance but to explore its application to pest management. The reader is referred to previously published reviews for additional coverage of the literature (Table 1, Pappas et al. 2017, Karban et al. 2020, Pereira et al. 2021, Poelman et al. 2023).

2 Constitutive and inducible resistance traits

The morphological and biochemical traits that underlie plant resistance are generally understood to include both constitutive and herbivore-inducible elements (Wang et al. 2023). Thus, Wu and Baldwin (2010) distinguish between constitutive traits that plants possess 'regardless of the presence of herbivores' and inducible defenses that 'are mounted only after plants are attacked' by herbivores. Similarly, according to Poelman et al. (2023), plants 'perceive and respond to risks of herbivore attack' in addition to possessing a 'constitutive first layer of chemical and morphological defences'. These constitutive and inducible resistance-related traits include traits that directly affect the biology and behavior of herbivores (direct resistance) and traits whose effects are mediated by the third trophic level (indirect resistance). Although the distinction between herbivore-inducible and constitutive traits has heuristic value, in broader context patterns of expression of most if not all plant traits that contribute to resistance are shaped not just by prior herbivory but by multiple other biotic and abiotic factors. As a result, the resistance phenotype expressed by a plant at any given moment in time is likely the outcome of the unique combinations of biotic and abiotic stresses to which the plant was subject over its growth. This is supported by many studies that show, in both the presence and absence of herbivory, variation in expression of plant resistance with

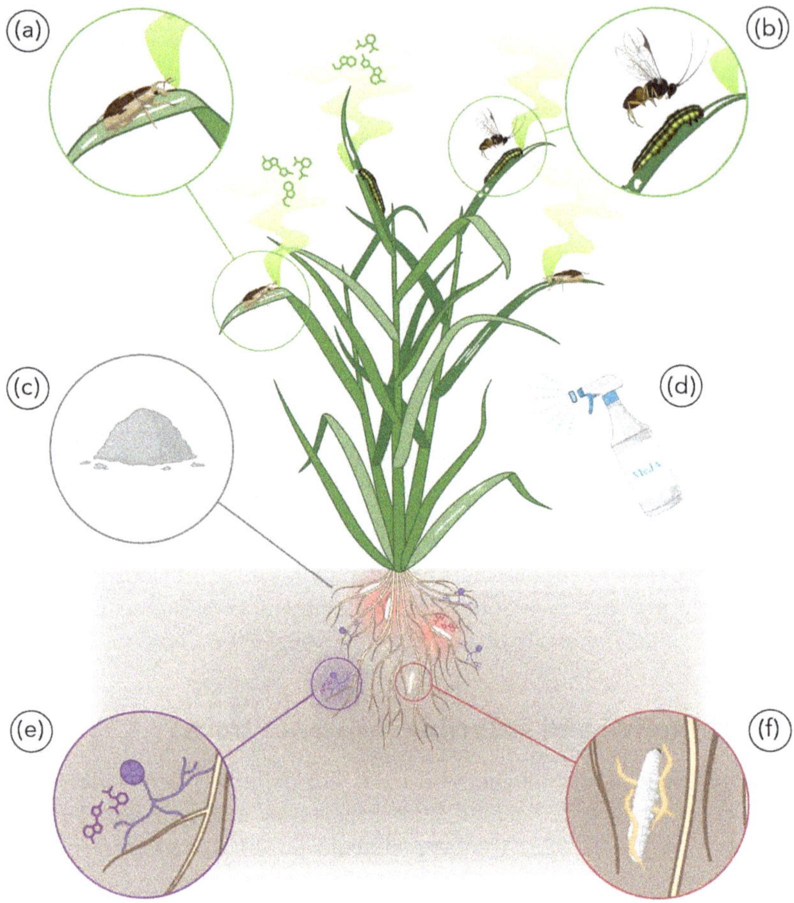

Figure 1 Schematic overview of phenomena discussed in this chapter: (a) Feeding by herbivores (e.g. rice water weevil adult) results in induction of resistance-related traits (e.g. toxic/repellent secondary metabolites), which may directly affect herbivores. (b) Feeding by caterpillars and other herbivores (e.g. fall armyworm) on above-ground portions of the plant can result in the release of volatiles that are attractive to parasitoids and predators indirectly affecting herbivores. (c) Soil amendments (e.g. silicon) can induce direct and indirect resistance against herbivores. (d) Elicitors (e.g. methyl jasmonate, MeJA) have been used to induce or prime crop resistance traits in the absence of herbivores. (e) Soil microorganisms (e.g. arbuscular mycorrhizal fungi) influence plant–insect interactions by inducing resistance traits. (f) Root herbivore larvae induce the release of signaling compounds, which attract entomopathogenic nematodes, another example of induced indirect resistance traits. (a–f) Each of these elements may influence not only crop growth and yield but also herbivore performance. Created with BioRender.com.

Table 1 Examples of agents that induce or strengthen plant resistance to herbivores

Type of strengthener/ biostimulant	Examples	Presumptive mechanisms	References
Elicitors			
Jasmonates	Jasmonic acid (JA); methyl jasmonate	Activate JA-related signaling	Sobhy et al. (2022); Stout et al. (2002)
Salicylic acid and analogues	Salicylic acid (SA); S-methyl 1,2,3-benzothiadiazole-7-carbothiate (BTH)	Activate SA signaling; may suppress JA-associated responses	Yassin et al. (2021)
Herbicides	2,4-D	Interference with phytohormone signaling or biosynthesis of secondary metabolites (e.g. shikimate or mevalonate pathways)	Norris and Kogan (2000)
Amino acids	γ-Aminobutyric acid (GABA); β-amino-butyric acid (BABA)	Modulation of reactive oxygen signaling	Tarkowski et al. (2020); Yassin et al. (2021)
Priming agents	*Z*-3-hexenol and other green leaf volatiles	Prime or sensitize plants to respond more quickly or strongly to subsequent herbivory; JA- or SA involved	Frost et al. (2008)
Soil microorganisms			
Endophytic fungi	Arbuscular mycorrhizal fungi (Glomeromycota)	Increase in plant growth and vigor; activation or priming of JA and SA signaling	Jung et al. (2012); Wang et al. (2023)
	Trichoderma afroharzianum; Trichoderma harzianum	Elicitor-induced activation of JA, SA, and reactive oxygen signaling	Nawrocka and Malolepsza (2013)
	Beauveria bassiana; Metarhizium anisopliae	Increase in seedling emergence and overall plant growth; induction of signaling pathways and production of metabolites detrimental to generalist herbivores	Pineda et al. (2017); Jaber and Enkerli (2016); Gange et al. (2019)

(*Continued*)

Table 1 (*Continued*)

Type of strengthener/ biostimulant	Examples	Presumptive mechanisms	References
Plant growth-promoting rhizobacteria	*Pseudomonas* sp., *Bacillus* sp.	JA- and ethylene-dependent signaling (induced systemic resistance)	Pereira et al. (2021)
Plant nutrients and soil amendments			
Essential macronutrients	Nitrogen, potassium	Changes in plant nutritional quality for herbivores; changes in plant compensatory ability	Singh and Sood (2017)
Silicon	Calcium silicate, slag, wollastonite	Increase in plant tissue toughness (biomechanical); induction/priming of JA signaling	Suganthy et al. (2023); Alhousari and Greger (2018)
Biochar	Various feedstock types (e.g. grass residues)	Activation of JA and SA signaling and alteration of soil microbiomes	Iacomino et al. (2022)
Humic substances	Humic and fulvic acids	Poorly understood; alterations in phytohormone levels and secondary metabolites	Pereira et al. (2021)

changes in environmental factors, including light, herbivory, water stress, soil fertility, salinity, and pathogen attack as well as plant ontogeny (Kulbat 2016; Mertens et al. 2021; Lindroth et al. 2023; Park et al. 2023).

Moreover, plants rarely express their resistance-related traits at the highest levels or concentrations possible, as demonstrated by numerous studies in which expression of resistance factors is induced to higher levels following administration of a stressor or elicitor (Thaler 1999; Thaler et al. 2001). Full consideration of the question of why plants do not continually express resistance (and resistance-related traits) to herbivores at a maximal level is beyond the scope of this chapter, but two general (non-exclusive) answers have been given by Karban (2020), Poelman et al. (2023), and others: (1) constitutive expression of resistance traits is costly; and (2) phenotypically plastic resistance is more effective.

The hypothesis most invoked to explain inducibility is that expression of resistance-related traits entails costs to the plant, and thus plants benefit from expressing resistance only when needed (i.e. when attacked by an herbivore). There is robust experimental evidence supporting the hypothesis that resistance can be costly (Heil 2002; Zust and Agrawal 2017; Karban 2020). Convincing evidence for 'allocation' costs of resistance (metabolic costs associated with synthesizing and expressing resistance-related traits) comes from studies showing reductions in plant fitness or reproduction when normally inducible defenses are activated such that they are expressed constitutively (Guo et al. 2018; Zavala and Baldwin 2004). Moreover, there are ecological and other non-allocation costs of resistance, including autotoxicity from expression of secondary metabolites, reduced competitiveness, interference with pollinators and other mutualists, and delayed flowering or seed set (Heil 2002; Karban 2020), and these may be as important or more important than allocation costs.

Several potential advantages to plasticity in the expression of plant resistance have been hypothesized (Karban 2020). One hypothesis centers on the need for plants to effectively resist diverse and changing combinations of stresses over the course of their lifetimes. According to this hypothesis, phenotypic plasticity in general and herbivore-inducible resistance, in particular, allows a plant to adjust or tailor its resistance phenotype to cope with the array of threats facing a plant at a particular moment in time (Karban 2020; Mertens et al. 2021). Furthermore, induced responses lead to increased spatial and temporal variability in the expression of resistance-related traits, and this increased variability ('moving target') may present challenges to insect physiology and behavior over both ecological and evolutionary time scales (Poelman et al. 2023). According to this hypothesis, it is not the reduced cost of induced resistance that has driven its evolution but its greater effectiveness.

As a further consideration, the resistance phenotypes of plants, including the equilibria of constitutive and inducible elements, are thought to be the outcomes of reciprocal evolutionary processes in which the risks of herbivory and the costs and benefits of resistance-related traits are balanced to produce more-or-less optimal patterns of expression. Inducible resistance is thought to be favored over constitutive resistance when resistance is costly, when herbivory is unpredictable in space and time, and when reliable cues for herbivory exist. Constitutive resistance, in contrast, is favored when risk of herbivory is constant and high (Karban 2020; Poelman et al. 2023; Pappas et al. 2017). The process of domestication and selection for improved yield, quality, and other desirable agronomic traits in crop plants, however, may have disrupted this optimization of resistance expression, and as a consequence crop plants may not express their constitutive or inducible resistance-related traits optimally (Bernal and Medina 2018; Stenberg et al. 2015). Notably, domesticated plants are thought to express lower levels of resistance-related traits in general because of their putative yield costs (Szcepaniec et al. 2012; Stenberg et al. 2015; Bernal and Medina 2018; Poelman et al. 2023). Moreover, interactions between crop plants and herbivore pests involve, from an evolutionary perspective, novel associations (i.e. plants and herbivores that have not co-evolved) and simplified environments that lack many of the ecological relationships present in natural environments.

These general considerations have three major implications for the use of plant resistance in pest management. First, the role and importance of induced resistance in crop plants may differ from their role and importance in non-domesticated plants. Second, if phenotypic plasticity is an important dimension of crop resistance to herbivorous pests, then adjustments to management practices may be needed to account for plasticity. Third, if substantial plasticity has been retained in crop plants, it may be possible to manipulate the expression of plant resistance via various production practices to optimize the expression of resistance by crop plants.

3 Relative importance of constitutive and inducible resistance in plants

Many studies have documented variation in both constitutive and inducible resistance among genotypes or populations of a plant species under controlled conditions. Onran et al. (2019) showed in a screenhouse common garden study that the (constitutive) expression of three resistance-related traits – trichomes, trypsin inhibitors, and glucosinolates – differed between two geographically distinct populations of wild arugala, *Eruca vesicaria*, that all three traits were inducible by exogenous jasmonic acid (JA), and, importantly, that the extent to which each trait was inducible differed between the two populations. Likewise, Lanka et al. (2021) demonstrated that levels of both constitutive resistance and inducibility differed among 18 soybean, *Glycine max*, genotypes and presented some evidence for a tradeoff between the two modes of resistance. What studies like these two leave largely unresolved, however, are the relative contributions of constitutive and inducible resistance traits to overall plant resistance under natural conditions – that is, are communities of herbivores and associated organisms structured more by differences in constitutive traits or induced responses under natural conditions?

One body of research that provides insight into the relative importance of inducible resistance in structuring communities of organisms on plants is research on so-called plant-mediated interactions, in which initial attack by an herbivore changes the expression of resistance-related traits by plants and consequently impacts other herbivores and organisms associated with the plant, even those spatially or temporally separated from the initial herbivore (Fig. 1a). These plant-mediated interactions have been shown in some cases to have substantial impacts on the communities of organisms centered on plants, and thus these studies provide *prima facie* evidence for an essential role for phenotypic plasticity in determining the strength and form of plant resistance in plants.

Plant-mediated interactions between initial and subsequent herbivores may operate at several time scales. At short time scales, rapid and local induced responses may affect the initial (inducing) herbivore and cause it to disperse from the site of damage, as was shown for the specialist beetle *Trirhabda pilosa* on sagebrush, *Artemisia tridentata*, which releases volatiles rapidly upon defoliation (Karban and Yang 2020). At a longer time scale, McArt et al. (2013) experimentally manipulated Japanese beetle, *Popillia japonica,* defoliation on leaves of common evening primrose, *Oenothera biennis*, and showed that previously injured plants were less attractive to, and less damaged by, three native Lepidopteran seed predators later in the growing season. Beetle-induced plants were shown to have higher levels of JA and polyphenolic compounds in the reproductive portions of the plant. Plant-mediated interactions may even extend beyond a single growing season. Chokecherry, *Prunus virginiana*, plants attacked by forest tent caterpillars, *Malacosoma californicum*, the previous year were six times less likely to have new tent caterpillar egg masses. This difference in egg-laying was correlated with lower insect performance on previously injured trees (Barnes and Murphy 2018).

Plant-mediated interactions among herbivores may operate at systemic spatial scales and thus induced resistance or susceptibility may extend to plant organs not injured by the initial herbivore, as was the case in the McArt et al. (2013) study described above in which injury to leaves resulted in increased resistance in reproductive organs in primrose. Many studies have shown that systemic plant responses can lead to plant-mediated interactions even between aboveground and belowground herbivores (Erb et al. 2011). Wan et al. (2021) demonstrated that feeding by adult flea beetles, *Bikasha collaris*, or weevils, *Heterapoderopsis bicallosicollis*, on above-ground portions of tallow tree, *Triadica sebifera*, resulted in reduced survival of *B. collaris* larvae on tallow roots, and that patterns of larval mortality were at least partly explained by changes in concentrations of tannins and carbohydrates. Similarly, in maize, *Zea mays*, leaf feeding by fall armyworm, *Spodoptera frugiperda,* resulted in reduced colonization, weight gain, and emergence of the root-feeding beetle the western corn rootworm, *Diabrotica virgifera virgifera* (Erb et al. 2011).

In most cases, the effects of an initial herbivory event on subsequent herbivores are negative (induced resistance), but not always: sometimes herbivory has little to no effect, and sometimes a positive effect, on subsequent herbivores. As an example of the former, artificially imposed, pre-flood defoliation by fall armyworm in rice had no effect on population densities of rice water weevil, *Lissorhoptrus oryzophilus*, after flooding in five of seven experiments, although in two experiments rice water weevil densities were lower in previously defoliated plots than in non-defoliated plots (Kraus and Stout 2019c). Specialist herbivores are often unaffected or positively affected by prior herbivory because they are adapted to the changes in secondary metabolism induced by herbivory. Poelman et al. (2008), for example, showed that small field plots of cabbage, *Brassica oleracea*, artificially infested with imported cabbageworm, *Pieris rapae,* caterpillars early in the growing season later experienced higher abundance of the specialist diamondback moth, *Plutella xylostella,* but lower abundance of the generalist *Mamestra brassicae* than previously uninfested plots. The effects of chewing insects and other herbivores that induce jasmonate-related responses in plants are often more strongly negative than the effects of sucking insects. Ali and Agrawal (2014) found asymmetry in plant-mediated interactions in milkweed, *Asclepias syrica*, that was probably related to the feeding guilds of herbivores: milkweeds first damaged by caterpillars, which induce higher levels of JA-related traits such as cardenolides and latex, were more resistant to aphids, but plants first infested by aphids, which induced salicylic acid (SA)-related responses, were more susceptible to subsequent caterpillars. In the same study, no plant-mediated interactions between aphids and caterpillars were detected on a related species of milkweed, *Asclepias tuberosa*. Clear patterns that would allow predictions about the direction and strength of plant-mediated interactions have not yet emerged, but, as the above examples make clear, the degree of herbivore specialization, order of attack, and feeding guild are factors (Lu et al. 2016).

Attack by an herbivore may impact not only other herbivores but also a range of other organisms associated with the plant. The best-studied of these types of effects is indirect resistance. Indirect resistance involves the facilitation of the activities of natural enemies of herbivores (predators and parasitoids) following herbivore attack (Pappas et al. 2017; Poelman et al. 2023). Facilitation often occurs via emission of herbivore-induced plant volatiles (HIPVs) attractive to predators and parasitoids but may also involve induction of extrafloral nectaries, refuges for natural enemies, or other mechanisms (Stenberg et al. 2015). Emission of HIPVs is critical to the success of natural enemies because HIPVs (as opposed to constitutive volatiles) provide reliable information to natural enemies on the presence of their herbivorous prey. Importantly, plant genotype (crop variety) can have a substantial impact on emission of HIPVs and consequent attraction of natural enemies. HIPVs emitted by cabbage following attack by *Pieris brassicae* differed among three varieties of the crop. These differences were correlated with attractiveness of the damaged plants to two parasitoids of *P. brassicae* in both the lab and the field (Poelman et al. 2009). Similarly, emission of the sesquiterpene caryophyllene from roots of maize attacked by Western corn rootworm differed substantially among genotypes, and the emission of caryophyllene was strongly implicated in the attraction of entomopathogenic nematodes in both the laboratory and field (Rasmann et al. 2005). Volatile emission may also be induced by other insect activities such as oviposition (Penaflor et al. 2011) or by pathogens (Sun et al. 2021). For example, when rice plants were infected by the fungus *Rhizoctonia solani*, volatile production increased, resulting in the attraction of two insects, the brown planthopper, *Nilaparvata lugens*, and its parasitoid *Anagrus nilaparvatae* (Sun et al. 2021)

The combination of herbivore-induced effects on other herbivores and natural enemies of herbivores may result in complex changes in arthropod communities (Pappas et al. 2017). For example, in natural populations of lima bean, *Phaseolus lunatus*, early-season herbivory by both a defoliator, *Cerotoma ruficornis*, and a green-pod feeder, *Apion godmani*, resulted in reduced abundance of a late-season, seed-feeding herbivore, the Mexican bean weevil, *Zabrotes subfasciatus*, but also reduced emergence by the parasitoid *Stenocorse bruchivora* from *Z. subfasciatus*. These effects were associated with herbivore-induced reductions in both protein and cyanogenic glycosides (Hernandez-Cumplido et al. 2016). In cabbage, attack by an early season aphid, *Brevicoryne brassicae,* but not by diamondback moth resulted in higher populations of aphid parasitoids later in the season and changed the later community of herbivores on flowers (Stam et al. 2018). In rice, *Oryza sativa*, silencing of linalool or caryophyllene emission had various effects on arthropod community structure under field conditions, with higher brown planthopper abundance on lines with reduced linalool production, reduced abundance of brown planthoppers on lines with reduced caryophyllene emission, and reduced parasitism of planthopper eggs and lower spider abundance on lines with reduced caryophyllene or linalool emission (Xiao et al. 2012).

Plant pathogenic microorganisms and pollinators may also interact with herbivores via induced changes in the shared host plant (Stout et al. 2006; Poelman et al. 2023). Herbivore-induced changes in plant resistance may result in increased resistance or increased susceptibility to pathogenic microorganisms; likewise, infection by a pathogen may increase or decrease plant resistance to subsequent herbivores. For example, maize plants infested with the root-feeding beetle Western corn rootworm were more resistant to the aboveground necrotrophic pathogen *Setosphaeria turcica* (Erb et al. 2009), and rose plants, *Rosa hybrida,* infected with the necrotrophic fungus *Botrytis cinerea* became more resistant to the aphid *Rhodobium porosum* (Mouttet et al. 2011). Regarding pollinators, nectar often contains secondary metabolites toxic or deterrent to non-adapted pollinators, and herbivory can sometimes alter levels of these compounds (Adler 2000; Poelman et al. 2023).

Despite an abundance of studies showing that plant responses to herbivore feeding can mediate interactions with subsequent herbivores and other organisms, only a few of these studies have been conducted in a way that allow a comparison of the strength of constitutive vis-à-vis inducible resistance. In a manipulative study in oak, *Quercus robur*, experimental augmentation of an early-season leaf-chewing caterpillar, *Acrobasis consociella,* led to a slightly lower abundance of late-season galls, *Andricus quercuscalicis,* and an increase in the diversity of sessile leafminers and gallers, but the effects of prior herbivory were weaker than the effects of tree location, a factor which incorporated tree genotype and other factors such as soil fertility. Thus, induced responses in this system appear to be a weaker force in structuring late-season arthropod communities than tree-dependent factors (Visakorpi et al. 2019). In contrast, Poelman et al. (2010) used a factorial design involving a susceptible and resistant variety of *B. oleracea* artificially infested or not with the early-season pest *P. rapae* to show that the effects of early-season herbivory (induced resistance) had larger effects on the late-season community of herbivores than did variety (constitutive resistance). In a study using maize and several wild relatives of maize, overall resistance to fall armyworm was correlated with inducibility of several resistance-related genes, including genes for a proteinase inhibitor and a pathogenesis-related (PR) protein, suggesting that induced responses were an important component of overall resistance (Szcepaniec et al. 2012).

4 Implications of phenotypic plasticity in crop plants for pest management practices

The existing literature thus makes it clear that plant resistance is fluid and that the activities of herbivores can change plants in ways that influence the communities of herbivores and other organisms associated with them. Unfortunately, relatively few of these studies have been conducted in agroecosystems, and for the most part, the implications of plasticity in resistance for the management of pests in crop plants remain insufficiently explored (Lu et al. 2016; Karban 2020).

The question that must be first addressed concerns the prevalence and importance of induced resistance in crop plants in the field. The methodologies commonly employed to assess crop varietal resistance in the field involve measuring differences in insect populations among crop varieties at a specific time point or points during the growing season, with little regard to (or experimental control of) the possible influence of prior herbivory or other biotic or abiotic stresses. Thus, these assessments of varietal resistance likely conflate the contributions of constitutive and induced resistance to overall resistance. Future assessments of varietal resistance should be designed to parse the contributions of constitutive and inducible resistance by employing factorial designs that couple manipulation of (inducing) herbivore populations with varieties known to differ in resistance, similar to the design employed by Poelman et al. (2010) and Stam et al. (2018) to demonstrate that herbivore-induced responses have a substantial influence on herbivore community structure in *B. oleracea*.

In crops or crop varieties in which induced resistance is a substantial component of overall plant resistance, it may be necessary to adjust management recommendations to better account for phenotypic plasticity. If, for example, early-season herbivory changes the composition of insect communities later in the season, adjustments in scouting recommendations or sampling protocols may be desirable. Such appears to be the case in white cabbage, *B. oleracea* var. *alba* plants, in which early season attack by *P. rapae* results in higher abundance of specialists but lower populations of generalists later in the season (Poelman et al. 2008; Stam et al. 2018), and in maize, in which aboveground defoliation by caterpillars results in reduced preference of *Diabrotica* spp. larvae for roots, potentially resulting in lower populations of rootworms (Lu et al. 2016). In both agroecosystems, the effects of early-season herbivory on later populations of herbivores may be consistent enough to warrant changes to sampling and scouting recommendations.

Furthermore, changes in plant resistance resulting from early attack by an herbivore may lead to reduced (or increased) need for management interventions such as insecticide applications later in the season and thus may alter economic or action thresholds. Grapevines inoculated early in the season with Willamette mites, *Eotetranychus willametti*, a non-damaging species, show increased resistance to subsequent mites, resulting in reduced populations of damaging Pacific mites, *Tetranychus pacificus*, later in the season and reducing the need for applications of miticides (Karban et al. 1997). In some cases, responses to an early season herbivory may result in compensatory or overcompensatory responses that change the vulnerability of the crop to yield losses. Some potato, *Solanum tuberosum*, cultivars, for example, overcompensate for tuber damage by tuber moths, *Tecia solnivora*, and potato weevils, *Premnotrypes* spp., by producing higher yields of (undamaged) tubers. In such cases, treatment following low levels of tuber damage would be counterproductive (Poveda et al. 2018).

When herbivory induces the emission of volatiles that attract predators and parasitoids (Fig. 1b), increased populations of natural enemies in the crop may reduce both the need for insecticides, because higher populations of natural enemies presumably help suppress pest populations, and the desirability of insecticide applications, because populations of natural enemies are often reduced by insecticide applications (Poelman et al. 2009). In addition, herbivore-induced resistance may enhance the success of natural enemies in other ways, such as slowing herbivore growth and thereby extending the period of time during which herbivores are present in smaller, more vulnerable stages (Kaplan et al. 2007). On the other hand, herbivore-induced responses are known in some instances to interfere with the activities of natural enemies and reduce biological control, such as when direct defenses have negative impacts on predator or parasitoid searching behavior or survival (Pappas et al. 2017). Some volatiles with roles in attracting natural enemies may be inhibited by herbivory, preventing beneficial indirect interactions. This may ultimately increase herbivore feeding if natural enemies and predators are unable to locate the host plant. However, as these volatiles play a pivotal role in host plant location, insects may not find host plants compromised in their ability to emit these compounds (Ramos et al. 2023).

Current management recommendations in some crops account for changes in crop susceptibility as the crop progresses through its ontogeny, such as when pests only attack certain stages of a crop plant. Beyond this, however, most management recommendations wrongly assume that resistance in crop plants is static. Unfortunately, insufficient data exist in any crop to determine whether phenotypic plasticity is strong or consistent enough to warrant modifications to sampling or scouting recommendations, or to factor into pest forecasting models, and crop-specific management recommendations that account for phenotypic plasticity have not been developed for any crop.

5 Exploiting phenotypic plasticity for pest management: use of plant elicitors to induce crop resistance to insect pests

In addition to necessitating adjustments in management practices such as sampling and thresholds, phenotypic plasticity may enable pest managers to proactively stimulate plant resistance in anticipation of herbivore attack. Substances that stimulate or prime plant resistance or tolerance to biotic or abiotic stresses have been variously termed biostimulants, elicitors, or plant strengtheners (Pereira et al. 2021; Sobhy et al. 2014). A wide variety of agents have been explored for such effects, including plant growth promoting bacteria and humic acids, silicon, mycorrhizae, and synthetic hormones such as JA (Fig. 1, Table 1) (Pineda et al. 2010; Pereira et al. 2021; Yassin et al. 2021; Annamalai et al. 2022). The mechanisms of action of some of these substances are well characterized (e.g. synthetic hormones), others less so. The potential for use of some of the most common of these biostimulants is explored below (Table 1).

Feeding, oviposition, and other activities of herbivores generate mixtures of molecules at the site of attack that allow plants to perceive the threat of herbivory. These cues may be of herbivore, plant, or mixed herbivore/plant origin (Erb 2012; Mittal et al. 2024). Some of the cues generated during herbivore or pathogen attack result in the upregulation of plant defense responses (e.g. herbivore-associated molecular patterns [HAMPs] and damage-associated molecular patterns [DAMPs]), while other cues suppress plant defenses (e.g. effectors). Perception of this mixture of cues initiates local and systemic signal cascades that have been partly elucidated. An interactive network of plant hormones mediates plant responses to herbivory, with JA and SA playing key roles. Plant responses to abiotic stresses are also mediated in part by these same signaling and hormonal networks. A full consideration of plant perception and signaling and hormonal mediation of plant responses is beyond the scope of this chapter and can be found elsewhere (Erb et al. 2012; Wang et al. 2021; Campos et al. 2014; Wu et al. 2024; Yassin et al. 2021; Mittal et al. 2024). This section will focus instead on the utility of molecules that activate these defensive responses (elicitors) and that have been identified as potentially effective tools for the management of herbivores.

The use of an elicitor to induce resistance traits is analogous to vaccination, meaning the plant's immunity is elevated and resistance elicited in the absence of an actual threat (Quintana-Rodriguez et al. 2018). This may be exhibited as immediate escalations in resistance traits in the absence of an herbivore or may involve priming plant defenses in the plant such that defenses are more quickly or strongly mobilized after attack is initiated. There are now dozens of examples in agriculture of elicitors (or primers) being used to stimulate plant resistance to insect pests. A large array of elicitors has been identified and they fall into several classes including proteins, carbohydrates, peptides, lipids, nucleotides, and synthetic chemicals (Yang et al. 2022a). Several elicitors have garnered more attention than others for their potential value and efficacy in agricultural systems. Thus, further examples are provided here for the elicitors JA, SA, S-methyl 1,2,3-benzothiadiazole-7-carbothiate (BTH), aminobutyric acid, and select herbicides with the ability to induce defensive cascades.

6 Types of plant elicitor: jasmonates

JA has been identified as the hormone primarily responsible for mediating responses to the DAMPs and HAMPs generated during attack by chewing herbivores (Campos et al. 2014). JA and its derivatives jasmonic acid isoleucine (JA-Ile) and methyl jasmonate (MJ) are collectively referred to as jasmonates. Synthetic jasmonates have been used as elicitors of induced plant resistance to insect pests in a variety of crop systems (Fig. 1d). Treatments of seed and foliage with jasmonates have been shown to induce crop resistance, as manifested by reductions in pest survival, growth, and population density. There are, however, potential drawbacks to the use of jasmonates as a pest management tool, the most important of which is the negative effect on plant growth or crop yield that may accompany elicitation – i.e. upregulation of defenses may be accompanied by a simultaneous decrease in growth and yield (Kraus and Stout 2019a). There may also be 'crosstalk' between hormonal pathways, in which the upregulation of one hormonal pathway may cause downregulation or inhibition of another pathway. Probably the most important example of crosstalk in plant defense is the antagonism between JA- and SA-mediated responses in many plants (Costarelli et al. 2020). This may result in elevated susceptibility to pests within the same system not targeted by the elicitor application.

A series of experiments using JA and MJ as seed treatments under both field and greenhouse conditions demonstrated that jasmonates can induce resistance in rice to the rice water weevil (Kraus and Stout 2019a; Bhavanam and Stout 2021). Seed treatment with MJ significantly reduced weevil densities under greenhouse and field conditions, with the resistance subsiding over time (Kraus and Stout 2019a). In the field, there appeared to be negative effects of MJ treatment on plant growth; however, due to heavy natural weevil infestation, this was difficult to untangle from the effects of weevil feeding on growth (Kraus and Stout 2019a). Subsequent experiments under greenhouse conditions (Bhavanam and Stout 2021) using both JA and MJ treatments in the absence of weevils were conducted to isolate the effects of jasmonates on growth and ascertain the potential tradeoff between growth and resistance. This study supported results from the field experiment with respect to reductions in growth and reproductive characteristics from MJ seed treatments. However, JA seed treatments showed no significant lasting effect on plant growth. The studies concur that the effects of seed treatments on defense and growth subside over time (Bhavanam and Stout 2021). Overall, these studies support the idea that jasmonate seed treatments offer a novel method of weevil control in rice.

Jasmonates have also shown promise as foliar applications in both cotton, *Gossypium hirsutum L.* (Yang et al. 2022b) and soybean (Lanka et al. 2021; Gordy et al. 2015; Shikano et al. 2018). Experiments in cotton show when JA-treated leaves are fed to cotton bollworm, *Helicoverpa armigera,* growth was reduced, resulting in a lower relative growth rate than controls. Additionally, both humoral immunity and activities of detoxification enzymes were decreased. The authors reported application of this elicitor, even at low concentrations (0.1 mM) induced plant resistance and may reduce the need for synthetic pesticides in the system (Yang et al. 2022b). In soybeans, exogenous treatments of JA reduced foliar consumption by fall armyworms by as much as 65% and reduced larval weight by as much as 76% (Lanka et al. 2021). Thus, jasmonate-induced direct resistance shows potential for use in pest management in multiple crop systems. Jasmonates may also be useful in activating the emission of HIPVs and other forms of indirect resistance thereby increasing the effectiveness of natural enemies in cropping systems (Sobhy et al. 2014). In an early example of the use of an elicitor to stimulate indirect resistance, Thaler (1999) showed that parasitism of beet armyworm, *Spodoptera exigua* by *Hyposoter exiguae* almost doubled in small field plots of tomato treated with MJ. The effects were most likely attributable to increased attraction or retention of *H. exiguae* to MJ-treated plots. Similarly, treatment of rice plants with 1 mM or 10 mM JA increased parasitism of brown planthopper eggs by *A. nilaparvatae* in the greenhouse and field, with a concomitant increase in volatile emission (Lou et al. 2005).

7 Types of plant elicitor: salicylic acid and its analogs

SA is associated with responses to biotrophic pathogens and piercing-sucking insects. Synthetic SA has been studied as an elicitor of resistance against herbivores in several agricultural systems with examples in wheat, *Triticum aestivum* (Feng et al. 2021), tomato, *Solanum lycopersicum* (Pulga et al. 2020), and cotton (El-Sherbeni et al. 2022). In each of these crops, foliar applications induced resistance to one or more arthropod pests. As with jasmonates, exogenous treatments with SA can induce resistance to herbivores resulting in reductions in insect growth or density or other insect activities such as oviposition, movement, or consumption. This was the case in tomatoes with two common arthropod pests, *Tetranychus urticae,* the two-spotted spider mite, and *Tuta absoluta*, the pinworm (also known as tomato leafminer). Foliar SA applications reduced the oviposition of both arthropods. Additionally, there were reductions in the distance traveled by the spider mites and the severity of damage from the pinworm (Pulga et al. 2020). This was especially interesting as mites utilize piercing-sucking mouthparts, while the pinworm is a chewing herbivore. Thus, the resistance-inducing effects of SA are not always restricted to piercing-sucking herbivores, as sometimes suggested.

The SA analog BTH has also been shown to induce resistance to plant pests. Commercial formulations of BTH (Actigard® [Syngenta]) have been registered for over 20 years against pathogenic fungi in several crops. BTH has also been tested against arthropods in at least five cropping systems. One study used foliar treatments with BTH to assess induced responses in melon, *Cucumis melo L., to* the cotton aphid, *Aphis gossypii*. Aphids produced fewer nymphs, had reduced intrinsic rates of growth, and experienced higher mortality on

BTH-treated plants (Moreno-Delafuente et al. 2020). Another study compared the effects of BTH and JA on plant resistance to fall armyworm, in four different crops (Gordy et al. 2015), soybean, corn, cotton, and rice. JA applications reduced the weight of fall armyworm in soybean and cotton, but not in rice or corn. However, armyworms that fed on BTH-treated cotton and soybean gained more weight compared to controls. The authors posit several hypotheses for these results. BTH, being an SA analog, is expected to crosstalk with JA. Thus, the plants may have become more susceptible to chewing caterpillars. This would explain why insects on the BTH-treated leaves gained weight rather than losing weight. When considering exogenous elicitors as a mode of pest management, the crop characteristics, pest traits, and the elicitor used will likely play a part in the exhibition of induced responses in individual plant–insect interactions.

SA and its analogs can also affect the emission of volatiles. Several experiments have shown that when treatment of maize seedlings with BTH (by soil drenches or foliar sprays) is combined with infestation by cotton leafworm, *Spodoptera littoralis,* plant attractiveness to three parasitic wasps (*Microplitis rufiventris, Cotesia marginiventris*, and *Campoletis sonorensis*) was increased (Rostas and Turlings 2008; Sobhy et al. 2012; Sobhy et al. 2014). Additionally, emissions of indole, (E)-b farnesene, and (E)-b-caryophyllene were reduced. Elicitor application to plants that did not receive subsequent herbivory did not affect wasp behavior. Indole has been noted to interfere with parasitoid attraction, thus its reduction may allow parasitoids to detect otherwise undetectable and as yet unknown volatiles. The authors hypothesized that BTH-mediated crosstalk with the JA pathway may have resulted in suppression of these volatiles (Sobhy et al. 2012). These results were very promising for pest management as the BTH applications increased parasitoid attraction and induced direct defense against cotton leafworm.

8 Types of plant elicitor: aminobutyric acids and synthetic herbicides

Amino acids have also been shown to induce plant defenses against insects and pathogens (Ghazanfar et al. 2020). Specifically, aminobutyric acid in its forms g-aminobutyric acid (GABA) and DL-β-aminobutyric acid (BABA) have been reported to induce resistance to insects in several studies. Notably, the induction of resistance by these non-protein amino acids has not always been accompanied by expected reductions in plant growth. Exogenous applications of GABA were shown to both increase the biomass of tomatoes and to induce defenses to the oriental fruit fly, *Bactrocera dorsalis,* represented by decreased pupal weights (Li et al. 2023). In a separate study, BABA was used as a root drench in five species of legumes. The application inhibited both growth and reproduction of *Acyrthosiphon pisum,* the pea aphid (Hodge et al. 2005). In this study, the effects on plant growth were not assessed.

Applications of other synthetic chemicals such as herbicides have been noted to induce plant defensive responses to insects and alter plant and insect communities (Xin et al. 2012; Bohnenblust et al. 2013; Eigenbrode et al. 1993). Herbicides may directly toxify herbivores or may affect them indirectly via reductions in plant diversity and host plant availability (Kraus and Stout 2019b). Additionally, herbicides by their very nature can stress plants and alter plant physiology in ways that inhibit insect establishment and successful reproduction (Ramos et al. 2023). This may occur, for example, by herbicide interference with the shikimate or mevalonate pathways, which are involved in the production of volatiles. Thus, synthetic herbicides may be responsible for the induction of indirect defenses to herbivores. For example, rice treated with the auxin homolog 2,4-D emitted more volatiles and was more attractive to *A. nilaparvatae*, an egg parasitoid of the brown planthopper (Xin et al. 2012). This latter study also provided an example of crosstalk, as 2,4-D-treated plants became more resistant to the striped stem borer (also commonly known as Asiatic rice borer), *Chilo suppressalis*, but also became more attractive to brown planthoppers as well as their parasitoid, *A. nilaparvatae*.

9 Plant defense priming

Finally, chemical substances may also be used to prime plant defense. Induced defenses may benefit plants by allowing them to invest in resources as needed rather than constitutively; however, the time lag between attack and induction of resistance may leave plants vulnerable to significant damage (Frost et al. 2008). Priming is a physiological state in which plants are sensitized or prepared for herbivore attack without the investment of significant resources. Primed plants respond to low levels of a stimulus more rapidly and/or more strongly when challenged by a stressor (Quintana-Rodriguez et al. 2018; Conrath et al. 2015). While our understanding of the phenomenon of priming is incomplete, true priming is sometimes difficult to distinguish from simple elicitation, and its importance of it is unclear. Balmer et al. (2015) suggest several states of priming through which plants may proceed. These are, first, the 'priming phase' which can be triggered by an exogenous application of an primer; the 'post-challenge primed stage', in which upregulation of resistance-related traits occurs, and the 'transgenerational primed state' in which epigenetic changes alter parental plants such that plants grown from the seed of primed parental plants exhibit greater resistance (Balmer et al. 2015). This has been termed 'transgenerational defense priming' (Martinez-Medina et al. 2016). SA, JA, and HIPVs (Balmer et al. 2015) are thought to be involved in the priming phase. Ultimately, plants may combine induced resistance with priming for the most effective defense strategy.

Frost et al. (2008) differentiate between induced resistance and priming and clarify that a single plant may prime portions of itself not yet under attack and thereby reduce vulnerability to overall damage. Three studies (Engleberth et al. 2004, 2007; Ton et al. 2007) are described by Frost et al. (2008) in which priming of maize plants by exposure to six-carbon green leaf volatiles (GLVs) resulted in greater induction of JA and HIPVs in response to applications of caterpillar regurgitant. Cotton leafworm larvae feeding on primed plants had lower growth rates and the parasitic wasp *Cotesia marginiventris* was preferentially attracted to primed plants. Additional examples involve the priming of wild tobacco, *Nicotiana attenuata*, by VOCs such that primed plants show accelerated induction of proteinase inhibitor activity when exposed to tobacco hornworm, *Manduca sexta,* herbivory (Kessler et al. 2006), and increased production of extrafloral nectar in lima beans exposed to VOCs from neighboring damaged plants (Choh et al. 2006). In the latter example, extrafloral nectar is an indirect defense that attracts predatory mites and defends against two-spotted spider mites.

In conclusion, various chemical agents have been successfully used to induce or prime direct and indirect resistance to piercing-sucking and chewing herbivores in half a dozen crop species. Various concentrations of elicitors have been assessed, from 0.1 mM to 5 mM, and various application methods (seed treatments, soil drenches, foliar applications) have been used. The potential utility of elicitors in pest management is thus clear, but there is a dearth of practical research on approaches for integrating elicitors into management programs. Further progress in the use of elicitors to induce resistance in crops will depend first on answers to two major questions: (1) whether the benefits of activating induced resistance, in terms of increased resistance to yield-reducing pests, outweigh the direct costs to growth and yield under field conditions; and (2), whether crosstalk among signaling pathways will interfere with the management of multiple pests within a system. One approach to addressing the former question is to compare pest populations and yields in induced and non-induced crops in both the presence and absence of herbivores under field conditions, but to our knowledge, this approach has not been implemented with complete success (Kraus and Stout 2019a; Zavala and Baldwin 2004). To address the latter issue, specific cropping systems, their array of pests, and effective elicitors must be assessed using a whole-system approach to evaluate whether crosstalk-associated susceptibilities compromise the benefits of elicitor use.

Beyond these two major questions, there are a host of other system-specific conditions that must be met for the use of elicitors in a crop to be successful (Stout et al. 2002). Because elicitor-induced resistance rarely rises to the level of complete immunity, elicitors are likely to be more useful against resident pests (those that build up to damaging levels after several generations) and other pests for which thresholds are high. Elicitors will also be more useful when pest infestations can be anticipated by pest managers, when the plant stage attacked by the herbivore is highly inducible, and when induced resistance is long-lasting. Elicitor-induced indirect resistance relies on populations of effective natural enemies attracted by induced plants. These conditions have not been fully explored in any crop-herbivore-elicitor system.

10 Microorganisms as inducers of crop resistance to insects

In agricultural systems, the soil microbiome and its various constituents, such as arbuscular mycorrhizal fungi (AMF), endophytic fungi, bacteria, and nematodes, influence not only crop growth and yield but also herbivore performance via effects on plant signaling pathways and resistance-related traits (Mariotte et al. 2018). Soil microbiomes are thus a major generator of phenotypic plasticity in plants (Fig. 1e and f). The effects of soil microbiome, however, tend to be species-specific and strongly context-dependent, and thus a better understanding of these interactions at a fundamental level will be needed before they can be effectively manipulated as a method for pest management.

AMF form symbiotic associations with the roots of many plant systems, including crops. Literature from the last two decades provides examples of this symbiosis affecting both the direct and indirect resistance of plants to above- and belowground herbivores (Pineda et al. 2010). Symbiosis between AMF and crop plants often involves direct activation of plant hormonal signaling. In maize, Stratton et al. (2022) reported that colonization of plants by AMF resulted in the upregulation of genes involved in JA (*mpi*) and SA (*pr5*) pathways, thereby providing resistance against the generalist fall armyworm and pathogens. Maize-AMF symbiosis may also improve plant growth, although improvements in growth may be dependent on nutrient availability. In Ramírez-Serrano et al. (2022), AMF symbiosis promoted maize growth and reduced the performance of the generalist beet armyworm. These effects were not tied to beet armyworm's gut microbiota but instead, its performance declined as N availability decreased.

Additionally, plant genotypes within the same species can show variable responsiveness to AMF (Berger and Gutjahr 2021). In strawberries, *Fragaria x ananassa*, Whyle et al. (2022) reported that plant genotype affected responses to AMF inoculation and herbivore performance in a field-based experiment. Strawberry genotype was the key factor in shaping physiological responses (stomatal conductance and photosynthesis); however, AMF-inoculated strawberry plants showed improvements in reproductive traits (more flowers and fruit) and phytochemical traits (floral volatiles). Also, strawberry genotype and AMF inoculation mitigated the negative effects of feeding by painted lady caterpillars, *Vanessa cardui*, on the reproductive success of the plant. Therefore, developing a better understanding of the effects of crop genotype on AMF-induced responses will be critical to leveraging AMF to help shape phenotypic plasticity and achieve their potential to mediate species interactions and enhance crop production.

Conversely, microbes can sometimes reduce crop plant resistance. For instance, in potato, Wang et al. (2023) reported that belowground herbivory by potato tuberworm, *Phthorimaea operculella*, on potato plants colonized by AMF facilitated subsequent feeding by conspecific larvae on potato tubers, even though there was induced resistance to the adult *P. operculella* aboveground. These tripartite interactions resulted from alterations in the primary and secondary metabolites in the whole plant. Similarly, in rice, Bernaola et al. (2018) reported AMF-induced susceptibility in field and greenhouse settings to rice water weevil, fall armyworm, and rice sheath blight, *Rhizoctonia solani*, all of which specialize in different plant tissues. This outcome was not based on nutrient availability and was possibly due to attenuation of resistance-related traits.

Furthermore, examples of direct induced resistance against crop pests resulting from plant associations with other mutualistic endophytic fungi have been investigated. Paradza et al. (2021) described how root inoculation with *Trichoderma* sp. caused a systemic increase in defense responses of tomato and bean crops against greenhouse whiteflies, *Trialeurodes vaporariorum*, by reducing oviposition preference and adult longevity. In a similar study on the interaction between endophytic *Trichoderma* and tomato, Agbessenou et al. (2022) explored the mechanism involved in reducing tomato leafminer damage via activation of the JA pathway, leading to systemic release of methyl salicylate. In addition, in onions, Muvea et al. (2014) found that colonization by endophytic fungi resulted in reduced feeding sites and oviposition preference of onion thrips, *Thrips tabaci*, with the underlying mechanisms possibly involving the release of metabolites and/or volatiles affecting thrips.

In addition, plant growth-promoting rhizobacteria (PGPR) are beneficial soil bacteria that can contribute to direct plant defenses by way of volatile organic compound (VOC) emissions and other mechanisms. In lima bean plants, association with rhizobia changed the VOC blend emitted by plants, thereby leading to a stronger olfactory preference for non-induced plants by the specialist insect herbivore Mexican bean beetle, *Epilachna varivestis*, (Ballhorn et al. 2013). In peppers, *Capsicum annum* L., Pappas et al. (2021) found that green peach aphid (*Myzus persicae*) numbers and two-spotted spider mite oviposition were negatively affected by inoculating with beneficial bacteria (*Bacillus amyloliquefaciens* and *Pseudomonas* spp.) and fungi (*Trichoderma* spp. and *Cordyceps fumosorosea*). Furthermore, a functional multitrophic framework for understanding the complex interactions among herbivores, crops, and their microbiota was provided by Di Lelio et al. (2023) in tomato plants. These authors showed that tomato plants colonized by the soil fungus, *Trichoderma afroharzianum*, were poorer hosts for the insect pest *Spodoptera littoralis,* and that the negative effects on *S. littoralis* resulted from *T. afroharzianum*-mediated alterations of the gut microbiota of *S. littoralis*, changing its nutritional uptake, larval developmental and survival.

Moreover, plant inoculation with beneficial microbes can be involved in indirect resistance to above- or belowground herbivores by changing plant physiology in ways that affect plant-volatile profiles (Vega 2018). For example, Hoffmann et al. (2011) studied common bean plants, *Phaseolus vulgaris*, colonized by AMF. AMF symbiosis enhanced populations of the herbivorous two-spotted spider mite, but this direct effect was mitigated by increased attractiveness to an acarine predator *Phytoseiulus persimilis*. Similarly, in melon and cotton plants, González-Mas et al. (2021) showed that endophytic colonization by the entomopathogenic fungus *Beauveria bassiana* influenced the emission of plant volatiles in the presence or absence of leaf-chewing caterpillars, *S. littoralis* and fall armyworm, and sap-sucking cotton aphids. Some of those emitted volatile compounds are known to be released in response to herbivore attack and have been implicated in natural enemy attraction. Likewise, in tomato, Papantoniou et al. (2022) demonstrated that inoculation with AMF and *Trichoderma harzianum* negatively influenced the performance of beet armyworm caterpillars by attracting its zoophytophagous mirid predator *Macrolophus pygmaeus*. Furthermore, when the predator was given a choice in Y-tube olfactometer assays between AMF- or *Trichoderma*-inoculated plants, the predator preferred HIPVs of the *Trichoderma*-inoculated plants. This indirect defense response resulted from higher expression of genes involved in JA and SA biosynthetic pathways that induce plant volatiles, such as green leaf volatiles, methyl salicylate, and several others, preferred *by M. pygmaeus*.

In addition, another role of volatiles is to mediate interactions of plants with undamaged neighboring plants with herbivores (Turlings and Erb 2018; Papantoniou et al. 2022). Regarding these induced volatiles, Song et al. (2014) showed that belowground interplant communication of herbivore-induced defense signals in tomato plants was made possible by networks formed by AMF associations. After AMF formed a belowground network, infestation by the common cutworm, *Spodoptera litura*, on one plant led to increased resistance in neighboring tomato plants not yet infested.

Plant tolerance, the ability of plants to recover from or compensate for injury caused by insect pests, is another aspect of crop plant resistance in which soil microbes play a role (Stout et al. 2024). In rice, the susceptibility of some varieties to yield losses from rice water weevil injury depends on symbiotic associations with AMF. Despite the presence of higher densities of weevil larvae on rice plants inoculated with AMF in a three-year field study, rice plants inoculated with AMF showed greater tolerance to root herbivory damage as measured by increased plant biomass and yield (Bernaola and Stout 2021). This suggests that AMF could potentially be a sustainable tool in IPM as a soil amendment or seed treatment. In addition, combinations of plant-beneficial soil microorganisms, such as AMF and entomopathogenic fungi (EPF), constitute another alternative to traditional pest management strategies with the caveat that deep understanding of the ecology is needed. In Zitlalpopoca-Hernandez et al. (2017), the combined interactions of EPF *Beauveria bassiana* and AMF with maize crops were investigated using the root-feeding larvae of *Phyllophaga vetula*. In this study, the combination of native AMF populations and *B. bassiana* induced tolerance in maize to root herbivory by increasing plant growth and N shoot content. In potatoes, Wang et al. (2023) observed that interaction between AMF and tuber herbivory by the potato tuberworm led to overcompensation in tuber growth and quality, as measured by starch, soluble proteins, and carbon concentration in the tubers.

As these examples show, the impact of soil microbiomes on pests of crops is variable, and there is still work to be conducted at a fundamental level before the effects of these potential management tools can be predicted with any accuracy. As Poveda et al. (2022) pointed out, more field studies on the efficacy of these microorganisms are necessary, since many of the greenhouse studies referenced herein may not accurately reflect conditions in agroecosystems. Additional layers of complexity beyond host nutrition, growth promotion, and pest resistance may be important to understanding the benefits of AMF symbiosis and vital to the integration of AMF into agricultural systems. For example, Pozo de la Hoz et al. (2021), reported that light intensity was a key regulator of AMF colonization and modulated the benefits of symbiosis, thereby highlighting a potentially important parameter overlooked under controlled conditions. Thus, applications of beneficial microbes in agriculture will require a nuanced understanding of their multifaceted effects under varying field conditions. An additional hurdle may be a lack of capacity of industry stakeholders to develop the bioinoculants at sufficient levels (Poveda et al. 2022). Finally, in terms of utilizing phenotypic plasticity in pest management, proper consideration of plant responses, not only to soil microbiomes but to different environmental conditions is needed. As Howard et al. (2020) mention, plant responses to microbiomes are species-specific, therefore a better understanding of these complex interactions and how they can be manipulated is needed to enable applications in different agricultural settings. This will be valuable for crop-specific management practices that account for pest resistance.

11 Silicon soil amendments as inducers of crop resistance to insects

The presence, abundance, and availability of nutrients in the soil have impacts on plant physiology that can influence resistance against insect herbivores. The availability of nitrogen, one of the most important nutrient inputs in agricultural systems, has received the most attention in this regard. Nitrogen availability can alter plant chemistry and nutritional quality and as a consequence can influence the suitability of host plants for insect herbivores (Throop and Lerdau 2004). Findings by Rashid et al. (2017) showed that variation in fertilizer inputs to rice plants had a substantial impact on the fitness traits of brown planthoppers with higher nitrogen levels leading to increased survival rates, greater fresh body weight, and enhanced overall development of planthoppers. However, the effects of nitrogen fertility interacted with rice variety to determine the effects on brown planthopper. Horgan et al. (2021) indicated that, although the fitness traits of brown planthoppers increased in nitrogen-treated rice plants, resistant and tolerant rice varieties were able to overcome this effect to a greater extent than were susceptible varieties. Thus, crop responses to insect herbivores can be manipulated by manipulating soil fertility. Here, we focus on the role of soil silicon amendment as a tool to manipulate crop resistance against insect pests.

The amendment of agricultural soils with silicon has been proposed as a strategy to enhance plant resistance against phytophagous insect pests (Fig. 1c). Silicon, one of the most abundant elements on Earth, is widely distributed in soil. However, it primarily exists in forms that are not readily available for plant use (Tubana and Heckman 2015). For silicon to be accessible to plants, it must be in the bioavailable form of silicic acid. Plants take up silicic acid via the roots and translocate it through the xylem to the shoots and leaves, where it condenses into amorphous silica deposits known as phytoliths (Ma and Yamaji 2006). Once silicon is deposited in plant tissues, it remains immobile and is not redistributed within the plant (Coskun 2019). There are two general hypotheses regarding the role of silicon in mediating plant resistance against phytophagous insects: the mechanical resistance hypothesis and the physiological resistance hypothesis (Teixeira et al. 2017). The mechanical resistance hypothesis posits that insoluble silicon, when deposited in the epidermal cells of plants, acts as a physical barrier or defense against insect pests. In contrast, the physiological resistance hypothesis suggests that silicon plays a more dynamic role by enhancing or inducing resistance by activating plant biochemical and physiological processes (Coskun et al. 2019).

Silicon is involved in strengthening plant tissues, thereby delaying penetration and feeding of host plants by insect pests. In sugarcane, silicon accumulation in the stem epidermal tissue enhances resistance to the lepidopteran stem borer *Eldana saccharina* by reducing larval stalk penetration, leading to reduced survival and feeding efficiency (Keeping et al. 2009). Previous studies have also documented silicon-associated reductions in boring success of several stem-boring pests in rice, including yellow stem borer, *Scirpophaga incertulas* (Annamalai et al. 2022), sugarcane borer, *Diatraea saccharalis* (Sidhu et al. 2013), and Asiatic rice borer (Hou and Han 2010). The delay in host plant penetration by stem borers extends the period during which these insect pests remain vulnerable to natural enemies, unfavorable environmental conditions, and chemical interventions.

Furthermore, the abrasiveness of silicified leaves and plant tissues results in greater wear of mouthparts during insect feeding, thereby acting as a deterrent for chewing insects. Massey and Hartley (2008) observed that mandibles of nutgrass armyworm larvae, *Spodoptera exempta*, experienced increased wear when fed with a high-silicon grass diet. This rapid wear on mandibles was also correlated with reduced larval growth rates and nitrogen absorption. Similar results were observed by Acevedo et al. (2021) where diets supplemented with silicon led to increased mandibular wear in fall armyworms. In contrast, the incisor teeth of the mandibles in tomato leafminer showed no damage when fed with silicon-treated tomato leaves; lack of effect was likely attributable to the distinct feeding strategy of this insect, which targets soft tissues between epidermal cell layers (Alhousari and Greger 2018; Dos Santos et al. 2015). The abrasiveness and toughness of plant tissues due to silicon deposition can also contribute to reductions in food intake and digestibility in insect pests. Biological assays conducted by Yang et al. (2017) demonstrated that brown planthopper feeding was reduced in silicon-amended rice plants compared to non-amended plants. In sugarcane, the application of silicon led to a reduction in relative growth rates and root consumption by the greyback canegrub, *Dermolepida albohirtum* (Frew et al. 2016). Results of a greenhouse experiment showed reductions in the relative growth rates of the sugarcane borer when fed with rice plants amended with silicon slag (Sidhu et al. 2013). In addition, results from a study conducted by Wang et al. (2021) indicated a reduction in feeding and growth of oriental armyworm, *Mythimna separata*, when fed with silicon-treated maize. Considerable literature therefore supports the role of silicon as a physical defense mechanism in crop systems.

There is, moreover, an increasing body of published research that highlights the role of silicon-mediated biochemical defense in plants. Ye et al. (2013) demonstrated that silicon can prime JA-mediated antiherbivore responses in rice against rice leaffolder, *Cnaphalocris medinalis*, making the plants physiologically prepared to respond more effectively to subsequent attacks. They utilized rice mutant lines in which genes responsible for jasmonate synthesis or jasmonate perception were silenced and observed strong interaction between soil-applied silicon and JA-mediated defense. Their study ultimately concluded that soil silicon amendment had a stimulatory effect on JA activity after insect damage, resulting in increased levels of defensive enzymes. In a study conducted by Wang et al. (2021) involving maize, the application of silicon significantly increased the expression of genes involved in phenolic synthesis and enhanced the accumulation of chemical defense substances, particularly chlorogenic acid. Chlorogenic acid is a polyphenol compound found in many plants that is effective against a wide array of insect herbivores (Kundu and Vadassery 2018). Findings by Acevedo et al. (2021) revealed that silicon treatment enhances biochemical defenses triggered by fall armyworm herbivory. They observed that silicon treatment in tomato plants upregulated the activities of defensive and oxidative enzymes including polyphenol oxidase (PPO) and peroxidase (POX) after exposure to fall armyworm larval feeding. In addition, results from their soybean experiment showed that the combination of silicon and fall armyworm herbivory led to increased trypsin protease inhibitor (PI) activities and greater accumulation of total phenolics compared to plants not supplemented with silicon. In wheat, silicon treatment led to elevated levels of PPO, POX, and phenylalanine ammonia-lyase (PAL) when plants were exposed to greenbug, *Schizaphis graminum* (Ranger et al. 2009). These studies offer evidence that silicon amendment improves crop resistance against insect pests by enhancing biochemical plant defense responses.

One of the least explored areas in literature is the role of silicon in tri-trophic interactions and indirect resistance. The first published study to elucidate the effects of silicon on plant defense, with a specific emphasis on HIPV-mediated effects, was conducted in cucumber, *Cucumis sativus,* plants (Kvedaras et al. 2010). Results from this study revealed an increased attraction of the predator red and blue beetle, *Dicranolaius bellulus*, to silicon-treated cucumber plants infested with Old World bollworm, *Helicoverpa armigera*. Moreover, a small-scale field experiment employing bollworm eggs attached to potted cucumbers and placed in the field indicated a significant increase in biological control by natural predators in silicon-treated plants compared to control plants. A study by Liu et al. (2017) went a step further and identified the volatiles in rice plants grown with and without silicon supplementation and infestation by rice leaffolder. Their study showed that the HIPV blend emitted by silicon-treated rice plants infested with rice leaffolder enhanced the attraction of adult female parasitoids *Trathala flavo-orbitalis* and *Microptilis mediator* compared to non-treated rice infested with rice leaffolder. Liu et al. (2017) also demonstrated that the increased attractiveness of HIPV blends in silicon-treated plants was JA-dependent. More studies are needed in this area to fully elucidate the extent of interaction between silicon amendment and plant defense mechanisms.

These examples highlight how soil silicon amendment influences plant responses and insect pest behavior and thus shows management potential in specific crop systems. Further research, particularly under field conditions, across diverse agricultural systems will clarify the efficacy, applicability, and cost-effectiveness of silicon amendment in the context of insect pest management.

12 Conclusion and future trends

Past pest management practices in general have failed to account for the ecological complexity of agroecosystems and in particular strategies for incorporating plant resistance into management programs have failed to account for the complexity engendered by phenotypic plasticity in resistance. There is thus untapped potential for increasing the efficacy of plant resistance as a tactic by leveraging phenotypic plasticity. However, exploitation of phenotypic plasticity will necessitate addressing several unanswered questions. These questions have been mentioned throughout the chapter but bear repeating here.

One important set of questions relates to the driving force behind the evolution of plasticity in plant resistance and the retention of plasticity in crop plants. Is plasticity in plant resistance merely a 'cost-saving' measure or is plasticity *per se* an important contributor to effective resistance? Does the variation in resistance traits created by plasticity suppress herbivore performance by interfering with the ability of herbivores to adjust physiologically to their diets, as has been suggested (Pearse et al. 2018), or is some other mechanism at work? Importantly, has the process of domestication and breeding for desirable agronomic traits attenuated the mechanisms responsible for generating plasticity in crop plants, which are bred for uniformity (Stenberg et al. 2015; Poelman et al. 2023)? Does plasticity have the same importance and role in crop plants as in non-domesticated plants? If the mechanisms that generate plasticity have been attenuated in crop plants, it may be necessary to breed crop plants for inducibility (Poelman et al. 2023).

The possibility of using elicitors or other types of plant strengtheners to activate inducible resistance in crop plants raises important questions related to the cost of inducible resistance. Activation of induced resistance over prolonged periods can reduce plant growth and yield (Guo et al. 2018). However, it may be possible to use elicitors more strategically to stimulate resistance for short periods of time at key points during crop development such that the benefits associated with enhanced resistance outweigh any costs to growth and reproduction (yield). As noted above, the conditions under which the benefits of elicitors outweigh their costs have not been resolved for any crop–pest system. Furthermore, recent evidence suggests that reductions in plant growth and reproduction following activation of resistance may result not so much from allocation costs of plant resistance as from hormonal networks that are 'hardwired' to attenuate growth when resistance is activated (Campos et al. 2016; Guo et al. 2018). If this is the case, it may be possible to circumvent the costs of resistance (i.e. through the co-application of plant hormones or genetic modification).

One major hurdle to the use of soil microbes and other plant strengtheners such as silicon is the idiosyncrasy and context-specificity of their effects. As noted above, increased characterization of the ecological, physiological, biochemical, and molecular conditions under which soil engineering promotes plant resistance as well as a fuller understanding of the mechanisms by which plant strengtheners stimulate resistance should lead to greater predictability in the effects of plant strengtheners (Lutz et al. 2023). Further research should focus not only on bi-trophic interactions between plants and herbivores but also on the response of the natural enemies of herbivores (tri-trophic interactions); ultimately, this broad approach will allow optimization of the use of these substances in the context of sustainable pest management programs. Furthermore, the potential for plant strengtheners may increase when used in combination (e.g. the combined use of silicon and soil microorganisms) to improve plant performance, growth, and yield (Verma et al. 2022). Finally, the potential for soil microorganisms to improve the tolerance of plants to herbivory – i.e. to help plants compensate or recover from herbivore injury – deserves further attention.

Despite the many questions that remain to be answered, there are unquestionably major benefits to exploiting plasticity in plant resistance. As regulation surrounding the use of synthetic chemicals in pest management intensifies, some products previously relied upon for pest management have become or will become unavailable, and there is an increased need for novel tools with minimal non-target effects. Developing products with plant elicitors as active ingredients may reduce the overall use of broad-spectrum insecticides and contribute to a more targeted crop pest management approach. In addition, it may be possible to use plant phenotypic plasticity in a manner that would align with organic farming principles. Elicitors can be sourced from natural materials which would allow them to be labeled as organic pesticides under EPA and USDA regulations. This may provide a pest management tool currently out of reach for these agricultural producers. Finally, the production of additional registered products will increase options for rotation in insecticide resistance management programs. This may further protect existing products by reducing pest exposure and decreasing selection pressure for the development of resistance to these products.

13 Where to look for further information

For further reading, Schoonhoven et al. (2005) is a helpful older textbook on plant-insect interactions. Karban and Baldwin (1997) is a classic introduction to the phenomenon of herbivore-induced resistance, and Walters (2011) provides a concise overview of plant defense, albeit primarily from a plant pathology perspective. Many reviews of various aspects of plant-insect interactions and plant defense have recently been published, but the review by Poelman et al. (2023), Karban (2020), and Stout et al. (2024) are noteworthy for their discussion of the potential importance of induced resistance in pest management.

For key journals/conferences, several scientific societies are dedicated to the study of plant strengtheners, including the International Society for Silicon in Agiculture (https://issag.org). The International Plant Resistance to Insects working group holds a biennial conference dedicated to the use of plant resistance in pest management. The *Journal of Chemical Ecology,* the official journal of the *International Society of Chemical Ecology,* publishes many articles related to plant resistance, as does the journal *Entomologia Experimentalis et Applicata*.

14 References

Acevedo, F., et al. (2021), 'Silicon-mediated enhancement of herbivore resistance in agricultural crops', *Front. Plant Sci.*, 12, 631824.

Adler, L. (2000), 'The ecological significance of toxic nectar', *Oikos*, 91, 409–420.

Ali, J.G. and Agrawal, A.A. (2014), 'Asymmetry of plant-mediated interactions between specialist aphids and caterpillars on two milkweeds', *Funct. Ecol.* https://doi.org/10.1111/1365-2435.12271.

Agbessenou, A., et al. (2022), 'The endophyte *Trichoderma asperellum* M2RT4 induces the systemic release of methyl salicylate and (Z)-jasmone in tomato plant affecting host location and herbivory of *Tuta absoluta*', *Front. Plant Sci.*, 13. https://doi.org/10.3389/fpls.2022.860309.

Agrawal, A.A. (2001), 'Phenotypic plasticity in the interactions and evolution of species', *Science,* 294, 321–326.

Alhousari, F. and Greger, M. (2018), 'Silicon and mechanisms of plant resistance to insect pests', *Plants,* 7, 33.

Annamalai, M., et al. (2022), Silica amendment enhances resistance of rice yellow stem borer, *Scirpophaga incertulas* (Walker) with no detrimental effect on non-target organism *Eisenia fetida* (Savigny)', *Silicon,* 14, 11939–11949.

Ballhorn, D.J., et al. (2013), 'Induced plant defense via volatile production is dependent on rhizobial symbiosis', *Oecologia*, 172(3), 833–846.

Balmer, A., et al. (2015), 'The 'prime-ome': towards a holistic approach to priming', *Trends Plant Sci., 20*(7), 443–452.

Barnes, E.E. and Murphy, S.M. (2018), 'Time-lagged intraspecific competition in temporally separated cohorts of a generalist insect', *Oecologia,* 186, 711–718.

Berger, F. and Gutjahr, C. (2021), 'Factors affecting plant responsiveness to arbuscular mycorrhiza', *Curr. Opin. Plant Biol.*, 59, 101994.

Bernal, J.S. and Medina, R.F. (2018). 'Agriculture sows pests: how crop domestication, host shifts, and agricultural intensification can create pests from herbivores', *Curr. Op. Ins. Sci.,* 26, 76–81.

Bernaola, L., et al. (2018), 'Belowground inoculation with arbuscular mycorrhizal fungi increases local and systemic susceptibility of rice plants to different pest organisms', *Front. Plant Sci.*, 9, 747.

Bernaola, L. and Stout, M. J. (2021), 'The effect of mycorrhizal seed treatments on rice growth, yield, and tolerance to insect herbivores', *J. Pest Sci.*, 94(2), 375–392.

Bhavanam, S. and Stout, M.J. (2021), 'Seed treatment with jasmonic acid and methyl jasmonate induces resistance to insects but reduces plant growth and yield in rice, *Oryza sativa*', *Front. Plant Sci.*, 12, 691768. https://doi.org/10.3389/fpls.2021.691768.

Bohnenblust, E., et al. (2013), 'Direct and indirect effects of the synthetic-auxin herbicide dicamba on two lepidopteran species', *Environ. Entomol.*, 42(3), 586–594.

Campos, M.L., et al. (2014), 'Jasmonate-triggered plant immunity', *J. Chem. Ecol.*, *40*, 657–675.

Campos, M.L., et al. (2016), 'Rewiring of jasmonate and phytochrome B signalling uncouples plant growth-defense tradeoffs', *Nat. Comm.*, 7, 12570. https://doi.org/10.1038/ncomms12570.

Choh, Y., et al. (2006), 'Induced production of extrafloral nectar in intact lima bean plants in response to volatiles from spider mite-infested conspecific plants as a possible indirect defense against spider mites', *Oecologia* 147(3), 455–460.

Conrath, U., et al. (2015), 'Priming for enhanced defense', *Annu. Rev. Phytopathol.*, 53, 97–119.

Coskun, D., et al. (2019), 'The controversies of silicon's role in plant biology', *New Phytol.*, 221, 67–85.

Costarelli, A., et al. (2020), 'Salicylic acid induced by herbivore feeding antagonizes jasmonic acid mediated plant defenses against insect attack', *Plant Sig. Behav.*, 15(1), 1704517.

Di Lelio, I., et al. (2023), 'A soil fungus confers plant resistance against a phytophagous insect by disrupting the symbiotic role of its gut microbiota', *Proc. Natl. Acad. Sci.*, 120(10), e2216922120.

Dos Santos, M., et al. (2015), 'Effect of silicon on the morphology of the midgut and mandible of tomato leaf miner *Tuta abosulta* (Lepidoptera: Gelechiidae) larvae', *Invert. Surviv. J.*, 12, 158–165.

Ehrlich, P.R. and Raven, P.H. (1964), 'Butterflies and plants: a study in coevolution', *Evolution*, 18, 586–608.

Eigenbrode, S.D., et al. (1993), 'Managing lepidopteran pests in cabbage with herbicide-induced resistance, in combination with a pyrethroid insecticide', *Entomol. Exp. Appl.*, 69(1), 41–50.

El-Sherbeni, A.E., et al. (2022), 'Salicylic Acid Enhances the Activity of Some Insecticides Against Bollworms-Infested Cotton Plants', *Egypt. Acad. J. Biol. Sci. Toxicol. Pest Control*, 14(1), 247–257.

Engelberth, J., et al. (2004), 'Airborne signals prime plants against insect herbivore attack', *Proc. Natl. Acad. Sci.*, *101*(6), 1781–1785.

Engelberth, J., et al. (2007), 'Insect elicitors and exposure to green leafy volatiles differentially upregulate major octadecanoids and transcripts of 12-oxo phytodienoic acid reductases in *Zea mays*', *Mol. Plant Microbe Interact.*, 20(6), 707–716.

Erb, M., et al. (2009), 'Signal signature of aboveground-induced resistance upon belowground herbivory in maize', *Plant J.*, 59, 292–302.

Erb, M., et al. (2011), 'Sequence of arrival determines plant-mediated interactions between herbivores', *J. Ecol.*, 99, 7–15.

Erb M., et al. (2012), 'Role of phytohormones in insect-specific plant reactions', *Trends Plant Sci.*, 17(5), 250–259.

Feng, J.L., et al. (2021), 'Exogenous salicylic acid improves resistance of aphid-susceptible wheat to the grain aphid, *Sitobion avenae* (F.) (Hemiptera: Aphididae)', *Bull. Entomol. Res.*, 111(5), 544–552.

Fraenkel, G.S. (1959), 'The raison d'être of secondary plant substances', *Science*, 129, 1466–1470.

Frew, A., et al. (2016), 'Increased root herbivory under elevated atmospheric carbon dioxide concentrations reversed by silicon-based plant defences'. *J. Appl. Ecol.*, 54, 1310–1319.

Frost, C.J., et al. (2008), 'Plant defense priming against herbivores: getting ready for a different battle', *Plant Physiol.*, 146(3), 818–824.

Gange, A.C., et al. (2019), 'Meta-analysis of the role of entomopathogenic and unspecialized fungal endophytes as plant bodyguards', *New Phytol.*, 223, 2002–2010.

Ghazanfar, M.U., et al. (2020), 'Management of late blight and sucking insect pests of potato with application of salicylic acid and β-aminobutyric acid under greenhouse conditions', *Sarhad J. Agric.*, 36, 646–654.

Glassmire, A., et al. (2016), 'Intraspecific phytochemical variation shapes community and population structure for specialist caterpillars' *New Phytol.*, 212, 208–219.

González-Mas, N., et al. (2021), 'Endophytic Colonization by the Entomopathogenic Fungus *Beauveria bassiana* Affects Plant Volatile Emissions in the Presence or Absence of Chewing and Sap-Sucking Insects', *Front. Plant. Sci.*, 12, 660460.

Gordy, J.W., et al. (2015), 'Comparative effectiveness of potential elicitors of plant resistance against *Spodoptera frugiperda* (JE Smith) (Lepidoptera: Noctuidae) in four crop plants', *PLoS One*, 10(9), e0136689.

Gratani L. (2014), 'Plant phenotypic plasticity in response to environmental factor', *Adv. Bot.*, 2014, 208747.

Guo, Q., et al. (2018), 'Resolution of growth-defense conflict: mechanistic insights from jasmonate signaling', *Curr. Opin. Plant Biol.*, 44, 72–81.

Hairston, N.G., et al. (1960), 'Community structure, population control, and competition', *Am. Nat.*, 94, 421–425.

Heil, M. (2002), 'Ecological costs of induced resistance', *Curr. Opin. Plant Biol.*, 5, 345–350.

Hernandez-Cumplido, J., et al. (2016), 'Cascading effects of early-season herbivory on late-season herbivores and their parasitoids', *Ecology*, 97(5), 1283–1297.

Hodge, S., et al. (2005), 'Application of DL-β-aminobutyric acid (BABA) as a root drench to legumes inhibits the growth and reproduction of the pea aphid *Acyrthosiphon pisum* (Hemiptera: Aphididae)', *Bull. Entomol. Res.*, 95(5), 449–455.

Hoffmann, D., et al. (2011), 'Mycorrhiza-induced trophic cascade enhances fitness and population growth of an acarine predator', *Oecologia*, 166(1), 141–149.

Horgan, F., et al. (2021), 'Nitrogenous fertilizer reduces resistance but enhances tolerance to the brown planthopper in fast-growing, moderately resistant rice', *Insects*, 12, 989.

Hou, M. and Han, Y. (2010), 'Silicon-mediated rice plant resistance to the Asiatic rice borer (Lepidoptera: Crambidae): effects of silicon amendment and rice varietal resistance', *J. Econ. Entomol.*, 103, 1412–1419.

Howard, M.M., et al. (2020), 'Soil microbiomes from fallow fields have species-specific effects on crop growth and pest resistance', *Front. Plant Sci.*, 11, 1171.

Iacomino, G., et al. (2022), 'The suppressive effects of biochar on above- and belowground plant pathogens and pests: A review', *Plants*, 11, 3144. https://doi.org/10.3390/plants11223144.

Jaber, L.R. and Enkerli, J. (2016), 'Effect of seed treatment duration on growth and colonization of *Vicia faba* by endophytic *Beauveria bassiana* and *Metarhizium brunneum*', *Biol. Control,* 103, 187–195. https://doi.org/10.1016/j.biocontrol.2016.09.008.

Jung, S.C., et al. (2012), 'Mycorrhiza-induced resistance and priming of plant defenses', *J. Chem. Ecol.*, 38(6), 651–664. https://doi.org/10.1007/s10886-012-0134-6.

Kant, M.R., et al. (2015), 'Mechanisms and ecological consequences of plant defence induction and suppression in herbivore communities', *Ann. Bot.,* 115, 1015–1051.

Kaplan, I., et al. (2007), 'Lefhopper-induced plant resistance enhances predation risk in a phytophagous beetle', *Oecologia*, 152, 665–675.

Karban, R. (2020), 'The ecology and evolution of induced responses to herbivory and how plants perceive risk', *Ecol. Entomol.*, 45, 1–9.

Karban, R. and Baldwin, I.T. (1997) *Induced responses to herbivory*. Chicago, IL: The University of Chicago Press.

Karban, R., et al. (1997), 'Mite vaccinations for sustainable managemetn of spider mites in vineyards', *Ecol. Appl.*, 71, 183–193.

Karban, R. and Yang, L.H. (2020), 'Feeding and damage-induced volatile cues make beetles disperse and produce a more even distribution of damage for sagebrush', *J. Anim. Ecol.*, 89, 2056–2062. https://doi.org/10.1111/1365-2656.13270.

Keeping, M., et al. (2009), 'Epidermal silicon in sugarcane: cultivar differences and role in resistance to sugarcane borer *Eldana saccharina*', *Environ. Exp. Bot.*, 66, 54–60.

Kogan, M. (1986). Plant defense strategies and host-plant resistance. In: Kogan, M. (ed.), *Ecological theory and integrated pest management Practice*. New York: Wiley. 362 p.

Kessler, A., et al. (2006), 'Priming of plant defense responses in nature by airborne signaling between *Artemisia tridentata* and *Nicotiana attenuata*', *Oecologia,* 148, 280–292.

Kraus, E.C. and Stout, M.J. (2019a), 'Seed treatment using methyl jasmonate induces resistance to rice water weevil but reduces plant growth in rice', *PLoS One,* 14(9), e0222800. https://doi.org/10.1371/journal.pone.0222800.

Kraus, E.C. and Stout, M.J. (2019b), 'Direct and indirect effects of herbicides on insect herbivores in rice, *Oryza sativa*', *Sci. Rep.*, 9, 6998.

Kraus, E.C. and Stout, M.J. (2019c), 'Plant-mediated interactions among above-ground and below-ground life stages of a root-feeding weevil', *Ecol. Entomol.*, 44, 771–779. https://doi.org/10.1111/een.12757.

Kulbat, K. (2016), 'The role of phenolic compounds in plant resistance', *Biotechnol., Food Sci.*, 80, 97–108.

Kundu, A. and Vadassery, J. (2018), 'Chlorogenic acid-mediated chemical defence of plants against herbivores', *Plant Biol.,* 2019(2), 185–189.

Kvedaras, O., et al. (2010), 'Silicon enhances natural enemy attraction and biological control through induced plant defenses', *Bull. Entomol. Res.,* 100, 367–371.

Lanka, S.K., et al. (2021), 'Jasmonic acid-induced resistance to fall armyworm in soybeans: Variation among genotypes and tradeoffs with constitutive resistance', *Basic Appl. Ecol.*, 56, 97–109.

Li, H., et al. (2023), 'The nitrogen-dependent GABA pathway of tomato provides resistance to a globally invasive fruit fly', *Front. Plant Sci.,* 14, 1252455.

Lindroth, R.L., et al. (2023), 'Phenotypic variation in phytochemical defense of trembling aspen in western north america: Genetics, development, and geography', *J. Che.m Ecol.*, 49, 235–250.

Liu, J., et al. (2017), 'Silicon supplementation alters the composition of herbivore induced plant volatiles and enhances attraction of parasitoids to infested rice plants', *Front. Plant Sci.,* 8, 1265.

Lou, Y.G., et al. (2005), 'Exogenous application of jasmonic acid induces volatile emissions in rice and enhances parasitism of *Nilaparvata lugens* eggs by the parasitoid *Anagrus nilaparvatae*', *J. Chem. Ecol.*, 31(9), 1985–2003.

Lu, J., et al. (2016), 'A conserved pattern in plant-mediated interactions between herbivores', *Ecol. Evol.*, 6, 1032–1040.

Lutz, S., et al. (2023), 'Soil microbiome indicators can predict crop growth response to large-scale inoculation with arbuscular mycorrhizal fungi', *Nat. Microbiol.,* 8, 2277–2289.

Ma, J. and Yamaji, N. (2006), 'Silicon uptake and accumulation in higher plants', *Trends Plant Sci.*, 11, 392–397.

Mariotte, P., et al. (2018), 'Plant-soil feedback: bridging natural and agricultural sciences', *Trends Ecol. Evol.*, 33(2), 129–142.

Martinez-Medina, A., et al. (2016), 'Recognizing plant defense priming', *Trends Plant Sci.*, 21(10), 818–822.

Massey, F. and Hartley, S. (2008), 'Physical defences wear you down: progressive and irreversible impacts of silica on insect herbivores', *J. Anim. Ecol.*, 78, 281–291.

McArt, S.H., et al. (2013), 'Leaf herbivory increases plant fitness via induced resistance to seed predators', *Ecology,* 94, 966–975.

Mertens, D., et al. (2021), 'Plant defence to sequential attack is adapted to prevalent herbivores', *Nat. Plants,* 7, 1347–1353.

Mittal, D., et al. (2024), 'External jasmonic acid isoleucine mediates amplification of plant elicitor peptide receptor (PEPR) and jasmonate-based immune signalling', *Plant Cell Environ.*, 47, 1397–1415.

Moreno-Delafuente, A., et al. (2020), 'Effects of a salicylic acid analog on *Aphis gossypii* and its predator *Chrysoperla carnea* on melon plants', *Agronomy,* 10(11), 1830.

Mouttet, R., et al. (2011), 'Phytophagous arthropods and a pathogen sharing a host plant: evidence for indirect plant-mediated interactions', *PLoS One,* 6(5), e18840.

Muvea, A.M., et al. (2014), 'Colonization of onions by endophytic fungi and their impacts on the biology of *Thrips tabaci*', *PLoS One*, 9(9), e108242.

Nawrocka, J. and Małolepsza, U. (2013), 'Diversity in plant systemic resistance induced by *Trichoderma*', *Biol. Control,* 67(2), 149–156. https://doi.org/10.1016/j.biocontrol.2013.07.005.

Norris, R.F. and Kogan, M. (2000), 'Interactions between weeds, arthropod pests, and their natural enemies in managed ecosystems', *Weed Sci.*, 48, 94–158.

Onran, A., et al. (2019), 'Evolution of phenotypic plasticity: Genetic differentiation and additive genetic variation for induced plant defence in wild arugula *Eruca sativa*', *J. Evol. Biol.*, 33, 237–246.

Painter, R.H. (1951), *Insect resistance in crop plants*. Lawrence, KS: The University of Kansas Press.

Pappas, M.L., et al. (2021), 'Beneficial soil microbes negatively affect spider mites and aphids in pepper', *Agronomy*, 11(9), 1831.

Pappas, M.L., et al. (2017), 'Induced plant defences in biological control of arthropod pests: a double-edged sword', *Pest Manag. Sci.*, 73, 1780–1788.

Papantoniou, D., et al. (2022), 'Root symbionts alter herbivore-induced indirect defenses of tomato plants by enhancing predator attraction', *Front. Physiol.*, 13, 1003746.

Paradza, V.M., et al. (2021), 'Endophytic colonisation of *Solanum lycopersicum* and *Phaseolus vulgaris* by fungal endophytes promotes seedlings growth and hampers the reproductive traits, development, and survival of the greenhouse whitefly, *Trialeurodes vaporariorum*', *Front. Plant Sci.*, 12, 771534.

Park, Y.J., et al. (2023), 'Identification of drought-responsive phenolic compounds and their biosynthetic regulation under drought stress in *Ligularia fischeri*', *Front. Plant Sci.*, 14. https://doi.org/10.3389/fpls.2023.1140509.

Pearse, I.S., et al. (2018), 'Variation in plant defense suppresses herbivore performance', *Curr. Biol.*, 28, 1981–1986.

Penaflor, M.F., et al. (2011), 'Oviposition by a moth suppresses constitutive and herbivore-induced plant volatiles in maize', *Planta*, 234, 207–215.

Pereira, R.V., et al. (2021), 'The effects of biostimulants on induced plant defense', *Front. Agron.*, 3, 630596. https://doi.org/10.3389/fagro.2021.630596.

Pineda, A., et al. (2017), 'Steering soil microbiomes to suppress aboveground insect pests', *Trends Plant Sci.*, 22, 770–778.

Pineda, A., et al. (2010), 'Helping plants to deal with insects: the role of beneficial soil-borne microbes', *Trends Plant Sci.*, 15(9), 507–14.

Poelman, E.H., et al. (2008), 'Early season herbivore differentially affects plant defence responses to subsequently colonizing herbivores and their abundance in the field', *Mol. Ecol.*, 17, 3352–3365.

Poelman, E.H., et al. (2009), 'Field parasitism rates of caterpillars on *Brassica oleracea* plants are reliably predicted by differential attraction of *Cotesia* parasitoids', *Funct. Ecol.*, 23, 951–962. https://doi.org/10/1111/j.1365-2435.2009.01570.x.

Poelman, E.H., et al. (2010), 'Herbivore-induced plant responses in *Brassica oleracea* prevail over effects of constitutive resistance and result in enhanced herbivore attack', *Ecol. Entomol.*, 35(2), 240–247.

Poelman, E.H., et al. (2023), 'Bringing fundamental insights of induced resistance to agricultural management of herbivore pests', *J. Chem. Ecol.*, 49, 218–229.

Poveda, K., et al. (2018), 'Can overcompensation increase crop production?', *Ecology,* 90, 270–280.

Poveda, J., et al. (2022), 'Editorial: Beneficial effects of fungal endophytes in major agricultural crops', *Front. Plant Sci.*, 13, 1061112.

Pozo de la Hoz, J., et al. (2021), 'Mycorrhiza-induced resistance against foliar pathogens is uncoupled of nutritional effects under different light intensities', *J. Fungi (Basel)*, 7(6), 402.

Pulga, P.S., et al. (2020), 'Salicylic acid treatments induce resistance to *Tuta absoluta* and *Tetranychus urticae* on tomato plants', *Hortic. Bras.*, 38, 288–294.

Quintana-Rodriguez, E., et al. (2018), 'Damage-associated molecular patterns (DAMPs) as future plant vaccines that protect crops from pests', *Sci. Hortic.*, 237, 207–220.

Ramírez-Serrano, B., et al. (2022), 'Mycorrhizal benefits on plant growth and protection against *Spodoptera exigua* depend on N availability', *J. Plant Interac.*, 17(1), 940–955.

Ramos, S.E., et al. (2023), 'Herbicides and their potential to disrupt plant–insect chemical communication', *J. Syst. Evol.*, 61(3), 550–560.

Ranger, C., et al. (2009), 'Influence of silicon on resistance of *Zinnia elegans* to *Myzus percicae* (Hemiptera: Aphididae)', *Environ. Entomol.*, 38, 129–136.

Rashid, M., et al. (2017), 'Higher fertilizer inputs increase fitness traits of brown planthopper in rice', *Sci. Rep.*, 7, 4719.

Rasmann, S., et al. (2005), 'Recruitment of entomopathogenic nematodes by insect-damaged maize roots', *Nature*, 434, 732–737.

Rostas, M. and Turlings, T.C.J. (2008), 'Induction of systemic acquired resistance in Zea mays also enhances the plant's attractiveness to parasitoids', *Biol. Control*, 46, 178–186. https://doi.org/10.1016/J.Biocontrol.2008.04.012.

Schoonhoven, L.M. (2005), Insect-plant biology.Oxford, UK: Oxford University Press.

Szcepaniec, A., et al. (2012), 'Higher expression of induced defenses in teosintes (*Zea* spp.) is correlated with greater resistance to fall armyworm, *Spodoptera frugiperda*', *Entomol. Exp. Applic.*, 146, 242–251.

Shikano, I., et al. (2018), 'Jasmonic acid-induced plant defenses delay caterpillar developmental resistance to a baculovirus: Slow-growth, high-mortality hypothesis in plant–insect–pathogen interactions', *J. Invertbr. Pathol*, 158, 16–23.

Sidhu, J., et al. (2013), 'Effect of silicon soil amendment on performance of sugarcane borer, *Diatraea saccharalis*, (Lepidoptera: Crambidae) on rice', *Bull. Entomol. Res.*, 103, 1–9.

Singh, V. and Sood, A.K. (2017), 'Plant nutrition: A tool for the management of hemipteran insect pests – a review', *Agric. Rev.*, 38, 260–270.

Smith, C.M. (2021), 'Conventional breeding of insect-resistant crop plants: Still the best way to feed the world population', *Curr. Opin. Insect Sci.*, 45, 7–13.

Sobhy, I.S., et al. (2012), 'Less is more: treatment with BTH and laminarin reduces herbivore-induced volatile emissions in maize but increases parasitoid attraction', *J. Chem. Ecol.*, 38, 348–360. https://doi.org/10.1007/S10886-012-0098-6.

Sobhy, I.S., et al. (2014), 'The prospect of applying elicitors and plant strengtheners to enhance biological control of crop pests', *Philos. T. Roy. Soc. B.*, 369, 20120283.

Sobhy, I.S., et al. (2022), 'Editorial: Inducing plant resistance against insects using exogenous bioactive chemicals: Key advances and future perspectives', *Front. Plant Sci.*, 13, 890884. https://doi.org/10.3389/fpls.2022.890884.

Song, Y.Y., et al. (2014), 'Hijacking common mycorrhizal networks for herbivore-induced defence signal transfer between tomato plants', *Sci. Rep.*, 4, 3915.

Stam, J.M., et al. (2018), 'Order of herbivore arrival on wild cabbage populations influences subsequent arthropod community development', *Oikos*, 127, 1482–1493.

Stenberg, J.A., et al. (2015), 'Optimizing crops for biocontrol of pests and disease', *Trends Plant Sci.*, 20, 698–712.

Stout, M.J. (2023), 'Preface: Induced resistance from theory to practice', *J. Chem. Ecol.*, 49(5), 215–216.

Stout, M.J., et al. (2002), 'Potential for the use of elicitors of plant resistance in arthropod management programs', *Arch. Insect Biochem. Physiol.*, 51, 222–235.

Stout, M.J., et al. (2006), 'Plant-mediated interactions between pathogenic microorganisms and arthropod herbivores', *Annu. Rev. Entomol.*, 51, 663–689.

Stout, M.J., et al. (2024), 'Recent history and future trends in host–plant resistance', *Ann. Entomol. Soc. Am.* https://doi.org/10.1093/aesa/saae006.

Stratton, C.A., et al. (2022), 'Nutrition vs association: plant defenses are altered by arbuscular mycorrhizal fungi association not by nutritional provisioning alone', *BMC Plant Biol.*, 22(1), 400.

Suganthy, M., et al. (2023), 'Silicon – A potential alternative in insect pest management for sustainable agriculture', *Silicon*. https://doi.org/10.1007/s12633-023-02825-5.

Sun, Z., et al. (2021), 'Phytopathogenic infection alters rice–pest–parasitoid tri-trophic interactions', *Pest Manag. Sci.*, 77(10), 453–4538.

Syngenta US "Actigard 40 WG Fungicide" Retrieved 2024-03-08.

Tarkowski, L.P., et al. (2020), 'G-Aminobutyric acid and related amino acids in plant immune responses: Emerging mechanisms of action', *Plant Cell Environ.*, 43, 1103–1116.

Thaler, J.S. (1999), 'Jasmonate-inducible plant defences cause increased parasitism of herbivores', *Nature*, 399, 686–687.

Thaler, J.S., et al. (2001), 'Jasmonate-mediated induced plant resistance affects a community of herbivores', *Ecol. Entomol.*, 26, 312–324.

Teixeira, N., et al. (2017), 'Silicon-mediated resistance against specialist insect in sap-sucking and leaf-chewing guilds in Si non-accumulator collard', *Entomol. Exp. Appl.*, 165, 94–108.

Ton, J., et al. (2007), 'Priming by airborne signals boosts direct and indirect resistance in maize', *Plant J.*, 49(1), 16–26.

Tubana, B. and Heckman, J. (2015), 'Silicon in soils and plants'. In F. Rodrigues and L. Datnoff (eds), *Silicon and plant diseases*, Cham: Springer International Publishing, pp. 7–52.

Throop, H. and Lerdau, M. (2004), 'Effects of nitrogen deposition on insect herbivory: implications for community and ecosystem processes', *Ecosystems*, 7, 109–133.

Turlings, T.C.J. and Erb, M. (2018), 'Tritrophic interactions mediated by herbivore-induced plant volatiles: Mechanisms, ecological relevance, and application potential', *Annu. Rev. Entomol.*, 63, 433–452.

Vega, F.E. (2018), 'The use of fungal entomopathogens as endophytes in biological control: a review', *Mycologia*, 110, 4–30.

Verma, K.K., et al. (2022), 'Silicon and soil microorganisms improve rhizospheric soil health with bacterial community, plant growth, performance and yield', *Plant Signal. Behav.*, 17, 2104004. https://doi.org/10.1080/15592324.2022.2104004.

Visakorpi, K., et al. (2019), 'Insect community structure covaries with host plant chemistry but is not affected by prior herbivory', *Ecology,* 100(8), e02739. https://doi.org/10.1002/ecy.2739.

Walters, D.R. (2011), Plant defences: Warding off attack by pathogens, herbivores, and parasitic plants. Chichester, UK: Wiley-Blackwell.

Wan, J., et al. (2021), 'Species-specific plant-mediated effects between herbivores converge at high damage intensity', *Ecol.*, 103, e3647. https://doi.org/10.1002/ecy.3647.

Wang, X., et al. (2023), 'Phenotypic plasticity plays an essential role in the confrontation between plants and herbivorous insects', *CABI Agric. Biosci.*, 4, 58.

Wang, D., et al. (2023), 'Arbuscular mycorrhizal fungi and belowground herbivores interact to determine plant productivity and performance of subsequently feeding conspecifics by changing plant metabolites', *J. Pest Sci.* https://doi.org/10.1007/s10340-023-01704-7.

Wang, M., et al. (2023), 'Effects of arbuscular mycorrhizal fungi on plant growth and herbivore infestation depend on availability of soil water and nutrients', *Front. Plant Sci.,* 14, 1101932. https://doi.org/10.3389/fpls.2023.1101932.

Wang, Z., et al. (2021), 'Nanosilicon enhances maize resistance against oriental armyworm (*Mythimna separata*) by activating the biosynthesis of chemical defenses', *Sci. Total Environ.*, 778, 146378.

Wang, Y., et al. (2021), 'Function and mechanism of jasmonic acid in plant responses to abiotic and biotic stresses', *Int.J. Mol. Sci.*, *22*(16), 8568.

Whyle, R.L., et al. (2022), 'Genotype, mycorrhizae, and herbivory interact to shape strawberry plant functional traits', *Front. Plant Sci.*, 13, 964941.

Wimp, G.M., et al. (2005), 'Plant genetic determinants of arthropod community structure and diversity', *Evolution*, 59, 61–69.

Wu, J. and Baldwin, I.T. (2010). New insights into plant responses to the attack from insect herbivores. *Annu. Rev. Genet.*, 44, 1–24.

Wu, S., et al. (2024), 'The MYC2-PUB22-JAZ4 module plays a crucial role in jasmonate signaling in tomato', *Mol. Plant*, 17, 598–613.

Xiao, Y., et al. (2012), 'Specific herbivore-induced volatiles defend plants and determine community composition in the field', *Ecol. Lett.*, 15, 1130–1139.

Yang, S., et al. (2022), 'Jasmonic acid-treated cotton plant leaves impair larvae growth performance, activities of detoxification enzymes, and insect humoral immunity of cotton bollworm', *Neotrop. Entomol.*, 51(4), 570–582.

Yang, B., et al. (2022b), 'Plant immunity inducers: From discovery to agricultural application', *Stress Biol.*, 2(1), 5.

Yang, L., et al. (2017), 'Silicon amendment to rice contributes to reduced feeding in a phloem-sucking insect through modulation of callose deposition', *Ecol. Evol.*, 8, 631–637.

Yassin, M., et al. (2021), 'The rise, fall and resurrection of chemical-induced resistance agents', *Pest Managmt Sci.*, 77(9), 3900–3909.

Ye, M., et al. (2013), 'Priming of jasmonate-mediated antiherbivore defense responses in rice by silicon', *Proc. Nat. Acad. Sci.*, 110, 3631–3639.

Xin, Z., et al. (2012), 'The broad-leaf herbicide 2,4-dichlorophenoxyacetic acid turns rice into a living trap for a major insect pest and a parasitic wasp', *New Phytol.*, 194(2), 498–510.

Zavala, J.A. and Baldwin, I.T. (2004), 'Fitness benefits of trypsin proteinase inhibitor expression in *Nicotiana attenuata* are greater than their costs when plants are attacked', *BMC Ecol.*, 4, 11. https://doi.org/10.1186/1472-6785-4-11.

Zitlalpopoca-Hernandez, G., et al. (2017), 'Multitrophic interactions between maize mycorrhizas, the root feeding insect *Phyllophaga vetula* and the entomopathogenic fungus *Beauveria bassiana*', *Appl. Soil Ecol.*, 115, 38–43.

Züst, T. and Agrawal A.A. (2017), 'Trade-offs between plant growth and defense against insect herbivory: an emerging mechanistic synthesis', *Annu. Rev. Plant Biol.*, 68, 513–534.

Part 2

Improving biocontrol product development and use

Chapter 7

Key issues in formulation of biocontrol agents for plant protection

Linda C. Muskat, Hochschule Geisenheim University, Germany

1 Introduction

Synthetic chemical pesticides have become widely criticized for their adverse impacts on non-target organisms and humans. As a result, there is a significant paradigm shift, marked by an increasing demand for pest control alternatives that are more sustainable and environmentally friendly. The European Parliament even recommends prioritizing low-risk pesticides and biological control measures, stating that appropriate risk management measures should be established when pesticides are used (European Parliament, 2009).

http://dx.doi.org/10.19103/AS.2024.0149.09

Biocontrol agents (BCAs) are recognized as a favorable and environmentally friendly substitute for synthetic chemical pesticides (Glare et al., 2012; Lacey et al., 2015). Biocontrol agents, when available or traded as a commercial product, are often referred to as 'biopesticides.' Biopesticides offer advantages over synthetic chemical pesticides, such as higher target specificity, reduced toxicity to non-target organisms, a lower likelihood of insect resistance, and a common origin from renewable resources (Seiber et al., 2014; Thakore, 2006; Villaverde et al., 2016).

Biological control is often defined with the commonly accepted definition: 'The use of living organisms to suppress the population of a specific pest organism, making it less abundant or less damaging than it would otherwise be' (Eilenberg et al., 2001). A similar definition of biological control was earlier given by DeBach (1964): 'the action of parasites, predators, or pathogens in maintaining another organism's density at a lower average than would occur in their absence.'

There are discrepancies in the terminology of BCAs. Therefore, Eilenberg et al. (2001) postulated with the aim of clarifying and standardizing terminology, that the term 'biological control' should be limited to the use of living organisms. The living organisms, termed 'biological control agents' or shorter 'biocontrol agents' (Eilenberg et al., 2001), can include microorganisms, namely fungi, bacteria, protozoa and nematodes, and also viruses (Lacey et al., 2015) and microalgae, and macroorganisms, such as insect predators or parasitoids (Waage and Hassel, 1982).

Other authors also include non-living natural products such as extracts or metabolites of plants or microorganisms and further include semiochemicals as 'biocontrol agent' (Chandler et al., 2011; Villaverde et al., 2016). However, in the present chapter, the focus is set on the formulation challenges and solutions of living biological control agents.

Among the BCAs that are on the market as bioinsecticides, *Bacillus thuringiensis* (*Bt*) products, constitute the largest market share, followed by entomopathogenic fungi (EPFs), other bacteria, viruses, nematodes (EPNs), and others (Lacey et al., 2015). Although the global biopesticide market is anticipated to surpass US$10 billion by 2025, with a compounded annual growth rate of 15–18%, biopesticides are still projected to constitute only a 10% share of the global crop protection market in 2025 (DunhamTrimmer®, 2022). This relatively low share is attributed to non-economic mass-production methods of biopesticides, limited stability, and inconsistent efficacy under field conditions (Glare et al., 2012).

Consequently, the main challenges of the development of BCA products are:

1 the cost-efficient production of viable biomass;
2 the stabilization of the produced propagules; and
3 improved performance of the BCA after application under the present environmental conditions.

Therefore, there is a need to explore new formulation materials, methods, and technologies that address the specific requirements of living BCAs for their formulation to unlock their potential as effective plant protection agents and to offer bioproducts suitable for integration into commercial integrated or organic plant protection strategies. Nevertheless, developing a safe and efficient BCA product is a multi-step approach that cannot be solved through formulation alone (Jones and Burges, 1998).

2 Preliminary steps in biocontrol agents product development: strain selection and fermentation

The formulation of BCAs cannot be regarded as a stand-alone issue (Jones and Burges, 1998; Ravensberg, 2011). Before a suitable formulation can be developed, two crucial steps must be taken (Bejarano and Puopulo, 2020; Ravensberg, 2011; Jackson et al., 2010):

- First, the selection of a potential BCA candidate; and
- Second, its cultivation and cost-effective mass production.

An overview of the essential steps of the development of a living BCA is illustrated in Fig. 1.

2.1 Strain selection

For the first step, the selection of a potential BCA species and strain, various strategies and related methods can be found in the literature. As most BCAs are known for their comparably high host specificity, a more targeted screening is affordable to find the most appropriate BCA candidates for a specific pest insect. The best-case scenario in the search for insect pest control agents is to find and isolate a naturally occurring entomopathogen, e.g. that can cause epizootics or pandemics within the pest insect's population (Quesada-Moraga et al., 2023; Meyling and Eilenberg, 2007). Alternatively, the isolation of potential candidates from the environment where the eventual product should be active warrants the adaptation of the candidate to the conditions at the place of application (Butt and Goettel, 2000; Glare et al., 2012). A classical established method to find and isolate entomopathogens, particularly entomopathogenic nematodes and fungi from soil for the control of soil-borne larval stages, is the so-called

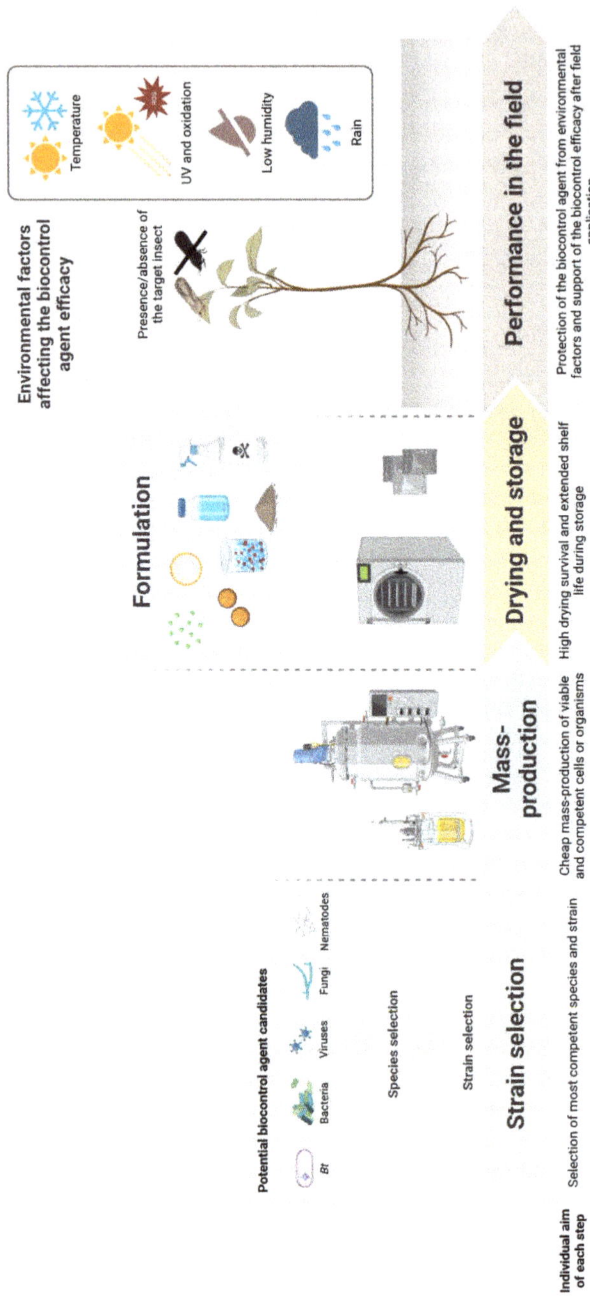

Figure 1 The development of biocontrol agents for plant protection is a multi-step approach

'insect bait method.' In this method, larvae that can easily be reared and are susceptible to various entomopathogens, commonly *Galleria mellonella*, are used as 'bait' in collected soil to catch and isolate the entomopathogenic organisms from the infected larvae afterward. The methodology was well explained and described by Zimmermann (1986) and Meyling (2007).

Alternatively, molecular methods can be employed to explore the presence of genes associated with potential entomopathogenism (Guerrero, 2023; Raymaekers et al., 2020). In bacteria, the search for common virulence factors, such as the genes for toxin expression, is a screening issue, as reviewed by Castagnola and Stock (2014). In EPFs the presence of insecticidal enzymes, such as chitinolytic enzymes that can support the penetration of the insect's cuticle, is a selection parameter of interest (Shubha et al., 2015). In the case of nematodes, selective breeding is a further option (Abd-Elgawad, 2023).

Nevertheless, it is assumed that inappropriate screening procedures are involved in the failure of BCAs in the field, as they often do not result in the selection of the most suitable candidates for the application site and scenario (Merriman and Russell, 1990; Folman et al., 2003). Therefore, the screening process should consider the influence of biotic and abiotic factors present at the application site and involve the selection of candidates adapted to the target environment (Butt and Goettel, 2000; Pliego et al., 2011; Lacey et al., 2015).

For example, strains that are more tolerant, resistant, or already adapted to environmental factors, such as UV, temperature, or low humidity conditions, can be specifically selected. Fernandes et al. (2015) gives a detailed overview of what is known about UV-tolerant EPF.

Strain selection can also involve the ease and cost-efficient mass-producibility or the potential for high drying survival, e.g. by osmotic stress (Krell et al., 2018b). The success of considering such production aspects becomes apparent when looking at bacteria such as *Bacillus* and *Streptomyces*, which produce heat- and desiccation-tolerant spores (Emmert and Handelsman, 1999).

2.2 Fermentation

The prevailing notion is that the effectiveness of entomopathogenic BCAs in controlling insect pests is directly linked to the application of a significant amount of inoculum (Jackson et al., 2010). Hence, for inundative control strategies, a critical condition for establishing an entomopathogenic microorganism as a BCA is the large-scale production of biomass. This is crucial for supplying a sufficient quantity of inoculant for extensive field application (Shah and Pell, 2003; Jaronski, 2023).

The state of the art of the mass-production processes of the individual BCA groups has been reviewed in detail for *Bt* (Brar et al., 2006), bacteria

(Jackson, 2017), fungi (Jaronski, 2023), viruses (Reid et al., 2023), and nematodes (Ehlers, 2001; Shapiro-Ilan et al., 2012) and will be summarized here only briefly.

The most used bioinsecticides are based on *Bt*. The cells are produced in large-scale fermentation tanks (Chandler et al., 2011) and separation of the insecticidal crystals from the culture has been a subject of biotechnological and industrial research since the 1950s (Pendleton and Morrison, 1966; Brar et al., 2006). Nevertheless, besides the insecticidal endotoxin crystals, most commercial *Bt* products also contain a mixture of viable bacterial cells and spores, enzymes, and unknown virulent factors (Brar et al., 2006). Bacteria are commonly produced in large-scale fermentation processes, and they are not only dependent on the species and conditions, viable cells but also spores, enzymes, toxins, and other unknown substances are produced, which can be harvested and subsequently formulated separately or as a mixture. Insect viruses as agents for biological control have been used for many years (Smith, 1959; Erlandson, 2020). Specific viruses have been successfully used in the field, particularly against leaf-feeding larvae (Payne, 1982). Genetic engineering and high-throughput sequencing technologies can further enhance their specificity, safety, and efficacy (Chejanovsky, 1992; Sun et al., 2023). The major drawback of viruses is that they need elaborate in vivo production (Erlandson, 2020). Entomopathogenic fungi can be produced in solid or liquid fermentation processes, depending on the desired biomass form and the available equipment, but most used EPF products are based on conidia produced in solid-state fermentation processes (Ruiu, 2018). Entomopathogenic nematodes can be efficiently produced on a large scale through either in vivo methods, involving infected insects, or in vitro methods, utilizing solid or liquid fermentation (Ehlers, 2001; Shapiro-Ilan et al., 2012; Koppenhöfer et al., 2020).

The individually suitable fermentation process is mainly dependent on (1) the BCA species and (2) the desired biomass form (whole organisms or cells, (endo)toxins, vegetative cells, spores, mycelia, hyphal bodies, conidia, chlamydospores, microsclerotia, or others). This also requires sufficient knowledge of the morphology, biology, and physiology of the organism. A prominent example are baculoviruses, which can only be produced in vivo within their host insect or in vitro in insect cell culture (Reid et al., 2023), what makes the mass production more elaborative compared to microbial BCAs that can be produced in classical fermentation processes. Knowing about the different cell morphologies is of particular interest in the production of EPF, as commonly one species can produce different cell types and morphologies (Jackson et al., 2003; Krell et al., 2018b; Muskat et al., 2022b).

However, already at this point in the development of a BCA for practical field application, the formulation and the requirements that the organism places on the formulation are basically known. Therefore, the formulator must decide on a biomass form and cell morphology that is to be specifically produced in the fermentation process.

Various mass-production techniques have been developed and refined over the past decades to enhance the yield of viable and virulent cells while minimizing costs. Solid-state fermentation stands out as the most widely employed method for generating aerial conidia in numerous hypocrealean EPF, which are the naturally occurring infection units (Jaronski and Mascarin, 2017). On the other hand, submerged fermentation processes enable brief fermentation times, produce high yields, offer control over biomass morphology, and are readily scalable for mass production (Jackson, 1997; Ravensberg, 2011; Jaronski, 2023).

In all cases, one central aim during the fermentation and mass production of the selected BCA is keeping the production costs low. The use of inexpensive, sometimes waste by-products from the food, agricultural, or other industries as a growth medium is an option to reduce production costs of mass fermentation processes (Hofstein and Chapple, 1998; Jaronski, 2023; Mattedi et al., 2023). Various by-products, such as molasses, malt extract, beer extract, and so on, have been tested or established as media compounds for the cultivation and mass production of BCAs (Mussatto et al., 2012; Mishra et al., 2016).

Furthermore, the fermentation of the BCA is of essential interest for the development of the final BCA product, as treatments of the cells during growth can contribute to higher stability, viability, and fitness of the BCAs during the further processing steps, the formulation, drying, storage, and application (Bejarano and Puopulo, 2020). For example, implementing specific conditions during the growth process can influence the spore-forming capability of various microorganisms (such as *Bacillus* spp.), enabling the production of cells with extended storability (Minh et al., 2011). Likewise, Mascarin et al. (2023) demonstrated improved fitness of *Beauveria bassiana* blastospores when set under osmotic stress during the cultivation by adding PEG200 or NaCl to the media. Previously, Krell et al. (2018b) found an increased shelf life of *Metarhizium brunneum* to be dependent on the culture age in a submerged fermentation process due to the enrichment of endogenous polyols within the cells, similar to the findings of Hallsworth and Magan (1994a, 1994b) for *B. bassiana*, *Metarhizium anisopliae*, and *Paecilomyces farinosus* on solid media when set under osmotic stress.

3 Formulation: purpose and components

The bridge between the production of viable biomass and its successful application in the field is the formulation of the BCA. The overall objective of why a BCA needs a formulation is to achieve a final product suitable for application, with a favorable shelf life and cost competitiveness. Formulating living BCAs can have multiple reasons. A formulation can enhance applicability and shelf life, as well as support growth, infectivity, establishment, and persistence following field application and, thus, may reduce inconsistency in field performance (Leggett et al., 2011; Burges, 1998; Boyetchko et al., 1999).

Exploring new processes, materials, methods, and technologies is essential to bridge the gap between the identification of new BCAs, their production, and their effective application in biocontrol.

But what is a formulation? A formulation is commonly composed of three components:

- the active ingredient;
- a carrier material; and
- additional formulation additives/adjuvants.

The carrier material often makes up the highest proportion of the formulations. It frequently serves a dual role not only as a filling agent, which contributes to cost reduction in formulations but can also improve handling and protect the cells from compression. In solid formulations, solid substances such as clay or starch serve as filling agents. In liquid water-based formulations, the water can be seen as the filling agent.

Formulation additives, sometimes also called adjuvants (Foy, 1989; Rhodes, 1993), are defined 'as compounds which assist or modify the action of the active ingredient' (Foy, 1989). Additives for improved stability are dependent on what the BCA should be protected from. Before selecting formulation additives, it is recommended to first evaluate whether additional additives are necessary, as they can increase formulation costs without any benefit. A very long list of formulation additives that have been tested for various effects can be found in Appendix Table I of Burges (1998b). In this chapter, the additives will be considered in terms of the challenges they were intended to address.

4 How to formulate living biocontrol agents

The selection of a suitable formulation type, material, and additives is dependent on many factors (Brar et al., 2006): the BCA (species, cell type, characteristics, regeneration mechanism and factors, persistence potential), application place (above-ground/foliage; below-ground/soil), mode of application (spray, land), application rate (L/ha or kg/ha, CFU/ha), application equipment (viscosity, density, particle size), target organism (species and type, infection or infestation factors and life cycle, feeding habits or etymology, host range), mode of action (contact, transdermal, oral), and environmental factors at the place of application (temperature, C/N availability, humidity, pH, UV radiation). Furthermore, the formulation is also determined by the cell type produced in the fermentation process.

The main challenges a BCA formulation needs to address are (Brar et al., 2006; Teixidó et al., 2022):

1 the stabilization of the BCA during the formulation, drying and storage;
2 improved handling and successful application of the product;
3 protection of the BCA from environmental factors that negatively affect its efficacy after field application; and
4 enhanced virulence of the formulated BCA.

The key to develop tailor-made formulations is knowledge of the physico-chemical or biological properties of the active ingredient or BCA and the environment in which the BCA needs to be effective. In the case of living BCAs, this is a complex issue and much more complicated compared to chemical substances that can be characterized by their reactivity, hydrophobicity, and so on. In the case of BCAs, the main characteristic is that the active ingredients are living cells or organisms (Rhodes, 1993).

5 Formulation issues: encapsulation materials

In formulating BCAs, it must always be kept in mind that they are living organisms sensitive to harsh conditions and engaged in metabolic processes. This consideration needs to be taken into account throughout the entire formulation process; otherwise, there is a risk of loss of viability. It is well known that living cells, including spores, are sensitive to temperature, humidity, pH, and also O_2/CO_2 concentrations (Miller, 1995).

Researchers have recognized that the selection of suitable materials is of fundamental importance for the success of the use of living cells. Failure of the microbial preparation may be a result of insufficient biocompatibility of the materials, well known from therapeutic pharmaceutical human applications such as implants (Orive et al., 2005).

When formulating living microbial cells, special attention needs to be given to the formulation material. A frequently used keyword regarding a materials characteristic is 'biocompatibility.' But what lies behind this term? A totally biocompatible material should not interfere with cell function (Orive et al., 2005). Contrary definitions postulate that a biocompatible material should support cell function. However, the term is often misused, as was critically discussed by Williams (2014). Therefore, in the case of BCAs, the definition 'The ability of a material to perform with an appropriate host response in a specific application' by Angelova and Hunkeler (1999) is well fitting.

However, there are some points that should be mentioned during the selection of the formulation material for encapsulation of living cells, including living BCAs. It primarily should not be toxic and should maintain the long-term viability and function of the formulated cells. Furthermore, it is recommended to select a material that has high batch-to-batch reproducibility. The material itself should be produced through an easily scalable manufacturing process and is, thus, always available in large amounts at preferably low costs (Leuenberger, 2001). The final aspect is evidently most crucial for the carrier material, considering that the carrier typically constitutes the highest proportion of commercialized biopesticides.

A common carrier material group broadly used in wettable and dispersible powders and granules are inorganic minerals such as clays like talc, kaolin, or bentonite (Burges, 1998a; Vemmer and Patel, 2013). Another group of carriers is natural polysaccharides like agar/agarose, α- and k-carrageenan, gellan, guar or bean gum, arabic gum, starch and starch-based materials, cellulose and its derivatives, such as carboxymethylcellulose or methylcellulose, xanthan, pectin, and chitosan, and the very popular alginate. Other carrier groups are polypeptides (e.g. poly L-lysine), proteins (e.g. gelatin or whey), lipids like oils and waxes, and lignin. Furthermore, synthetic polymers, copolymers, and composites of natural and synthetic materials have also been used as carrier material for living BCAs (Vemmer and Patel, 2013).

It is generally accepted that biomaterials derived from natural sources exhibit a higher level of biocompatibility compared with synthetic ones (Angelova and Hunkeler, 1999). Therefore, a current trend is to use natural biobased formulation materials, which have other advantages such as their abundance, relatively low cost, and biodegradability (Nussinovitch, 1997). Many studies have shown that polymer gels are highly suitable for entrapping living fungal and bacterial cells for use as BCAs in agriculture (Dommergues et al., 1979; Jung et al., 1982; Fravel et al., 1985; Mugnuier and Jung, 1985). A popular example of such hydrogel formulations is calcium alginate capsules, which have been proven as a promising carrier for the encapsulation of various living BCAs (Connick, 1988; Patel and Vorlop, 1994; Nussinovitch, 2010; Vemmer and Patel, 2013).

Another important issue is the safety aspect of a formulation. During and after the application of biopesticides, humans, the protected plant, and other non-target organisms can come in contact with the formulation, thus, with the active ingredient itself and the formulation additives. Therefore, it must be ensured that no danger arises from them, which is why the formulation materials and additives must be approved and registered for use in plant protection. Formulation ingredients registered for use in agricultural applications in European agriculture are listed in the FiBL list for biopesticides (Mäder et al., 2019), in the EPPO, EFSA, etc., to ensure harmlessness. In Canada, they are listed in the PMRA List of Formulants, and in the United States, formulation information can be found in Inert Ingredients Permitted in Pesticide Products (Leggett et al., 2011).

Besides the formulation, the application place and timing should additionally be considered in order to ensure the best possible conditions. For example, the sensitivity of nematodes to desiccation and ultraviolet (UV) light makes applications in soil or concealed habitats more successful, especially when carried out during the early morning or evening (Shapiro-Ilan et al., 2012).

The efficiency of EPF formulations can be further enhanced by 'virulence-enhancing additives,' which include substances that support or accelerate the infection and colonization of the target organism. For example, Krell et al. (2018a) demonstrated improved endophytism of the EPF*M. brunneum* in tomato plants by adding cellulase to facilitate the fungus's entry into the plant through the roots. Another additive that may serve as a virulence-enhancing formulation additive is chitin. The presence of chitin can induce the production of chitinolytic enzymes necessary for entering the target insect percutaneously by EPF (Charnley and St. Leger, 1991; Gerding-Gonzalez et al., 2007; Małagocka et al., 2015; Shah et al., 1999; Muskat et al., 2022a).

6 Formulation issues: water activity

When working on the formulation of BCAs, the water activity is a parameter that plays a role in different developmental steps, namely drying, storage, rehydration, and growth conditions after field application. But what is water activity?

Water activity is defined as $a_W = p_{sample}/p_{water}$, where p_{sample} is the partial vapor pressure in equilibrium with the tested material or solution, and p_{water} is the partial vapor pressure of pure water at the same temperature. To express it more simply, water activity is a measure of the biologically available water that is accessible to and can be utilized by the living biological control agent. The relative humidity of air in equilibrium with a material or solution is also called the Equilibrium Relative Humidity, expressed in %. It is equal to water activity according to ERH = $a_W \times 100\%$. Water activity should not be confused with water content, which also includes bound, biologically non-available water.

Water activity is an important parameter during different formulation steps and has been applied to the formulation of bacteria, nematodes, and fungi (Mugnier and Jung, 1985; Connick et al., 1996; Peters, 2016; Muskat et al., 2023). In particular, the a_W value is important for drying and storage of microbial BCAs, as a low a_W is known to favor the long-term viability of bacterial and fungal cells during storage (Mugnier and Jung, 1985).

7 Formulation issues: drying and storage

When aiming for the commercialization of a BCA, it is inevitable that the product must be storable for a certain period of time without losing its effectiveness. The storability of living cells is typically achieved through drying (Jackson et al., 2006; Morgan et al., 2006). Therefore, the topics of drying and storability must be considered together.

In the context of the storage of BCAs, the term 'shelf life' is inseparably linked to it and is often mentioned as an important characteristic of BCA products. It is of utmost importance for commercialization. With regards to BCAs, the term 'shelf life' refers to the time the BCA formulation can be stored at a defined temperature and remain viable and in a usable condition (Wraight and Carruthers, 1999; Wraight et al., 2003). Thus, the shelf life of a product is commonly given in the datasheet of commercialized products with the information on which temperature the product can be stored and for how long. For commercialization, a formulation needs a shelf life of at least 18 months (Couch et al., 1981). But the statements vary in a range from a minimum of 2 years being acceptable to up to 4 years being preferred (Rhodes, 1993). And this shelf life should be achieved when the formulation is stored at usual storage temperature, which means room temperature. However, this stays a challenge (Wraight et al., 2003), as indicated by the many BCA product user instructions specifying that a storage under cooled conditions at 4°C is required.

The two main factors affecting shelf life or storability are temperature and moisture. It is well known from several biological systems and cells that a certain shelf life is generally achieved by drying (Moore et al., 1996; Dietsch et al., 2021). Appropriate drying can retain biological activity during storage, while just drying itself can be detrimental. Therefore, humidity, water activity, and moisture content are some of the most intensively studied parameters in the formulation of EPF (Moore et al., 1996; Wraight and Carruthers, 1999). The outcome of many studies was that storage under low water activity maintains the longest viability, as, e.g. demonstrated for *Colletotrichum truncatum* conidia formulated in the so-called 'Pesta' formulation (Connick et al., 1996) and several other EPFs, such as *M. anisopliae*, *B. bassiana*, and *Paecilomyces* spp. (Clerk and Madelin, 1965; Daoust et al., 1983).

Special attention is needed during the drying process and rehydration from physical and chemical stress due to desiccation and rehydration. During storage, the cells can undergo chemical, enzymatic, and physical degradation processes.

The insufficient desiccation tolerance of many BCA cell types produced in liquid or solid culture is one of the main problems for the commercialization of some BCAs, such as fungi (Wraight et al., 2003) and non-spore-forming bacteria (Berninger et al., 2018). During desiccation, cells undergo various changes, including (Teixidó et al., 2022):

1 morphological alterations (cell deformation and decrease of cell volume resulting from internal water loss);
2 changes in osmotic pressure (precipitation of salts);
3 membrane damage (dehydration induces alterations in the membrane, leading to the total or partial loss of function);
4 biochemical modifications such as a reduction in phospholipid quantity, changes in the saturation level of fatty acids, peroxidation reactions, oxidation, and disruption of the tertiary structure of proteins; and
5 impairment of genetic material.

These changes can cause irreversible damage and result in the loss of viability of the cell. Therefore, this subject has been widely and intensively investigated, aiming to address it through:

1 the use of formulation additives in the form of drying protectants;
2 screening for the best drying parameters, such as the drying method, temperature, humidity, and speed (Stephan and Zimmermann, 1998; Jackson and Payne, 2007);
3 selection or targeted production of desiccation tolerant cell types (Schisler et al., 2004; Yánez-Mendizabal et al., 2012); and
4 conditioning of cells prior to drying, e.g. by osmotic stress (Teixidó et al., 2006; Krell et al., 2018b).

To understand the mode of stabilization of drying protectants, it is necessary to understand the mechanism of damage by drying. Decreases in cell volume, precipitation of salts, oxidation of compounds, and the rupture of the cell membrane can cause irreversible damage to cell structures and thus, loss of viability. Dehydration (and also freezing) can cause irreparable damage to the cell membrane and can cause the cell's death. The mechanism behind this damage was described by Burges (1998a) as follows: during the dehydration process, hydrogen-bonded water molecules that occupy the spaces among the hydrophilic tails of fatty acids in the membrane bilayer evaporate. The absence of these water molecules results in the fatty acid tails being drawn together by van der Waals forces, causing them to gel and transition into the liquid crystalline phase. This shift to a crystalline phase causes the membrane to lose its fluidity, along with some crucial transmembrane proteins, ultimately compromising its integrity. In other words, the lipid may undergo a transition from a liquid crystalline to a gel phase (Crowe et al., 1984; 1988). Upon rehydration, dry membranes, which are in the gel phase at room temperature, experience a shift from the gel to the liquid crystal phase. During this phase transition, certain regions exhibit packing defects, causing the membranes to become permeable (Blok et al., 1975).

Regarding the formulation additives in the drying context, there are two types of drying protectants. The first group of drying protectants can be referred to as membrane stabilizers (Burges, 1998a). They act by entering the membrane to replace the water molecules during dehydration, preserving the membrane's integrity (Crowe et al., 1984). Common membrane-stabilizing additives are non-reducing sugars, such as trehalose or sucrose. The addition of these sugars before drying reduces the transition temperature (T_m) of the dry membranes by displacing the water between the lipid headgroups. This prevents the occurrence of the phase transition and the resulting leakage upon rehydration (Crowe et al., 1988).

Drying protectants, particularly sugars, have been employed for an extended period to stabilize fungal propagules during desiccation, as documented by Crowe et al. (1987). Methods have ranged from the straightforward inclusion of sugars to more complex approaches such as encapsulation with starches, flours, and polysaccharide gels (Soper and Ward, 1981).

One of the most used formulation additives is skim milk (powder), which is a common carrier in spray-drying, fluidized bed-drying, and freeze-drying processes of BCAs (Teixidó, 2020), and has been extensively tested for probiotic bacteria (Wang et al., 2022). The mode of protection of skim milk is believed to be associated with the ingredient lactose, which interacts with the cell membrane and contributes to preserving membrane integrity, similar to the mechanism of non-reducing sugars (Fu and Chen, 2011). Skim milk can additionally serve as a nutritional additive in BCA formulations, e.g. in Ca-alginate capsules (Muskat et al., 2022a).

Another reason for damage, which is a big problem in hydrogel formulations such as Ca-alginate capsules, is cell damage due to the compression of the formulation matrix. When the hydrogel, which contains up to >90% water, is drying, it shrinks multifold from its initial size in the wet state. The encapsulated cells become compressed. So-called filling agents, which counteract this effect, represent the second group of drying protectants. The addition of filling agents can reduce shrinkage and prevent the compression. Common filling agents that not only act as drying protectants in polymer gel capsules but also in powder-type formulations are clay minerals, e.g. bentonite or kaolinite, or crystalline polymers, such as crystalline starch or cellulose (Przyklenk et al., 2017). The filling agents act as a non-shrinking spacer material, in which the cells are embedded. Thus, a good drying protectant with filling action is a material that does not absorb water to a great extent and does not swell and shrink during rehydration–dehydration events.

As a second function, the drying protective additives often improved the shelf life during storage. Common additives to enhance the shelf life of formulated microorganisms have been reviewed by Berninger et al. (2018) and include protective compounds such as polymers, dextran, polyethylene glycol, sugars, albumin, milk, salts, honey, polyols, or amino acids, which largely overlap with the conventional drying protectants.

There are some hypotheses on the protective mechanism of the substances and their positive effect on increased shelf life: the first mechanism lies in the physical protection to mitigate the heat and osmotic stress during the drying and desiccation process. The mode of drying protectants has been mentioned previously. The second is their function as humidity absorbents, leaving the dried formulation/maintaining the dried state to prevent the cells from rewetting during storage and support the rehydration process (Teixidó et al., 2022). This is one explanation for the improved shelf life of fungal conidia when formulated with clay due to its desiccating properties. This is also believed to be the mode of protection by non-reducing disaccharides. They probably protect the cells from rehydration during storage due to their water-binding capacities and the sugar glass formation that retard the molecular mobility and reaction rate (Santivarangkna et al., 2007; Fu et al., 2011). A third mode is based on oxygen-scavenging activity, preventing the cells from oxidation and, thus, degrading reactions (Dietsch et al., 2021). Hence, the beneficial effect of some clay minerals and silicates, such as kaolinite (Soper and Ward, 1981), is assumed to be based on their oxygen-scavenging properties (Busolo and Lagaron, 2012).

Besides the formulation, other strategies have been elaborated or established to further improve the shelf life during storage (Neale and Newton, 1999; Wraight et al., 2003):

1. Species or strain selection.
2. Conditioning of cells.
3. Packaging.
4. Low-temperature storage (≤4°C).

The first three are discussed further next.

Selection of strains, cells, or cell types with a natural tolerance against desiccation and natural stability during storage is important (Jaronski, 1997). In the case of bacteria, dormant states, such as resistant spores, are known to have a better shelf life compared to metabolically active states (Paau, 1988; Boyetchko et al., 1999; Glare et al., 2017).

Improved drying survival and storability can also be achieved by pretreatment of the cells. Osmotic stress conditioning and the influence of culture age and/or manipulation of culture conditions, especially nutrients, or the combination thereof, have been investigated and have been shown to improve the desiccation tolerance of BCAs (Hallsworth and Magan, 1994a; Jackson, 1997; Teixidó, 2006; Iwanicki et al., 2018; Cruz Barrera, 2020). This is achieved through the provocation of accumulation of cell-protecting substances, such as endogenous solutes like polyols or trehalose, by and within the cells themselves (Hallsworth and Magan, 1994a, 1994b; Krell et al., 2018b; Dietsch et al., 2021).

Packaging is crucial to maintain optimum storage environments, which means in most cases packaging regulates gas and moisture exchange, mainly to completely exclude oxygen and moisture from the package and the formulation. The most commonly used packaging material and method for excluding atmospheric gases is vacuum drying and packaging in polyethylene–aluminum foil laminates (Wraight et al., 2003; Jenkins et al., 1998).

8 Types of commercialized formulation for biocontrol agents

Numerous formulations for different BCAs have been developed and brought to market (Burges, 1998a; de Faria and Wraight, 2007). The classical and predominant formulation types used for BCAs can be categorized into dry and solid products (e.g. dusts, powders, granules, and briquettes), and suspensions or liquid preparations (oil- or water-based formulations, such as oil dispersions, soluble concentrates or suspension concentrates, and emulsions) (Rhodes, 1993; Jones and Burges, 1998; Burges, 1998a; Teixidó et al., 2022).

Formulation terminology adheres to a two-letter convention, as exemplified for granules (GR), listed in the Catalogue of Pesticide Formulation Types by CropLife International (formerly GIFAP, then GCPF) (CropLife International; GIFAP (2017)). It is worth noting that certain manufacturers deviate from these industry standards, potentially leading to confusion for users (Rhodes, 1993).

The dry preparations also include the so-called wettable powder (WP), soluble powder (SP), and water-dispersable granules (WG), which are formulated in a dry state designed to be added to a liquid carrier, normally water, just before application. The liquid suspension-type formulations include baits, suspension concentrates, oil-miscible concentrates, ultra-low volume suspensions, and oil dispersions (de Faria and Wraight, 2007).

In liquid formulations, entire cultures or cell suspensions are combined with additives. In many cases, the production and processing of these formulations are simpler, and their costs tend to be lower compared to solid-based formulations (Teixidó et al., 2022).

Solid formulations are typically subjected to drying. This process can enhance the shelf life of solid formulations during storage and, thus, has the capability to prolong the stability of the formulated BCA but needs costly drying processes (Teixidó et al., 2022; Dietsch et al., 2021). Common drying methods are freeze-drying, spray-drying, fluidized bed-drying, and fluidized bed-spray-drying (Teixidó et al., 2022). Drying mycelial mats of entomophthoralean fungi prepared by filtering fungal material produced in liquid fermentation processes and treated with a protective sugar solution prior to drying has been successfully tested as a less elaborate drying method (McCabe and Soper, 1985; Wraight et al., 2003).

An additional trend involves encapsulation within polymer beads or capsules. Vemmer and Patel (2013) have reviewed and illustrated various encapsulation methods. These methods have in common the entrapment of the BCA in a carrier matrix (called capsule or bead), often based on cross-linked polymer gels. Such capsule systems can, e.g. represent a 'microfermenter' enabling cell proliferation due to nutrient and shelter provision (Przyklenk et al., 2017) and enable the co-formulation of the BCA with other active ingredients (Vemmer et al., 2016).

A special type of capsule formulation is the so-called microcapsule. Microcapsules are characterized by their core-shell morphology and not by their size. The size of microcapsules can vary between only some nanometers and several millimeters. Microcapsules can be composed of a liquid or solid core surrounded by a solid shell. The functionality of the shell is dependent on the material or composite selection of the shell and allows for several functionalities, such as permeability, stability, surface properties, and so on (Scher, 1999). There are many different methods to prepare microcapsules (Sparks and Jacobs, 1999), dependent on the available equipment and the selected material. A well-known example of microcapsules in biocontrol applications is the encapsulation of semiochemicals, especially pheromones, with the aim of a sustained and long-term release of the behavior-manipulating substances through the shell material (Scher, 1999; Muskat and Patel, 2022).

The main active agent of *Bt* is a 'non-living' protein, which makes formulation development more similar to single substances with better characterizable physico-chemical properties compared to whole living organisms as BCAs. Several liquid and solid formulation types have been tested for *Bt* (Brar et al., 2006). In agricultural application scenarios, the final *Bt* formulations are commonly sprayable and easily applicable by common application equipment, similar to chemical insecticides, which makes *Bt* preparations more successful in replacing chemical insecticide applications than other biopesticides (Jones and Burges, 1998; Lacey et al., 2015). Other entomopathogenic bacteria have been formulated in several formulation types. The Gram-positive bacteria belonging to the Bacillales order stand out as the most commercially successful bacterial entomopathogens due to their ability to form endospores under challenging conditions. This is an advantage as they are more stable during drying and storage compared to non-spore-forming bacteria (Berninger et al., 2018). Most commercialized virus products are sold as suspension concentrates, WPs, or granules (Lacey et al., 2015). When applied to the leaves, the addition of UV protectants is a common measure to overcome the sensitivity of viruses to degradation by UV radiation (Erlandson, 2020). The majority of commercially available formulations for fungal-based insecticides are either emulsifiable suspensions or WPs containing fungal spores or mycelia or mixtures thereof. However, there are also sales of unformulated spores with a small, separate quantity of wetting agent (Jaronski, 2023; Jaronski and Mascarin, 2017). The majority of the commercialized mycoinsecticides consist of conidia (Ruiu, 2018). Another more innovative but elaborative attempt is the single-cell encapsulation of BCAs. This technique is very interesting for the use of sensitive cell types that are sensitive to external factors, such as desiccation or UV. The single-cell encapsulation was demonstrated by Yaakov et al. (2018) for *M. brunneum* spores and was achieved by a Pickering emulsion. Dietsch et al. (2024) developed a polyelectrolyte multilayering method for the single-cell encapsulation of *M. brunneum* blastospores in order to increase their desiccation tolerance. Entomopathogenic nematodes are highly sensitive to desiccation and low moisture conditions. Therefore, the focus of nematode formulation is commonly set on overcoming the humidity problem by formulation carriers and additives that create a moist microenvironment around the nematodes, such as water-retaining gels in order to maintain their viability and motility (Shapiro-Ilan et al., 2012; Wu et al., 2023).

9 Formulation to protect biocontrol agents from environmental variables in the field: ultraviolet radiation

The main challenge is to let the BCA work in the field. The inefficacy of disease-control by BCAs is frequently attributed to environmental factors, which can be challenging to precisely define (Renwick and Poole, 1989). Under field conditions, UV radiation, low and varying humidity conditions, temperature, and the limited availability of nutrients and carbon can contribute to restricting the effectiveness of the BCA (Burges, 1998).

Solar radiation is a common factor reducing cell viability in the field (Ignoffo and Garcia, 1992; Jaronski, 2010). Ultraviolet radiation is generally known for its cell-damaging action. But when talking about UV radiation, first, we have to define what UV radiation is. The wavelength ranges of UV light can be set as follows: UV light: nm = 100–380 nm; UV-A: 315–380 nm; UV-B: 280–315 nm; UV-C: 100–280 nm, which means that the wavelengths are shorter compared to the visible light and, thus, are more powerful and can more easily enter deeper into cells and cell structures. UV-C is not naturally present on Earth, as the ozone layer filters the short wavelengths before entering the atmosphere (Madronich, 1993; Madronich et al., 1998). UV-C have gained attention not only as a physical measure to control plant pathogenic fungi, e.g. in grapevine against powdery and downy mildew (Gadoury et al., 2023) and *Botrytis cinerea* (Klärner et al., 2015) but also as a pest management measure, e.g. against *Tetranychus urticae* in strawberries (Short et al., 2018).

UV-A and UV-B are the UV radiations present in the field that can damage and reduce the efficacy of beneficial BCAs (Yao et al., 2010). UV-B presents 5–10% of the total UV radiation, and UV-A constitutes 90–95% (Griffith et al., 1998; Madronich et al., 1998). UV-B with its shorter and more powerful wavelengths can thereby damage the membrane, proteins, and also the DNA and can cause the cell's death. UV-A is not that damaging but can have indirect effects by causing oxidative stress, meaning the direct damage is caused by reactive oxygen species (ROS). Even the wavelengths of visible light, in the range of 380–780 nm, can have negative effects on cells also due to ROS formation (Griffith et al., 1998).

Nevertheless, most BCA cells become inactivated within hours when directly exposed to UV radiation. This is well known for fungi, especially fungal spores (Quesada-Moraga et al., 2023; Wraight et al., 2003; Moore et al., 1993), viruses (El-Helaly et al., 2013), and nematodes (Gaugler et al., 1992). Hence, most BCAs need protection from UV and against light-induced oxidative stress when applied in the field and exposed to sunlight and radiation. Formulations designed for above-ground applications, such as foliar sprays, should contain UV protectants (Ravel et al., 1999).

Different UV protectants, sometimes called sunscreens or photoprotectants, have been evaluated and used as formulation additives for BCAs (Ignoffo and Garcia, 1992; Jackson et al., 2010). On the physical mode, UV protectants can reflect, scatter, or absorb the damaging wavelengths of UV radiation. Depending on the formulation type and carrier material, the formulation matrix itself can act as a UV protectant as it can reflect and scatter the waves or serve as a barrier to incoming waves. Additionally, filler materials have the potential to serve as UV protectant, as demonstrated by Zohar-Perez et al. (2003) who found that incorporating fillers such as kaolin, bentonite, or chitin into hydrogel matrices significantly reduced UV transmission by an average of 0.15–3.4% of the radiation.

UV protectants that reflect the waves are sometimes termed 'optical brighteners,' as these substances absorb energy from UV light and emit it as visible light, thereby prevent the UV light from entering the formulation and reaching the formulated cells. Common reflectants are fluorescent brighteners, stilbene brighteners, e.g. Calcofluor white (Shapiro, 1992), but also titanium dioxide (TiO_2) and zinc oxide (ZnO), which are common BCA formulation additives (Burges, 1998a).

Another group of protective agents that can be classified as indirect UV protectants are antioxidants, as they provide protection against oxidative stress in the form of ROS, mainly generated by the longer wavelengths of UV-A and visible light. Antioxidants can serve as (1) catchers for ROS by serving as a reaction partner or (2) as reduction–oxidation catalysts, and thus protecting the cell structures from those damaging reactions. Typical antioxidants are e.g. ascorbic acid or humic acid (Kaiser et al., 2019). In recent years, several natural substances and extracts have been examined as sources of photoprotective or antioxidative formulation additives for BCAs, such as green and black tea, eucalyptus, cacao, green coffee, red and green cabbage, and many others (El Salamouny et al. 2009; El-Helaly et al., 2013) and also some enzymes have antioxidant potential (Zhang and Feng, 2018).

The search for strong UV protectants has been a major obstacle in the formulation of baculoviruses since the 1980s (Wilson et al., 2020; Grzywacz 2017; Burges and Jones, 1998). Çakmak et al. (2021) tested different photoprotective substances for Alphabaculovirus of *Chrysodeixis chalcites* (ChchNPV-TF1) on the Canary islands and found no protection for benzopurpurine and benzimidazole, low protection for kaolinite, intermediate protection for green tea, moringa, and bentonite, and high protection for charcoal, cacao, and iron dioxide. In the field trial, the mortality rates were always higher in the UV-protective substance-treated parcels, with the highest protection observed with the charcoal (1%) treatment.

Oil-based formulations have been found to provide a good UV protection for fungal spores (Moore et al., 1993; Inglis et al., 1995; Behle et al., 2009; Fernandes et al., 2015; Kaiser et al., 2019). It has been observed that oil-based formulations can significantly increase the resistance of conidia to UV radiation, leading to more advanced germination compared to conidia suspended in water-based formulations (Alves et al., 1998).

Some microorganisms are able to synthesize their own photoprotectants in the form of pigments, e.g. some fungal spores are pigmented (Brags et al., 2006; Gao and Garcia-Pichel, 2011; Pagano and Dhar, 2015). Hence, when aiming to apply a BCA above-ground in a high sun exposure area, it could be part of the strain screening (see Section 2.1 'Strain selection') to select, e.g. a fungus as a BCA that has a higher self-protecting potential due to pigmentation (Fernandes et al., 2015).

10 Formulation to protect biocontrol agents from environmental variables in the field: humidity and temperature

Water availability and relative humidity play a crucial role in the success of a BCA application. Cells and organisms need available water for growth (Hallsworth and Magan, 1999). When investigating the humidity problem of BCAs, the water activity value (a_W value; see Section 6 Formulation issues: water activity') needs to be considered as an important parameter. Water activity value (a_W), which can be defined as the biologically available water, thus means water available to microorganisms for growth. As most cells need available water for growth and proliferation, a high water activity of the formulation material over a prolonged time can support the growth and release of infection units of the BCA after field application (Muskat et al., 2023).

Most attempts to overcome the humidity problem of BCAs through formulation have been examined with nematodes as active agents. In this context, the addition of water-retaining additives, such as superabsorbents has been investigated in field applications as an option to prolong high moisture conditions within the nematode formulation (Shapiro-Ilan et al., 2012). For example, in the screening for a suitable formulation composition for the nematode *Steinernema carpocapsae*, the co-application with Barricade®, a fire retardant, created preferable conditions for the living nematodes. This effect was assumed to be associated with the water retention properties of the gel, which maintained a high moisture level around the nematodes (Wu et al., 2023). Another option is a pre- and post-application wetting aimed at maximizing the humidity levels during and following application.

Also, for EPF, low water activity is a problem, but attempts to address this problem are rare. Co-application with synthetic polyacrylate superabsorbent polymers has also been tested for EPF as a granular formulation (Zhou and Feng, 2009). One drawback of synthetic polyacrylate-based superabsorbent materials is that they are manufactured in a toxic preparation process. Moreover, they are hardly biodegradable and therefore not an environmentally friendly solution when applied in the field. Recently, promising attempts have been made by Muskat et al. (2023) with the development of a biobased superabsorbent paste formulation providing high water activity over a prolonged period to enable sporulation of the entomophthoralean fungus *Pandora cacopsyllae* under field conditions.

Temperature is a critical environmental factor that cannot be resolved by the formulation. The options here include the selection and use of BCA candidates or strains that are adapted to the climatic zone and temperature conditions at the place of application (Köhl et al., 2011; Borisade and Magan, 2014; Sangorrín et al., 2014). The use of strain selection has been examined for some fungal entomopathogens. The first experimental step is to evaluate the sporulation and infection capability of different species or strains under different or varying temperature regimes under laboratory conditions to select and test the best-performing strain for the temperature conditions present in the field (Alali et al., 2019; Gava et al., 2022).

Recent attempts on this subject have been made by strain selection or strain conditioning in order to 'train' the EPF *M. brunneum* for relatively high-temperature applications in bee hives at 35°C for the control of *Varroa* mites

(Han et al., 2021). The fungus was treated with oxidative and nutritional stress during spore production to induce increased thermotolerance.

11 Ensuring adhesion to target areas such as leaves

The adhesion of the BCA formulation to the target area requires careful consideration of the mode of action of the BCA, the biology and behavior of the target organism, the surface properties of the target surface, the conditions at the place of application and the application technique. There are two typical sites to which the BCA needs to adhere to: (1) the plant surface or (2) the target organism.

Regarding the plant surface, adhesives, sticking, or spreading agents are common formulation additives for foliar spray formulations (Boyetchko et al., 1999; Santos et al., 2019). Improved adhesion to the plant surface can contribute to better rainfastness, increased uptake, reduction in active ingredient usage, and prolonged and thereby more sustainable efficacy duration, which improves the effectiveness, efficiency, and sustainability of applied plant protection products. Conventional chemical rainfastness enhancers are based on silicone surfactants, alkyl alkoxylate, and acryl polymers. Such petrochemical stickers or surfactants used for the application of chemical pesticides are not always compatible with living BCAs. Therefore, several biopolymers, peptides, and oils have been tested in combination with BCAs (Boyetchko, 2020).

The common parameter to be measured to analyze the wettability of surfaces, such as leaf surfaces, is the contact angle (Bejarano and Puopulo, 2020). A low contact angle indicates good spreading and, thus, well dispersion of a liquid on the (leaf) surface. Water, which is highly hydrophilic but a common liquid solvent or carrier of biopesticides, forms a high contact angle on plant leaves, which are highly hydrophobic. The contact angle of a liquid can be modified by the addition of surfactants. Surfactants are common spreading agents that are used as additives in order to modify the surface tension of a liquid formulation. Details about the physics behind surfactants can be found in Kronberg et al. (2014).

For larger particles that can also be applied to leaves, an analysis method from conventional sticker technology research, based on a tensile shear test in dynamic-mechanical shear experiment method, was adapted for the investigation of the adhesive properties of a newly developed oleogel formulation on apple leaves by Muskat et al. (2022c). This method is suitable at a laboratory level to investigate the relationship between the gel composition and the self-adhesion properties on plant leaves or other surfaces.

A method closer to practice is the analysis of the rainfastness of a formulation in a rain simulation laboratory. The effect of different intensities or angles of rainfall can be simulated, and the wash off of the formulation from the leaves can be quantified.

Besides solutions from the formulation side by adhesion agents or one of the novel technologies to improve adherence to the target area, the application technique, commonly spray application, is crucial for success. Several nozzle types and the choice between low-volume air-assist sprayers and high-pressure hydraulic spray application are common pesticide application techniques (Wraight and Carruthers, 1999). In the special case of living BCAs, the time of day can play an essential role in the success of the BCA's efficacy. Many instructions for living preparations recommend applying the product in the evening when the critical environmental factors, namely UV radiation, too low humidity and too high temperatures, are reduced (such information can be found in the manuals of the commercialized products).

In contrast to other entomopathogenic microorganisms, the infection caused by an EPF occurs through the cuticle (percutaneously) not via ingestion (Jaronski, 2023). Consequently, when EPFs are applied as BCAs, external contact with the infection units can be sufficient to initiate an infection. The adhesion of spores or conidia of insect-pathogenic fungi to the cuticle is assumed to be due to the hydrophobicity of the spores and hydrophobic interactions with the insect cuticle surface (Boucias et al., 1988; Holder et al., 2007; Zhang et al., 2011) and should be taken into account when formulating such cells. Thus, lipophilic oils are inherently compatible with hydrophobic conidia and have been successfully used as spray carriers for EPF's conidia (Prior et al., 1988; Jaronski, 1997; Wraight and Carruthers, 1999; Malsam et al., 2002). Oil-based formulations of fungal entomopathogens have the advantage of excellent adhesion not only to the hydrophobic cuticle of the pest insect but also to the leaf surface. Therefore, oil-based spray formulations are the first choice for spray application of spores of EPFs (Jaronski, 1997; Prior et al., 1988). Additionally, oil formulations have also been proven to improve the storage stability of fungal spores, as demonstrated for different *Metarhizium flavoviride*, *M. anisopliae*, and *B. bassiana* strains whose viability was found to be better when formulated in an oil carrier compared to dried powder when incubated for up to 266 days (Jaronski, 1997). Vegetable oils such as sunflower, rapeseed, or canola oil offer the benefit of being allowed in organic farming (Mäder et al., 2019).

12 Modes of release

When developing formulations, one important aspect is the release behavior of the active ingredient, including release rate and release duration (Akelah, 2013). The mode of release of the formulated BCA can be modified by the formulation material and the supplementation with release-modifying additives.

Typical modes of release that can be achieved by the formulation, in contrast to the unformulated agent, can be classified as slow, sustained, burst, or controlled/triggered release.

Slow release addresses the classical aim of preventing the release of the whole loading within a short time. Hence, with a slow release behavior, the depot of active ingredients can be maintained over a longer time, enabling a long-term release. Microcapsules can be designed as a slow release system as long as the shell material is designed to stay stable for a prolonged period, preventing the core from being released too fast, protecting the encapsulated BCAs from the environment, and maintaining a desirable microenvironment for the living BCA by the addition of nutrients. Bashan (1986) developed a slow release carrier based on core-shell alginate microcapsules in which a large reservoir of plant-growth-promoting bacteria was enclosed within the core surrounded by the Ca-alginate shell, releasing the bacteria at a slow and constant rate. Similar capsules were prepared by Kim et al. (2012) for the slow release of *Pantoea agglomerans* E325 for the control of *Erwinia amylovora*, the causative agent of fire blight disease in apple. A simple measure to slow down the release from any matrix or shell is increasing the material's density, e.g. by adjusting the gelling agent concentration (Kim et al., 2012) or by the addition of filling agents. This was demonstrated by Feng et al. (2019) who found the release of spores of a non-aflatoxigenic *Aspergillus flavus*, as a competitive BCA against aflatoxigenic *Aspergillus* strains, from Ca-alginate-matrix-type capsules was retarded when kaolin or rice husk powder was added as filling agents, and decrease correlated with filling agent content.

Sustained and slow often come along with each other, but sustained means that the active ingredient is released at a constant amount (zero order). Burst release stands in contrast to slow and sustained release, as in the best case, the whole active ingredient loading is released in one go. This mode of release is desirable when a high concentration or dose is needed to be effective. A burst release can be a controlled release at the same time if the burst is initiated by any external or internal stimulus. 'Controlled' release is often misused when actually 'slow' or 'sustained' release is realized by the formulation. Controlled release formulation systems can be divided into two main types: matrix and reservoir systems. In matrix-type formulations, the active agent is embedded and distributed homogeneously within the matrix material. Possible modes of release from matrix formulations are diffusion through, or dissolution, surface erosion, or bulk erosion of the formulation material (Esser-Kahn et al., 2011). Reservoir systems are commonly microcapsules, enclosing a reservoir of the active agent in the core surrounded by a shell material, often composed of a polymer (Nussinovitch, 2010).

The key characteristic of a formulation for it to be referred to as controlled release is that the release must be triggered or initiated by an external or internal stimulus (Esser-Kahn et al., 2011; Indermun et al., 2018). This can be achieved when the active agent is either dissolved, dispersed, or encapsulated within a matrix or entrapped by a coating composed of a stimuli-responsive material (Nussinovitch, 2010). A stimulus or trigger can be turned on by environmental conditions such as humidity, temperature, pH, or UV radiation, which cause a change in the material's physical or chemical state, or the breakdown of a formulation's shell or matrix material by UV light, redox reactions, or enzymatic activity, enabling the increased diffusion or the release of the active agent from the formulation (Camara et al., 2019).

The matrix material or the shell is prepared from a material or composition that can be modified by a stimulus, followed by the release of the core containing the active agent. Common formulation materials used as carriers or additives in controlled release systems are not only natural or synthetic polymers (Nussinovitch, 2010) but also clay materials, supplemented with disintegrating additives. Regulating the degradation rate ensures a controlled liberation of cells from the formulation microenvironment. The development of formulations that release the active ingredient under specific physico-chemical conditions is a common practice in pharmaceutical formulation technology, where certain active ingredients are intended to be released at a specific location in the body. Research on true 'controlled' release formulations of living BCAs is rare.

Recently, Hermann et al. (2021) developed a novel controlled release system for BCAs, with an accelerated mode of infection of click beetle larvae (*Agriotes* spp.) and a fast-swelling coating composed of disintegrants that enable the immediate disintegration of the coating layer to release infective blastospores of *M. brunneum*. This approach overcomes the *lag* time until the fungus encapsulated within the dried Ca-alginate capsule core begins to grow and produce aeroconidia on the bead surface (Hermann et al., 2023).

Another promising controlled release formulation developed by Machado et al. (2020) for the control of wood-decaying fungi of the Esca complex causing grapevine trunk disease is a lignin-based nanocapsule. Fungi of the disease consortium are able to degrade lignin. As a proof of concept, classical fungicides were encapsulated as active ingredients, which become released only in the presence of the pathogenic fungi and their enzymes within the vascular system of the plant.

13 Conclusion and future trends

Biopesticides represent promising substitutes for environmentally friendly plant protection. Hence, upcoming trends in biopesticide formulation should encompass the improvement of bioformulation methods, refinement of commercialization approaches, and the development of biopesticides suitable for crop systems and application techniques, in order to enable seamless integration into existing pest management programs. The future of biopesticide formulation is promising, with a focus on enhancing product performance and extending activity (Gašić, 2013). In this context, the development of a BCA formulation needs consideration of all developmental steps, beginning from strain selection, the production of competent cells or organisms, and the best performance under the environmental conditions at the place of application. The summary of the key challenges and solutions that can be addressed during the developmental process of BCAs are given in Table 1.

Despite limitations, such as slow killing and production difficulties, biopesticides are expected to play a crucial role in future pest management (Glare, 2016). Different strategies should be considered to overcome the inconsistencies in biocontrol effectiveness of BCAs when applied alone. Therefore, one current trend is the integration of living BCAs into biocontrol strategies, e.g. improving the efficacy of BCAs by implementing them in strategies in co-application or combination with other active agents. Attract-and-kill or push-pull-kill strategies or the application via entomovectoring are the key words that lead to a series of current publications dealing with the integration of BCAs into innovative pest control strategies (Nunes, 2010). These are interesting investigations when an already registered BCA could be further improved without the high registration costs for a new species or strain.

The use of natural chemical products, encouraged by new pesticide regulations, is also a significant trend (Villaverde, 2016). Also imaginable is the accelerated and improved development of biopesticides on artificial intelligence and deep-learning systems, as already under development and testing for next-generation fungicides (Bayer Crop Science). Nanotechnology is a current trend in various areas. In many cases, the terms 'nano,' 'nanoencapsulation,' 'nanomaterial,' and 'nanoformulation' are misused and do not mean the development of novel formulation types. The acceptable definition of 'nanomaterial,' according to European Union law, is 'a natural, incidental, or manufactured material containing particles, in an unbound state or as an aggregate or as an agglomerate, and where, for 50% or more of the particles in the number size distribution, one or more external dimensions is in the size range 1–100 nm' (European Commission, 2011). Converting a living cell of micrometer scale, as would be the case for most BCAs, into a nanoscale formulation is unlikely to be achievable, but it is a significant topic for formulation technology for non-living active agents, including biopesticides such as essential oils or single chemical compounds for improving bioactivities and bioavailability of active agents and, thus, their efficacy (Hernandez-Tenorio, 2022). Nevertheless, the small size of the so-called 'nanoformulations' can have advantages for the formulation of viruses or another current trend for dsRNA or RNAi as active agents (Baum et al., 2007) that needs to be ingested by the target insect or as a formulation that should diffuse through a plant cuticle.

Table 1 Challenges and solutions for the key issues of biocontrol agent development for plant protection

Challenges	Solutions	Materials/Additives	References
		Production factors	
High production costs	A) Use of cheap waste materials or by-products as nutrient sources; B) Support of cell multiplication after application by the formulation	A) Molasses, malt extract, beer extract, extracts from slaughter waste materials; B) Nutrient sources for the BCA as formulation additives	A) Mussatto et al., 2012; Mishra et al., 2016; Muskat et al., 2022a, 2022b; B) Przyklenk et al., 2017
Low drying survival and shelf life during storage	A) Protection of the cells from disruption due to compression during drying; B) Maintenance of cell membrane fluidity; membrane stabilization; C) Optimization of the drying process; D) Improved drying survival and storability by pre-conditioning of the cells	A) Filling agents, e.g. Starch, clay, silicate; B) Humectants/wetteners, e.g. sugar alcohols; other membrane-stabilizing agents, e.g. skim-milk; C) Modification of drying parameters (temperature, humidity, duration, pre-treatment); D) Conditioning of cells during the fermentation process, e.g., by osmotic stress	A) Soper and Ward, 1981; Przyklenk et al., 2017; Berninger et al. 2018; B) McCabe and Soper, 1985; Crowe et al., 1987; Connick Jr et al., 1996; Jackson et al., 2006; Fu and Chen, 2011; C) Stephan and Zimmermann, 1998; Jackson and Payne, 2007; Mascarin et al., 2016; Przyklenk et al., 2017; D) Hallsworth and Magan, 1994a, 1994b; Iwanicki et al., 2018; Krell et al., 2018b; Cruz Barrera et al., 2020
		Environmental factors and performance in the field	
Damage by UV radiation	A) Direct protection of the formulated agent from solar radiation by sunscreens; B) Protection from chemical degradation due to redoxreactions; indirect protection against solar radiation; C) Strain selection	A) TiO_2, ZnO, talc, synthetic sunscreens, optical brighteners, anionic dyes; B) Antioxidants, pigments, phenolic compounds, enzymes, oils (in particular, natural oils contain antioxidants, e.g. tocopherols); C) Selection of naturally phototolerant strains or cells	A) Shapiro, 1992; Webb et al., 1994; Cohen et al., 2003; Zohar-Perez et al., 2003; B) Inglis et al., 1995; Alves et al., 1998; Behle et al., 2009; El Salamouny et al., 2009; El-Helaly et al., 2013; Fernades et al., 2015; Zhang and Feng, 2018; Kaiser et al., 2019; A & B) Moore et al., 1993; Edgington et al., 2000; Çakmak et al., 2021; C) Ignoffo and Garcia, 1992; Brags et al., 2006; Fernandes et al., 2015

(Continued)

Table 1 (*Continued*)

Challenges	Solutions	Materials/Additives	References
Low humidity conditions in the field	Rehydration support and maintenance of a high water activity in order to provide water for a prolonged period to enable the growth of the BCA	Humectant/water absorbents/water-retaining substances; Acrylic acid-based superabsorbents; Natural-based materials or polymers with high water absorption capacities, e.g. xanthan, cellulose, starch; Copolymers	Zhou and Feng, 2009; Shapiro-Ilan and Goolsby, 2021; Muskat et al., 2023; Wu et al., 2023
No optimal or varying temperature conditions	A) Strain selection B) Conditioning of cells for a special temperature range and conditions	A) Selection of strains from an area with a similar/same temperature regime as at the place of application B) Conditioning by temperature treatment of the strain or cells	A) Köhl et al., 2011; Borisade and Magan, 2014; Sangorrín et al., 2014; Alali et al., 2019; Gava et al., 2022; B) Han et al., 2021
Lack of contact between the active BCA and the target insect	Increasing the attractiveness of the BCA formulation for the target insect; Implementation into 'attract/lure-and-arrest,' 'attract/lure-and-arrest,' or 'attract/lure-and-kill' strategies	Semiochemicals ((volatile) pheromones or allomones), Colors, Shape; Phagostimulants (sugars, amino acids, oils)	Vemmer et al., 2016; Gregg et al., 2018; Chow et al., 2018
A) Low persistence of the BCA formulation after leaf application; low rainfastness; B) Low or too short adhesion on the insect's cuticle	A) Improved and prolonged adhesion and coverage to and of the plant's surface; B) Enabled adhesion of infection units to the target insect's cuticle; creation of a suitable environment for infection	A and B) Sticker/spreader/adhesives: Oil (vegetable or mineral oil), xanthan, cellulose derivatives (methylcellulose, carboxymethylcellulose), gelatine, gums (latex), molasses, waxes, silica, synthetic adhesives	A) Angus, 1959; Malsam et al., 2002; Santos et al., 2019; B) Prior et al., 1988; Bateman et al., 1993; Jaronski, 1997
Low virulence and insufficient killing efficacy	A) Support of the virulence of the BCA against the target, e.g., by improved or enfastened penetration of the target organism; B) Selection of a virulent strain; C) Fast or triggered release of the infection units	A) Virulence-enhancing additives; enzyme-inducing agents (e.g. the enzyme's substrate); B) Strain selection: screening for the most competent strain with the highest virulence in regard to the target insect; C) Stimuli-responsive additives, e.g. desintegrants	A) Gerding-Gonzalez et al., 2007; Krell et al., 2018a; B) Meyling and Eilenberg, 2007; Meyling, 2007; Castagnola and Stock, 2014; Shuba et al., 2015; Abd-Elgawad, 2023; C) Hermann et al., 2023

Another requirement for the successful application and further development in the field of BCAs is the acceptance and successful application by the growers. In order to increase the implementation of BCAs in plant protection strategies, it needs (1) improved efficacy of BCA products, which is the responsibility of research and development to provide effective BCA products, and (2) increased acceptance by the growers through more education on the proper use of such living preparations, as well as knowledge transfer and promotion of these environmentally friendly alternatives.

14 Where to look for further information

- Formulation of Microbial Biopesticides: Beneficial microorganisms, nematodes, and seed treatments. Edited by H. D. Burges. Published in 1998 by Kluwer Academic Publishers, Dordrecht. ISBN 0 412 62520 2. → A standard book about the formulation of microbial BCAs.
- Rhodes, D. J. (1993). Formulation of biological control agents. In *Exploitation of microorganisms* (pp. 411–439). Dordrecht: Springer Netherlands.
- Glare, T. R., Moran-Diez, M. E. and Glare. (2016). *Microbial-based biopesticides*. New York: Springer Science+Business Media.
- Hajek, A. E. and Eilenberg, J. (2018). *Natural enemies: an introduction to biological control*. Cambridge: Cambridge University Press. → For more background information on the biology and interaction of naturally occurring entomopathogens and strategies on how to implement BCAs in insect pest control.
- Butt, T.M., et al. (1999). Microbial biopesticides. In: Hall, F.R. and Menn, J.J. (eds) *Biopesticides: use and delivery. methods in biotechnology*, vol 5. Totowa, NJ: Humana Press.
- Lacey, L. (2017). *Microbial control of insect and mite pests*. Boston, MA: FNP: Academic Press.
- Ravensberg, W.J. (2011). *A roadmap to the successful development and commercialization of microbial pest control products for control of arthropods* (Vol. 10). New York: Springer Science and Business Media.
- Mollet, H. and Grubenmann, A. (2008). *Formulation technology: emulsions, suspensions, solid forms*. Hoboken, NJ: John Wiley and Sons.
- Wijffels, R. H. (Ed.). (2001). *Immobilized cells*. New York: Springer Science and Business Media.
- Nussinovitch, A. (2010). Beads and special applications of polymers for agricultural uses. In: Nussinovitch, A. (ed.) *Polymer macro-and micro-gel beads: fundamentals and applications*, pp. 231–253. New York: Springer.

Journals/Where to find the publications:

- Publications on all aspects of biological control of insect pests:
- *BioControl, Biological Control, Biocontrol Science and Technology, Pest Management Science, Journal of Invertebrate Pathology, Crop Protection*; with the keywords 'fermentation', 'mass production,' 'formulation', 'encapsulation', 'delivery systems' etc. for 'biocontrol agents' in general or individual species or classes.
- Publications on formulation technology aspects:
- ACS journals on material science, physics, and chemistry, *Journal of Controlled Release*.

15 References

Abd-Elgawad, M.M. (2023), Optimizing entomopathogenic nematode genetics and applications for the integrated management of horticultural pests. *Horticulturae*, 9(8), 865.

Akelah A (2013) Polymers in the controlled release of agrochemicals: Functionalized Polymeric Materials in Agriculture and the Food Industry (ed. Springer, pp. 133–192.

Alali, S., et al. (2019), Thermotolerant isolates of *Beauveria bassiana* as potential control agent of insect pest in subtropical climates. *PLoS One*, 14(2), e0211457.

Alves, R.T., et al. (1998), Effects of simulated solar radiation on conidial germination of *Metarhizium anisopliae* in different formulations. *Crop Prot*, 17(8), 675–679.

Angelova, N. and Hunkeler, D. (1999), Rationalizing the design of polymeric biomaterials. *Trends Biotechnol*, 17(10), 409–421.

Angus, T.A. (1959), Potential usefulness of vinyl latices as stickers. *Can Entomol*, 91(4), 254–255.

Bashan, Y. (1986) Alginate beads as synthetic inoculant carriers for the slow release of bacteria that affect plant growth. Appl Environ Microbiol 51:1089–1098

Bateman, R.P., et al. (1993), The enhanced infectivity of *Metarhizium flavoviride* in oil formulations to desert locusts at low humidities. *Ann Appl Biol*, 122(1), 145–152.

Baum, J.A., et al. (2007), Control of coleopteran insect pests through RNA interference. *Nat Biotechnol*, 25(11), 1322–1326.

Behle, R.W., et al. (2009), Evaluation of soyscreen in an oil-based formulation for UV protection of *Beauveria bassiana* conidia. *J Econ Entomol*, 102(5), 1759–1766.

Bejarano, A. and Puopolo, G. (2020), Bioformulation of microbial biocontrol agents for a sustainable agriculture. In: De Cal, A., Melgarejo, P., Magan, N. (eds) *How research can stimulate the development of commercial biological control against plant diseases*, pp. 275–293. Cham: Springer. https://doi.org/10.1007/978-3-030-53238-3_16

Berninger, T., et al. (2018), Maintenance and assessment of cell viability in formulation of non-sporulating bacterial inoculants. *Microb Biotechnol*, 11(2), 277–301.

Blok, M. C., et al. (1975). The effect of chain length and lipid phase transitions on the selective permeability properties of liposomes. Biochimica et Biophysica Acta (BBA)-Biomembranes, 406(2), 187–196.

Borisade, O.A. and Magan, N. (2014), Growth and sporulation of entomopathogenic *Beauveria bassiana*, *Metarhizium anisopliae*, *Isaria farinosa* and *Isaria fumosorosea* strains in relation to water activity and temperature interactions. *Biocontrol Sci Techn*, 24(9), 999–1011.

Boucias, D.G., et al. (1988), Nonspecific factors involved in attachment of entomopathogenic deuteromycetes to host insect cuticle. *Appl Environ Microbiol*, 54(7), 1795–1805.

Boyetchko, S.M. (2020), Improving methods for developing new microbial biopesticides. In: Birch, N. and Glare, T. (eds) *Biopesticides for sustainable agriculture*, pp. 3–20. Cambridge: Burleigh Dodds Science Publishing.

Boyetchko, S.M., et al. (1999), Formulations of biopesticides. In: Hall, F.R. and Menn, J.J. (eds) *Biopesticides: use and delivery. methods in biotechnology*, Vol 5. Totowa, NJ: Humana Press.

Brags, G.U., et al. (2006), Conidial pigmentation is important to tolerance against solar-simulated radiation in the entomopathogenic fungus *Metarhizium anisopliae*. *J Photochem Photobiol*, 82(2), 418–422.

Brar, S.K., et al. (2006), Recent advances in downstream processing and formulations of *Bacillus thuringiensis* based biopesticides. *Process Biochem*, 41(2), 323–342.

Burges, H.D. (1998a), Appendix I - List of additives tested until 1998 with microorganisms. In: Burges, H.D. (ed.) *Formulation of microbial biopesticides*, pp. 33–127. Dordrecht: Kluwer Academic Publishers.

Burges, H.D. (1998b), Formulation of mycoinsecticides. In Burges, H.D. (ed.) *Formulation of microbial biopesticides: beneficial microorganisms, nematodes and seed treatments*, pp. 131–185. Dordrecht: Springer Netherlands.

Burges, H.D. and Jones, K.A. (1998), Formulation of bacteria, viruses and protozoa to control insects. In Burges, H.D. (ed.) *Formulation of microbial biopesticides*, pp. 33–127. Dordrecht: Kluwer Academic Publishers.

Busolo, M.A. and Lagaron, J.M. (2012), Oxygen scavenging polyolefin nanocomposite films containing an iron modified kaolinite of interest in active food packaging applications. *Innovative Food Sci Emerg Technol*, 16, 211–217.

Butt, T.M. and Goettel, M.S. (2000), Bioassays of entomogenous fungi. In: Navon, A. and Ascher, K.R.S. (eds) *Bioassays of entomopathogenic microbes and nematodes*, pp. 141–195. Wallingford, UK: CABI International Press.

Çakmak, T., et al. (2021), Effects of several UV-protective substances on the persistence of the insecticidal activity of the Alphabaculovirus of Chrysodeixis chalcites (ChchNPV-TF1) on banana (*Musa acuminata*, Musaceae, Colla) under laboratory and open-field conditions. *PLoS One*, 16(5): e0250217.

Camara, M.C., et al. (2019), Development of stimuli-responsive nano-based pesticides: emerging opportunities for agriculture. *J Nanobiotechnol*, 17, 100.

Castagnola, A. and Stock, S.P. (2014), Common virulence factors and tissue targets of entomopathogenic bacteria for biological control of lepidopteran pests. *Insects*, 5(1), 139–166.

Chandler, D., et al. (2011). The development, regulation and use of biopesticides for integrated pest management. Philosophical Transactions of the Royal Society B: Biological Sciences, 366(1573), 1987–1998.

Charnley, A.K. and St. Leger, R.J. (1991), *The role of cuticle-degrading enzymes in fungal pathogenesis in insects*. In: Cole, G.T. and Hoch, H.C. (eds) *The fungal spore and disease initiation in plants and animals*, pp. 267–286. Boston, MA: Springer.

Chejanovsky, N. and Tal, J. (1992), Genetic engineering of insect viruses for insect biological control. *Phytoparasitica*, 20, 25–31.

Chow, A., et al. (2018), Field efficacy of autodissemination and foliar sprays of an entomopathogenic fungus, *Isaria fumosorosea* (Hypocreales: Cordycipitaceae), for control of Asian citrus psyllid, *Diaphorina citri* (Hemiptera: Liviidae), on residential citrus. *J Econ Entomol*, 111(5), 2089–2100.

Clerk, G. C., & Madelin, M. F. (1965). The longevity of conidia of three insect-parasitizing hyphomycetes. Transactions of the British Mycological Society, 48(2), 193–209.

Cohen, E., et al. (2003), Photostabilization of an entomopathogenic fungus using composite clay matrices¶. *J Photochem Photobiol*, 77(2), 180–185.

Connick Jr, W.J. (1988), Formulation of living biological control agents with alginate. In pesticide formulations. *ACS Symposium Series*, 371, 241–250.

Connick Jr, W.J., et al. (1996), Water activity and other factors that affect the viability of *Colletotrichum truncatum* conidia in wheat flour-kaolin granules ('Pesta'). *Biocontrol Sci Techn*, 6(2), 277–284.

Couch, T. L., et al. (1981). Microbial control of pests and plant diseases 1970–1980. HD Burges, Ed, 621–634.

CropLife International; GIFAP (2017), *Catologue of pesticide formulation types and international coding system*. Brussels: GIFAP.

Crowe, J.H., et al. (1984), Preservation of membranes in anhydrobiotic organisms: the role of trehalose. *Science*, 223, 701–703.

Crowe, J.H., et al. (1987), Stabilization of dry phospholipid bilayers and proteins by sugars. *Biochem J*, 242(1), 1.

Crowe, J.H., et al. (1988), Interactions of sugars with membranes. *Biochim Biophys Acta*, 947, 367–384.

Cruz Barrera, M., et al. (2020), Anhydrobiotic engineering for the endophyte bacterium *Kosakonia radicincitans* by osmoadaptation and providing exogenously hydroxyectoine. *World J Microbiol Biotechnol*, 36, 1–16.

Daoust, R.A., et al. (1983), Effect of formulation on the viability of *Metarhizium anisopliae* conidia. *J Invertebr Pathol*, 41(2), 151–160.

de Faria, M. R., & Wraight, S. P. (2007). Mycoinsecticides and mycoacaricides: a comprehensive list with worldwide coverage and international classification of formulation types. Biological control, 43(3), 237–256.

DeBach, P. (1964), *The scope of biological control*. In: P. DeBach (ed.) *Biological control of insect pests and weeds*. London: Chapman and Hall Ltd.

Dietsch, R., et al. (2021), Desiccation-tolerant fungal blastospores: from production to application. *Curr Res Biotechnol*, 3, 323–339.

Dietsch, R., et al. (2024). Single-cell polymer coating improves the desiccation tolerance of *Metarhizium brunneum* blastospores. *Biotechnology for the Environment*, 1(1), 13.

Dommergues, Y. R., Diem, H. G., & Divies, C. (1979). Polyacrylamide-entrapped Rhizobium as an inoculant for legumes. Applied and Environmental Microbiology, 37(4), 779–781.

DunhamTrimmer® Global Biocontrol Market Overview Update 2022 (2022). https://dunhamtrimmer.com/products/biocontrol-global-market-report/. Accessed: 06/24/2022

EC-European Commission (2011), Commission Recommendation of 18 October 2011 on the definition of nanomaterial. *Off J Eur Union* 275, 38.

Edgington, S., et al. (2000), Photoprotection of *Beauveria bassiana*: testing simple formulations for control of the coffee berry borer. *Int J Pest Manag*, 46, 169–176.

Ehlers, R.U. (2001), Mass production of entomopathogenic nematodes for plant protection. *Appl Microbiol Biotechnol*, 56, 623–633.

Eilenberg, J., et al. (2001), Suggestions for unifying the terminology in biological control. *BioControl*, 46, 387–400.

El-Helaly, A., et al. (2013), Promising additives to protect the activity of baculovirus biocontrol agent under field-sunlight conditions in Egypt. *J Life Sci*, 7(5), 495.

El Salamouny, S., et al. (2009), Black tea and lignin as ultraviolet protectants for the beet armyworm nucleopolyhedrovirus. *J Entomol Sci*, 44(1), 50–58.

Emmert, E. and Handelsman, J. (1999), Biocontrol of plant disease: a (Gram-) positive perspective. *FEMS Microbiol Lett*, 171, 1–9.

Erlandson, M. (2020), Advances in the use of entomopathogenic viruses as biopesticides in suppressing crop insect pests. In: Birch, N. and Glare, T. (eds) *Biopesticides for sustainable agriculture*, pp. 167–194. Cambridge: Burleigh Dodds Science Publishing.

Esser-Kahn, A.P., et al. (2011), Triggered release from polymer capsules. *Macromolecules*, 44(14), 5539–5553.

European Parliament (2009), Directive 2009/128/EC of the European Parliament and of the Council of 21 October 2009 establishing a framework for Community action to achieve the sustainable use of pesticides. *Off J Eur Union*, 309: 71–86.

Feng, J., et al. (2019), Controlled release of biological control agents for preventing aflatoxin contamination from starch–alginate beads. *Molecules*, 24(10), 1858.

Fernandes, É.K., et al. (2015), Tolerance of entomopathogenic fungi to ultraviolet radiation: a review on screening of strains and their formulation. *Curr Genet*, 61, 427–440.

Folman, L.B., et al. (2003), Inability to find consistent bacterial biocontrol agents of *Pythium aphanidermatum* in cucumber using screens based on ecophysiological traits. *Microb Ecol*, 45, 72–87.

Foy, C. L. (1989). Adjuvants: terminology, classification, and mode of action. In Adjuvants and agrochemicals (pp. 1–15). CRC Press.

Fravel, D. R., et al. (1985). Encapsulation of potential biocontrol agents in an alginate-clay matrix. Phytopathology, 75(7), 774–777.

Fu, N. and Chen, X.D. (2011), Towards a maximal cell survival in convective thermal drying processes. *Food Res Int*, 44(5), 1127–1149.

Gadoury, D.M., et al. (2023), Effects of nighttime applications of germicidal ultraviolet light upon powdery mildew (*Erysiphe necator*), downy mildew (*Plasmopara viticola*), and sour rot of grapevine. *Plant Dis*, 107(5), 1452–1462.

Gao, Q. and Garcia-Pichel, F. (2011), Microbial ultraviolet sunscreens. *Nat Rev Microbiol*, 9(11), 791–802.

Gašić, S. M., & Tanović, B. (2013). Biopesticide formulations, possibility of application and future trends. Pesticides and Phytomedicine/Pesticidi i fitomedicina, 28(2).

Gaugler, R., et al. (1992), Ultraviolet inactivation of heterorhabditid and steinernematid nematodes. *J Invertebr Pathol*, 59(2), 155–160.

Gava, C.A.T., et al. (2022), Selecting thermal tolerant strains of entomopathogenic fungi to control *Ceratitis capitata* (Wiedeman) in tropical semi-arid conditions. *Biol Control*, 176, 105062.

Gerding-Gonzalez, M., et al. (2007), Use of chitin to improve a *Beauveria bassiana* alginate-pellet formulation. *Biocontrol Sci Techn*, 17(1), 105– 110.

Glare, T.R., et al. (2012), Have biopesticides come of age? *Trends Biotechnol*, 30, 250–258.

Glare, T.R., et al. (2016). *Microbial-based biopesticides*. New York: Springer Science+ Business Media.

Glare, T.R., et al. (2017), *Basic and applied research: entomopathogenic bacteria*. In: Lacey, L.R. (ed.) *Microbial control of insect and mite pests*, pp. 47–67. London: Academic Press.

Gregg, P.C., et al. (2018), Advances in attract-and-kill for agricultural pests: beyond pheromones. *Annu Rev Entomol*, 63, 453–470.

Griffiths, H. R., et al. (1998). Molecular and cellular effects of ultraviolet light-induced genotoxicity. Critical reviews in clinical laboratory sciences, 35(3), 189–237.

Grzywacz, D. (2017), Basic and applied research: baculovirus. In: Lacey, L.R. (ed.) *Microbial control of insect and mite pests: from theory to practice*, pp. 27–46. London: Academic Press.

Guerrero, M.S. (2023), Utilizing metagenomics and metabarcoding for assessing the diversity and abundance of entomopathogenic fungi in soil ecosystems. *J Appl Entomol*, 3(3), 1–7.

Hallsworth, J.E. and Magan, N. (1994a), Effect of carbohydrate type and concentration on polyhydroxy alcohol and trehalose content of conidia of three entomopathogenic fungi. *Microbiology*, 140(10), 2705–2713.

Hallsworth, J.E. and Magan, N. (1994b), Effects of KCl concentration on accumulation of acyclic sugar alcohols and trehalose in conidia of three entomopathogenic fungi. *Lett Appl Microbiol*, 18(1), 8–11.

Hallsworth, J.E. and Magan, N. (1999), Water and temperature relations of growth of the entomogenous fungi *Beauveria bassiana*, *Metarhizium anisopliae*, and *Paecilomyces farinosus*. *J Invertebr Pathol*, 74(3), 261–266.

Han, J.O., et al. (2021), Directed evolution of *Metarhizium* fungus improves its biocontrol efficacy against *Varroa* mites in honey bee colonies. *Sci Rep*, 11(1), 10582.

Hermann, K.M., et al. (2021), Formalin-casein enhances water absorbency of calcium alginate beads and activity of encapsulated *Metarhizium brunneum* and *Saccharomyces cerevisiae*. *World J Microbiol Biotechnol*, 37(9), 156.

Hermann, K.M., et al. (2023), Polyvinyl alcohol coating releasing fungal blastospores improves kill effect of attract-and-kill beads. *AMB Express*, 13(1), 72.

Hernandez-Tenorio, F., Miranda, A. M., Rodríguez, C. A., Giraldo-Estrada, C., & Sáez, A. A. (2022). Potential strategies in the biopesticide formulations: a bibliometric analysis. Agronomy, 12(11), 2665.

Hofstein, R. and Chapple, A. (1998), Commercial development of biofungicides. In: Frhajj, M. (ed.) *Biopesticides: use and delivery*, pp. 77–102. Totowa, NJ: Humana Press.

Holder, D.J., et al. (2007), Surface characteristics of the entomopathogenic fungus *Beauveria* (*Cordyceps*) *bassiana*. *Microbiology*, 153(10), 3448–3457.

Ignoffo, C.M. and Garcia, C. (1992), Influence of conidial color on inactivation of several entomogenous fungi (hyphomycetes) by simulated sunlight. *Environ Entomol*, 21, 913–917.

Indermun, S., et al. (2018), Stimuli-responsive polymers as smart drug delivery systems: Classifications based on carrier type and triggered-release mechanism. In: Makhlouf A.S.H. and Abu-Thabit N.Y. (eds) *Stimuli responsive polymeric nanocarriers for drug delivery applications*, Volume 1, pp. 43–58. Duxford, UK: Woodhead Publishing.

Inglis, G.D., et al. (1995), Influence of ultraviolet light protectants on persistence of the entomopathogenic fungus, *Beauveria bassiana. Biol Control*, 5, 581–590.

Iwanicki, N.S.A., et al. (2018), Modified Adamek's medium renders high yields of *Metarhizium robertsii* blastospores that are desiccation tolerant and infective to cattle-tick larvae. *Fungal Biol*, 122(9), 883–890.

Jackson, M. (1997), Optimizing nutritional conditions for the liquid culture production of effective fungal biological control agents. *J Ind Microbiol Biotechnol*, 19, 180–187.

Jackson, M.A., et al. (2003), Media and fermentation processes for the rapid production of high concentrations of stable blastospores of the bioinsecticidal fungus *Paecilomyces fumosoroseus*. *Biocontrol Sci Techn*, 13(1), 23–33.

Jackson, M.A., et al. (2006), Influence of formulation additives on the desiccation tolerance and storage stability of blastospores of the entomopathogenic fungus *Paecilomyces fumosoroseus* (Deuteromycotina: Hyphomycetes). *Biocontrol Sci Techn*, 16(1), 61–75.

Jackson, M.A., et al. (2010), Ecological considerations in producing and formulating fungal entomopathogens for use in insect biocontrol. *BioControl*, 55, 129–145.

Jackson, T.A. (2017), Entomopathogenic bacteria: mass production, formulation, and quality control. In: Lacey, L.R. (ed.) *Microbial control of insect and mite pests*, pp. 125–139. London: Academic Press.

Jackson, M.A. and Payne, A.R. (2007), Evaluation of the desiccation tolerance of blastospores of *Paecilomyces fumosoroseus* (Deuteromycotina: Hyphomycetes) using a lab-scale, air-drying chamber with controlled relative humidity. *Biocontrol Sci Techn*, 17(7), 709–719.

Jaronski, S.T. (1997), New paradigms in formulating mycoinsecticides. *ASTM Spec Tech*, 1328, 99–114.

Jaronski, S.T. and Mascarin, G.M. (2017), *Mass production of fungal entomopathogens*. In: Lacey, L.R. (ed.) *Microbial control of insect and mite pests*, pp. 141–155. London: Academic Press.

Jaronski, S.T. (2023), Mass production of entomopathogenic fungi–state of the art. In: McLaughlin, R.E. and Bell, M.R. (eds) *Mass production of beneficial organisms*, pp. 317–357. London: Academic Press.

Jenkins, N. E., Heviefo, G., Langewald, J., Cherry, A. J., & Lomer, C. J. (1998). Development of mass production technology for aerial conidia for use as mycopesticides. Biocontrol News and information, 19, 21N–32N.

Jones, K.A. and Burges, H.D. (1998), Technology of formulation and application. In Burges, H.D. (ed.) *Formulation of microbial biopesticides: beneficial microorganisms, nematodes and seed treatments*, pp. 7–30. Dordrecht: Springer Netherlands.

Jung, G., et al. (1982). Polymer-entrapped Rhizobium as an inoculant for legumes. Plant and Soil, 65, 219–231.

Kaiser, D., et al. (2019), Efficiency of natural substances to protect *Beauveria bassiana* conidia from UV radiation. *Pest Manag Sci*, 75(2), 556–563.

Kim, I.Y., et al. (2012), Controlled release of *Pantoea agglomerans* E325 for biocontrol of fire blight disease of apple. *J Control Release*, 161(1), 109–115.

Klärner, S., et al. (2015), Studies on mould prevention in viticulture by means of UV-C application of vines (Vitis vinifera L.). *Landtechnik–Agric Engin*, 70, 139–148.

Köhl, J., et al. (2011), Stepwise screening of microorganisms for commercial use in biological control of plant-pathogenic fungi and bacteria. *Biol Control*, 57(1), 1–12.

Koppenhöfer, A.M., et al. (2020), Entomopathogenic nematodes in sustainable food production. *Front Sustain Food S*, 4, 125.

Krell, V., et al. (2018a), Cellulase enhances endophytism of encapsulated *Metarhizium brunneum* in potato plants. *Fungal Biol*, 122(5), 373–378.

Krell, V., et al. (2018b), Endogenous arabitol and mannitol improve shelf life of encapsulated *Metarhizium brunneum*. *World J Microbiol Biotechnol*, 34, 1–9.

Kronberg, B., et al. (2014), Types of surfactants, their synthesis, and applications. In: Kronberg, B., Holmberg, K. and Lindman, B. (eds) *Surface chemistry of surfactants and polymers*, 1–47. Hoboken, NJ: John Wiley & Sons, Ltd.

Lacey, L.A., et al. (2015), Insect pathogens as biological control agents: back to the future. *J Invertebr Pathol*, 132, 1–41.

Leggett, M., et al. (2011). Formulation of microbial biocontrol agents–an industrial perspective. Canadian Journal of Plant Pathology, 33(2), 101–107.

Leuenberger, H. (2001), New trends in the production of pharmaceutical granules: the classical batch concept and the problem of scale-up. *Eur J Pharm Biopharm*, 52(3), 279–288.

Machado, T.O., et al. (2020), Bio-based lignin nanocarriers loaded with fungicides as a versatile platform for drug delivery in plants. *Biomacromolecules*, 21(7), 2755–2763.

Mäder, R., et al. (2019), *Betriebsmittelliste 2019 für den ökologischen Landbau in Deutschland*. Frick, Switzerland: FiBL.

Madronich, S. (1993), *UV radiation in the natural and perturbed atmosphere*. In: Tevini, M. (ed.) *UV-B radiation and ozone depletion*, pp. 17–69. Boca Raton: Lewis.

Madronich, S., et al. (1998), Changes in biologically active ultraviolet radiation reaching the Earth's surface. *J Photochem Photobiol B Biology*, 46(1–3), 5–19.

Małagocka, J., et al. (2015), Transcriptome of an entomophthoralean fungus (*Pandora formicae*) shows molecular machinery adjusted for successful host exploitation and transmission. *J Invertebr Pathol*, 128: 47–56.

Malsam, O., et al. (2002), Oils for increased efficacy of *Metarhizium anisopliae* to control whiteflies. *Biocontrol Sci Techn*, 12(3), 337–348.

Mascarin, G.M., et al. (2016), Improved shelf life of dried *Beauveria bassiana* blastospores using convective drying and active packaging processes. *Appl Microbiol Biotechnol*, 100, 8359–8370.

Mascarin, G.M., et al. (2023), Impact of osmotic stress on production, morphology, and fitness of *Beauveria bassiana* blastospores. *Appl Microbiol Biotechnol*, 107, 1–17.

Mattedi, A., et al. (2023), Solid-state fermentation: Applications and future perspectives for biostimulant and biopesticides production. *Microorganisms*, 11(6), 1408.

McCabe, D. and Soper, R.S. (1985), *U.S. Patent No. 4,530,834*. Washington, DC: U.S. Patent and Trademark Office.

Merriman, P. and Russell, K. (1990), Screening strategies for biological control. In: Campbell, R. (ed.) *Biological control of soil-borne plant pathogens*, 427–435. Oxford: CAB International.

Meyling, N.V. (2007), Methods for isolation of entomopathogenic fungi from the soil environment-laboratory manual. *Lab Manual*, 32, 818–890.

Meyling, N.V. and Eilenberg, J. (2007), Ecology of the entomopathogenic fungi *Beauveria bassiana* and *Metarhizium anisopliae* in temperate agroecosystems: Potential for conservation biological control. *Biol Control*, 43(2), 145–155.

Miller, D.W. (1995), *Commercial development of entomopathogenic fungi: formulation and delivery*. ACS Symposium Series, Vol. 595. Biorational Pest Control Agents. Chapter 15, 213–220.

Mishra, S., et al. (2016), Suitability of agricultural by-products as production medium for spore production by *Beauveria bassiana* HQ917687. *Int J Recycling Org*, 5, 179–184.

Moore, D., et al. (1993), Ultra-violet radiation damage to *Metarhizium flavoviride* conidia and the protection given by vegetable and mineral oils and chemical sunscreens. *Ann Appl Biol*, 122(3), 605–616.

Moore, D., et al. (1996), Effects of moisture content and temperature on storage of *Metarhizium flavoviride* conidia. *Biocontrol Sci Techn*, 6(1), 51–62.

Morgan, C.A., et al. (2006), Preservation of micro-organisms by drying; a review. *J Microbiol Methods*, 66(2), 183–193.

Mugnier, J. and Jung, G. (1985), Survival of bacteria and fungi in relation to water activity and the solvent properties of water in biopolymer gels. *Appl Environ Microbiol*, 50(1), 108–114.

Muskat, L.C. and Patel, A.V. (2022), Innovations in semiochemical formulations. *Entomol Gen*, 42(2): 231–249.

Muskat, L.C., et al. (2022a), Encapsulation of the psyllid-pathogenic fungus *Pandora* sp. nov. inedit. and experimental infection of target insects. *Pest Manag Sci*, 78(3), 991–999.

Muskat, L.C., et al. (2022b), Fermentation of the psyllid-pathogenic fungus *Pandora* sp. nov. inedit. (Entomophthorales: Entomophthoraceae). *Biocontrol Sci Techn*, 32(5), 564–585.

Muskat, L.C., et al. (2022c), Development of a self-adhesive oleogel formulation designed for the slow release of semiochemicals. *Macromol Mater Eng*, 307(10), 2200276.

Muskat, L.C., et al. (2023), A biobased superabsorbent formulation for above-ground application of a new entomophthoralean fungus for biological psyllid pest control. *BioControl*, 68(6), 629–641.

Mussatto, S.I., et al. (2012), *Use of agro-industrial wastes in solid-state fermentation processes*. In: Show, K.-Y. and Guo, X. (eds) *Industrial waste*, pp. 121–140. Rijeka: InTech-Open Access Publisher.

Neale, M. and Newton, P. (1999), Registration regulatory requirements in Europe. In: Hall, F.R., Menn, J.J. (eds) *Biopesticides: use and delivery*, Methods in Biotechnology, vol 5. Humana Press. pp. 453–471. https://doi.org/10.1385/0-89603-515-8:453

Nguyen Thi Minh, H., et al. (2011). Effect of sporulation conditions on the resistance of Bacillus subtilis spores to heat and high pressure. Applied microbiology and biotechnology, 90, 1409–1417.

Nunes, C. (2010), New developments in alternative methods to control postharvest fruit decay. In: Nunes, C. (ed.) *Environmentally friendly and safe technologies for quality of fruit and vegetables*, pp. 131–143. Brussels: COST Association.

Nussinovitch, A. (2010), Beads and special applications of polymers for agricultural uses. In: *Polymer macro-and micro-gel beads: fundamentals and applications*, pp. 231–253.

Orive, G., et al. (2005), Challenges in cell encapsulation. In: Nedović, V. and Willaert, R. (eds) *Applications of cell immobilisation biotechnology*, 185–196. Dordrecht: Springer.

Paau, A.S. (1988), Formulations useful in applying beneficial microorganisms to seeds. *Trends Biotechnol*, 6(11), 276–279.

Pagano, M.C. and Dhar, P.P. (2015), Fungal pigments: an overview. In: *Fungal biomolecules: sources, applications and recent developments*, pp. 173–181. https://doi.org/10.1002/9781118958308.ch1

Patel, A.V. and Vorlop, K.D. (1994), Entrapment of biological control agents applied to entomopathogenic nematodes. *Biotechnol Tech*, 8, 569–574.

Payne, C.C. (1982), Insect viruses as control agents. *Parasitology*, 84(4), 35–77.

Pendleton, I.R. and Morrison, R.B. (1966), Separation of the spores and crystals of *Bacillus thuringiensis*. *Nature*, 212(5063), 728–729.

Peters, A. (2016), Formulation of Nematodes. In: Glare, T. and Moran-Diez, M. (eds) *Microbial-based biopesticides methods in molecular biology*, vol 1477. New York: Humana Press.

Pliego, C., et al. (2011), Screening for candidate bacterial biocontrol agents against soilborne fungal plant pathogens. *Plant Soil*, 340, 505–520.

Prior, C., et al. (1988), Infectivity of oil and water formulations of Beauveria bassiana (Deuteromycotina: Hyphomycetes) to the cocoa weevil pest *Pantorhytes plutus* (Coleoptera: Curculionidae). *J Invertebr Pathol*, 52(1), 66–72.

Przyklenk, M., et al. (2017), A bioencapsulation and drying method increases shelf life and efficacy of *Metarhizium brunneum* conidia. *J Microencapsul*, 34(5), 498–512.

Quesada-Moraga, E., et al. (2023), Ecosystem services of entomopathogenic ascomycetes. *J Invertebr Pathol*, 201, 108015.

Ravel, D. (1999), Hurdles and bottlenecks on the road to biocontrol of plant pathogens. *Austral Plant Pathol*, 28, 53–56.

Ravensberg, W.J. (2011), Roadmap to success and future perspective. In: Ravensberg, W. J. (ed.) *A roadmap to the successful development and commercialization of microbial pest control products for control of arthropods*, pp. 357–376. Dordrecht: Springer.

Raymaekers, K., et al. (2020), Screening for novel biocontrol agents applicable in plant disease management–a review. *Biol Control*, 144, 104240.

Reid, S., et al. (2023), Production of entomopathogenic viruses. In: McLaughlin, R.E. and Bell, M.R. (eds) *Mass production of beneficial organisms*, pp. 375–406. London: Academic Press.

Renwick, A., & Poole, N. (1989). The environmental challenge to biological control of plant pathogens. Biotechnology of fungi for improving plant growth. Symposium of the British Mycological Society held at the University of Sussex, September 1989, 277–290 ref. 57.

Rhodes, D.J. (1993), Formulation of biological control agents. In: Jones, D. G. (ed.) *Exploitation of microorganisms*, pp. 411–439. London: Chapman and Hall.

Ruiu, L. (2018). Microbial biopesticides in agroecosystems. Agronomy, 8(11), 235.

Sangorrín, M.P., et al. (2014), Cold-adapted yeasts as biocontrol agents: Biodiversity, adaptation strategies and biocontrol potential. In: Buzzini, P. and Margesin, R. (eds) *Cold-adapted yeasts: biodiversity, adaptation strategies and biotechnological significance*, pp. 441–464. Heidelberg: Springer-Verlag Berlin.

Santivarangkna, C., et al. (2007), Alternative drying processes for the industrial preservation of lactic acid starter cultures. *Biotechnol Prog*, 23, 302–315.

Santos, C.A.M.D., et al. (2019), Effect of addition of adjuvants on physical and chemical characteristics of *Bt* bioinsecticide mixture. *Sci Rep*, 9(1), 12525.

Scher, H.B. (1999), *Controlled-release delivery systems for pesticides* (1st ed.). New York: Routledge.

Schisler, D.A., et al. (2004), Formulation of *Bacillus* spp. for biological control of plant diseases. *Phytopathology*, 94(11), 1267–1271.

Seiber, J.N., et al. (2014), Biopesticides: state of the art and future opportunities. *J Agr Food Chem*, 62, 11613–11619.

Shah, P.A., et al. (1999), Production factors involved in the formulation of *Erynia neoaphidis* as alginate granules. *Biocontrol Sci Techn*, 9(1), 19–28.

Shah, P.A. and Pell, J.K. (2003), Entomopathogenic fungi as biological control agents. *Appl Microbiol Biotechnol*, 61(5), 413–423.

Shapiro, M. (1992), Use of optical brighteners as radiation protectants for gypsy moth (Lepidoptera: Lymantriidae) nuclear polyhedrosis virus. *J Econ Entomol*, 85(5), 1682–1686.

Shapiro-Ilan, D.I., et al. (2012), Entomopathogenic nematode production and application technology. *J Nematol*, 44(2), 206–217.

Shapiro-Ilan, D.I. and Goolsby, J.A. (2021), Evaluation of Barricade® to enhance survival of entomopathogenic nematodes on cowhide. *J Invertebr Pathol*, 184, 107592.

Short, B.D., et al. (2018), UV-C irradiation as a management tool for *Tetranychus urticae* on strawberries. *Pest Manag Sci*, 74(11), 2419–2423.

Shubha, S., et al. (2015), Screening of entomopathogenic fungi for chitinolytic activity. *Trends Biosci*, 8(24), 6869–6872.

Smith, K.M. (1959), The use of viruses in the biological control of insect pests. *Outlook Agr*, 2(4), 178–184.

Soper, R.S. and Ward, M.G. (1981), *Production, formulation and application of fungi for insect control*. In: Biological control in crop production, 161–180.

Sparks, R.E. and Jacobs, I.C. (1999), Selection of coating and microencapsulation processes. In: Scher, H.B. (ed.) *Controlled-release delivery systems for pesticides*, pp. 3–29. New York: Routledge.

Stephan, D. and Zimmermann, G. (1998), Development of a spray-drying technique for submerged spores of entomopathogenic fungi. *Biocontrol Sci Techn*, 8(1), 3–11.

Sun, K., et al. (2023), Leveraging insect viruses and genetic manipulation for sustainable agricultural pest control. *Pest Manag Sci*, 80, 2515–2527.

Teixidó, N., et al. (2006), Improving low water activity and desiccation tolerance of the biocontrol agent *Pantoea agglomerans* CPA-2 by osmotic treatments. *J Appl Microbiol*, 101(4), 927–937.

Teixidó, N., et al. (2020), Formulations to improve biocontrol products shelf-life and/or ecosystem adaptation. In: *How research can stimulate the development of commercial biological control against plant diseases*, pp. 257–273. http://dx.doi.org/10.1007/978-3-030-53238-3

Teixidó, N., et al. (2022), Insight into a successful development of biocontrol agents: production, formulation, packaging, and shelf life as key aspects. *Horticulturae*, 8(4), 305.

Thakore, Y. (2006), The biopesticide market for global agricultural use. *Ind Biotechnol*, 2, 194–208.

Vemmer, M. and Patel, A.V. (2013), Review of encapsulation methods suitable for microbial biological control agents. *Biol Control*, 67(3), 380–389.

Vemmer, M., et al. (2016), Development of a CO_2-releasing coformulation based on starch, *Saccharomyces cerevisiae* and *Beauveria bassiana* attractive towards western corn rootworm larvae. *Pest Manag Sci*, 72(11), 2136–2145.

Villaverde, J.J., et al. (2016), Biopesticides from natural products: Current development, legislative framework, and future trends. *BioResources*, 11(2), 5618–5640.

Waage, J. K., & Hassell, M. P. (1982). Parasitoids as biological control agents–a fundamental approach. Parasitology, 84(4), 241–268.

Wang, N., et al. (2022), The extent and mechanism of the effect of protectant material in the production of active lactic acid bacteria powder using spray drying: A review. *Curr Opin Food Sci*, 44, 100807.

Webb, R.E., et al. (1994), Effect of optical brighteners on the efficacy of gypsy moth (Lepidoptera: Lymantriidae) nuclear polyhedrosis virus in forest plots with high or low levels of natural virus. *J Econ Entomol*, 87(1), 134–143.

Williams, D.F. (2014), There is no such thing as a biocompatible material. *Biomaterials*, 35(38), 10009–10014.

Wilson, K., et al. (2020), A novel formulation technology for baculoviruses protects biopesticide from degradation by ultraviolet radiation. *Sci Rep*, 10, 13301.

Wraight, S.P. and Carruthers, R.I. (1999), Production, delivery, and use of mycoinsecticides for control of insect pests on field crops. In: Hall, F.R. and Menn, J.J. (eds) *Biopesticides: use and delivery methods in biotechnology*, Vol. 5. Totowa, NJ: Humana Press.

Wraight, S.P., et al. (2003), *Zoophthora radicans* (Zygomycetes: Entomophthorales) conidia production from naturally infected *Empoasca kraemeri* and dry-formulated mycelium under laboratory and field conditions. *Biol Control*, 28(1), 60–77.

Wu, S., et al. (2023), Novel formulations improve the environmental tolerance of entomopathogenic nematodes. *Biol Control*, 186, 105329.

Yaakov, N., et al. (2018), Single cell encapsulation via pickering emulsion for biopesticide applications. *ACS Omega*, 3(10), 14294–14301.

Yánez-Mendizabal, V., et al. (2012), Endospore production allows using spray-drying as a possible formulation system of the biocontrol agent *Bacillus subtilis* CPA-8. *Biotechnol Lett*, 34, 729–735.

Yao, S.L., et al. (2010), In vitro and in vivo responses of fungal biocontrol agents to gradient doses of UV-B and UV-A irradiation. *BioControl*, 55, 413–422.

Zhang, S., et al. (2011), Two hydrophobins are involved in fungal spore coat rodlet layer assembly and each play distinct roles in surface interactions, development and pathogenesis in the entomopathogenic fungus, *Beauveria bassiana*. *Mol Microbiol*, 80(3), 811–826.

Zhang, L.B. and Feng, M.G. (2018), Antioxidant enzymes and their contributions to biological control potential of fungal insect pathogens. *Appl Microbiol Biotechnol*, 102, 4995–5004.

Zhou, X. and Feng, M.G. (2009), Sporulation, storage and infectivity of obligate aphid pathogen *Pandora nouryi* grown on novel granules of broomcorn millet and polymer gel. *J Appl Microbiol*, 107(6), 1847–1856.

Zimmermann, G. (1986), The *Galleria* bait method for detection of entomopathogenic fungi in soil. *J Appl Entomol*, 102, 213–215.

Zohar-Perez, C., et al. (2003), Structure of dried cellular alginate matrix containing fillers provides extra protection for microorganisms against UVC radiation. *Radiat Res*, 160(2), 198–204.

Chapter 8

Challenges in commercialising microbial-based biopesticides for controlling insect pests

Travis R. Glare, Lincoln University, New Zealand

1 Introduction

Biological products are often touted as a major tool in the future of insect pest management due to their utility, specificity and safety, both environmentally and for vertebrates. Biopesticides can be defined broadly to include all pesticidal bioproducts, including commercially available predators and parasites, as well as products from microorganisms or plant extracts. The definition of a biopesticide can even include genetically modified plants (Marrone 2019; Glare and Nollet 2023), or definitions can be more limited. For the purposes of this chapter, I will concentrate on microbial-based biopesticides, formulations of microbes, nematodes and their metabolites used in the control of pest insects, sometimes called bioinsecticides.

http://dx.doi.org/10.19103/AS.2024.0149.13

The use of microbes to control insects has a long and checkered history, with the first practical attempt occurring around 1888 in Russia, using *Metarhizium* against a wheat chafer (Zimmermann et al. 1995). From there, the use of microbes against insect pests continued to develop until the advent of chemical insecticides in the 1950s, which were cheap and efficacious. It was not until the 1980s that interest in the practical use and commercialisation of insect pathogenic microbes began to expand, driven by the need for more environmentally friendly pest management. The use of biopesticides in insect control has been steadily increasing, but there have been many 'false dawns' in the development of commercially available biopesticides, where their proposed rapid expansion failed to materialise.

Formulations of *Bacillus thuringiensis* have been commercially available since 1932 (Glare and O'Callaghan 2000) and for many years made up over 90% of all biopesticides commercially sold. The reasons for *B. thuringiensis*'s dominance of the biopesticide market likely reflect the chemical pesticide equivalence of the mode of action and products – the lethal action is due to toxins which are toxic to more than a single pest, do not need a live cell for activity, and can be produced cheaply with a consistent shelf life as a formulated product. Most other microbes (or their metabolites) have more specific requirements than *B. thuringiensis*, often leading to issues with production, shelf life or field efficacy. But many insect pathogens have shown great promise for development as biopesticides and a range of species/strains are now being exploited worldwide. However, the number of species and genera involved represents a small fraction of the thousands of insect pathogenic species. The reason for this shortfall is the subject of this chapter.

2 The current state of the biopesticide market

Estimates of the global biopesticide market vary with recent reports for 2023 at around $US6.13B and a compound growth rate of 19.8% (Reportlinker 2024). Growth rates around 15% have been reported over the past decade (Croplife 2022). Bioinsecticides reached around $US2.2B in 2020 (Qu et al. 2022). There is still a long way to go before biopesticides displace chemical pesticides for insect control, but Marrone (2023a) reported that biopesticides (including those that target plant disease and bioherbicides) now account for around 10% of the global pesticide market.

There are various estimates for how many biopesticides (targeting more than just insects) are available globally. Samada and Tambunan (2020) put the figure at 175. Marrone (2023a) reported around 340 registered microbial biopesticides with bioinsecticides making up 30% of sales among the biopesticides categories. In the United States (USA) alone, there are more than 200 products available, compared to 60 analogoues products in the

European Union (EU) market (Kumar and Singh 2015). In the United States, as of 2017, there were 125 biopesticide active ingredients registered based on microbes, with 57 labelled for use against nematodes and insects (Arthurs and Dara 2018). A recent review of mycopesticides (those based on fungi) put the number of registered mycoinsecticides in the United States, EU, China, Canada and Australia at only 27 (Jiang and Wang 2023).

An opportunity for biopesticides to take greater proportional market share is based on the regulatory restrictions that are being applied to synthetic pesticides, leading to a 2% decline per year in use in favour of a 10% increase in biopesticide use (Damalas and Koutroubas 2018). Environmental degradation, mammalian toxicity and the rise in pest resistance are driving some of this change. In the United States, Marrone (2023a, b) listed increased grower returns, especially when biopesticides are used in integrated pest management (IPM), increasing scientific knowledge, roles in resistance and residue management, new scientific tools (such as 'omics'), human and environmental safety, lower carbon footprint during manufacturing, shorter reentry times (typically 4 h) and the growth of precision agriculture as drivers of growth in biopesticide use. In response to issues around overuse of chemical pesticides, many governments have instituted policies intended to encourage the development of alternatives, including biopesticides. For example, the USA EPA has streamlined registration for biopesticides, the European Union has introduced legislation mandating synthetic pesticide reduction targets and the Ministry of Agriculture and Rural Affairs of China formulated and issued the Action Plan for Zero Growth in Pesticide Use by 2020 (Jiang and Wang 2023). In addition, most biopesticides can be incorporated into IPM systems or used in organic production (Marrone 2019). The growth in interest in regenerative agriculture, a loose term that encompasses changes in conventional agriculture to reduce inputs and promote the protection of biodiversity, is also stimulating the increased use of biopesticides in farming.

3 Issues holding back new biocontrol products: regulation and policy

Biopesticides are seen as having advantages, such as specific modes of action, short residual effect, few environmental impacts and consistent with sustainable production principles (Ayilara et al. 2023). As noted by various authors and agencies, despite the huge potential of biopesticides, that potential is far from being realised (Marrone 2023a; Fenibo et al. 2021). There is no doubting the advantages of effective biopesticides over traditional pesticides nor the place they can fill in IPM regimes. However, biopesticides still only make up under 10% of pest control products, and they have yet to fulfil the market need in most cases. Uptake of biopesticides is still limited due to the lack of products for many pests, the cost of the products available and often the slow action that sometimes results in damage by pests after application (Verma et al. 2021; Fenibo et al. 2021; Ayilara et al. 2023). In addition, limitations include variable quality control of products, susceptibility to environmental factors and lack of stability during storage (Fenibo et al. 2021).

There are several areas that are holding back the uptake of biopesticides. In 2012, we reviewed the field of microbial biopesticides and posed the question 'Have biopesticides come of age?' (Glare et al. 2012). In that work, we considered the advances in biopesticide development in light of market share and technological advances. At that time, the main areas that needed improvement were thought to be improvements in delivery and persistence, advancing the use of bioactives rather than the whole organism, strategic selection of target pests and markets, more investment in expertise and registration and legislative changes. In the intervening years, some of these needs have not changed, but the emphasis has shifted as biopesticides become mainstream pest control products with the potential of significant increased market share.

A major and continuing issue is the long and complex regulation regimes of biopesticides in some jurisdictions. In the European Union, for example, the average time for approval of a microbial biological control product is 1369 days (almost 4 years) (Jiang and Wang 2023). Fewer biopesticide-active substances are registered in the European Union than in the United States, India, Brazil or China, due to long and complex registration processes in the European Union, which follows the model for the registration of conventional pesticides. The balance of the precautionary principle with the need to bring new products to market rapidly is one that regulators are grappling with constantly. Microbial entomopathogens are particularly challenging, as while they are naturally occurring and show no mammalian toxicity historically, as research and knowledge increases, so does identification of potentially toxic compounds that could be produced. I say potentially because often the evidence is based on the presence of genes known to encode human toxins; however, expression at quantities that would cause mammalian toxicity may not naturally occur (Biggel et al. 2022). Nonetheless, for regulators, this must be demonstrated, especially during manufacturing.

Regulations that promote the registration of low-risk compounds with the provision of incentives could also facilitate the commercialisation and availability of biopesticides in the market (Damalas and Koutroubas 2018). As stated by Aoife Dillion (Fera 2023), finding the right balance in regulations to manage the risk of these new classes of active agents for pest control with the need to register them in a timely manner has been a challenge for all countries. Many countries have developed specific pathways for biologically based biopesticides, but this has not always worked to reduce the time it takes to register.

Policy changes targeted at reducing chemical pesticide use have been enacted in some regions, such as the 'Farm-to-Fork' strategy of the European Union in 2020, which was set to legislate a target to reduce chemical pesticide use by 50% by 2030. However, in November 2023, the 'Sustainable Use Regulation', as it was called, was vetoed by the EU parliament. Brazil made policy changes in 2014 that promoted biological approaches and has seen a 42% compound market growth since (Jenifer Lewis, quoted in Fera (2023)). In France, government policies that were intended to reduce synthetic pesticide use, implemented in 2008, 'Ecophyto', did not lead to a reduction over the subsequent 10 years (Stokstad 2018).

The effect of a regulatory framework that requires large, complex dossiers of information can be seen by the lack of microbial-based products registered in the European Union (Marrone 2023a). Croplife (2022) suggests that 'if the regulatory framework were better implemented and strengthened with appropriate regulatory guidance, the European Union could become a market more suitable for the development of biopesticides'. It is not only in the European Union where regulation needs are restricting progress on biopesticides, but it is a large market that needs to reduce synthetic pesticide use, so alternatives are needed. The OECD is working on a plan to provide testing guidelines for microbial biopesticides, which can reflect the unique properties of these microbe-led solutions. These efforts are aimed at reducing the time to approval for biopesticides, which in some areas have been long enough to deter industry investment. Finding that balance between risks and benefits that result in a clear regulatory framework that operates in a reasonable timeframe will greatly advance the use of biopesticides.

Another issue in some jurisdictions is the cost associated with registration. In Australia, for example, the long registration procedures coupled with the costs of development, the fees for registration and the small projected returns due to market size, are expected to limit the registration of biopesticides (Jiang and Wang 2023). This issue, of cost versus return, is even more evident in small economies such as New Zealand.

4 Issues holding back new biocontrol products: economic mass production

Biopesticides based on microbes or their actives require some form of mass production of the active agent, which can be an impediment to developing cost-effective pesticides. In some high-value niche markets, the cost of production is less of an issue, but for the majority of biopesticide markets, there are competing products or the farming system itself has limits on what can be spent on pest management. For many commodities, including large-acreage field crops, competition from less-expensive conventional pesticides remains a barrier to adoption. So, the cost of production is an issue.

Generally, production is based on either liquid fermentation or solid-state production or their combination. There are other approaches, such as the production of some insect viruses in host insects, but the majority of large-market targeted biopesticides will use a fermentation approach. The cost of each approach will vary depending on local factors, such as labour costs, availability and cost of media ingredients and power. Generally, liquid fermentation at scale is capital intensive but able to produce agents more cheaply in bulk than solid-state production. For this reason, there has been interest in using blastospores of entomopathogenic fungi such as *Beauveria* and *Metarhizium* produced in liquid fermentation rather than producing conidia using solid state (e.g. Kassa et al. 2004; Mascarin et al. 2015; Jackson and Mascarin 2016, 2017). Blastospores can be directly infective for insects and can be produced in higher quantities more quickly compared with conidia on grain substrates. However, there are still technical issues around formulation and survival to be overcome, and solid state remains the main method of production of entomopathogenic fungi. Bacteria are generally produced in large-scale liquid fermentation, contributing to the success of products such as those based on *B. thuringiensis*.

A ballpark figure for most fungal and bacterial applications is 10^{12}–10^{13} infective propagules/ha; producing this many propagules cheaply can be challenging. In addition, propagules or active compounds need to remain infective/toxic and have good storage characteristics, so the target of production includes stable, resilient cells and high biomass. The use of agricultural or industrial wastes as a substrate for the production of microbial insecticides has been explored to reduce production costs. This area was extensively reviewed recently (Astudillo et al. 2023). There are obvious benefits if costs can be reduced for the growth substrate, but a major issue with many agro-wastes is the lack of consistency in constituents, which affects predictability of the production process and can affect characteristics of the microbes or bioactives produced.

There has been a lot of focus on developing cheaper media, as that is often seen as the main driver of the cost of production. But in many commercial situations, depreciation of the capital investment on equipment and labour costs are more important for determining the final cost of production than the media. Equipment and labour costs are often overlooked during development.

5 Issues holding back new biocontrol products: formulation for long shelf life and survival after application

From a commercial viability viewpoint, shelf life can be a major determinant in biopesticide success. A long shelf life at ambient temperature both reduces the need for refrigeration during shipping and storage and, if long enough, allows stock to be used in multiple seasons, reducing waste. But shelf life over a year is unusual for biopesticides, especially those based on live organisms. Shelf life may be short, such as that of the New Zealand product, Invade™ (no longer sold in this form), which needed to be kept at 4°C with a shelf life of less than 7 days at 20°C (Jackson et al. 1992), although the product has subsequently been improved through various formulation changes (Johnson et al. 2001).

As that demonstrates, formulation can improve the efficacy of a biopesticide, shelf life and survival after application and is an area where advances are rapidly being made. For example, nanoformulations and microencapsulation technologies can improve the stability and residual action of biopesticide products, and this could improve their field use (Koul 2019). The field of nanobiopesticides, which includes various approaches to formulation such as nano-emulsions, nano-vesicles and nano-gels, has recently been extensively reviewed (Koul 2023).

Advances are also being made in formulation approaches that have been used for many years, such as the development of wettable powders based not on the conidia of entomopathogenic fungi but on liquid fermentation-produced blastospores, such as with *Metarhizium robertsii* where spray drying and air-drying techniques were used on the blastospores before formulating (Iwanicki et al. 2021). Another interesting example is the encapsulation approach that may allow the exploitation of Entomophthorales. *Pandora* and other Entomophthorales are notoriously difficult to use as applied biocontrols because the spores are environmentally very fragile, require high humidity to germinate and infect and are difficult to mass produce *in vitro*. Muskat et al. (2022) developed a novel formulation of *Pandora* sp. in Ca-alginate beads which encapsulated the fermenter-produced biomass and provided additional nutrients that promoted sporulation after application. Muskat et al. (2023) also developed a paste-type formulation containing bio-based super absorbency, and the co-application of the encapsulated beads and paste allowed sporulation of the fungus on plants.

There is a lot of research on various aspects of the process of formulating microbes, with knowledge increasing on the effects of additives, drying processes and the influence of growth media and conditions on the subsequent quality of the active agents. This knowledge is slowly making changes in the quality of the final products, which will hopefully lead to better outcomes from biopesticide use, such as producing more shelf-stable products. However, much of the research and development on formulation occurs within companies and is generally a trade secret or has other Intellectual Property (IP) protections. This proprietary aspect can limit the advancement of the field in general, as the lack of publicly available information can result in 'reinventing the wheel', where each development group must start without knowing the latest advances.

Of course, shelf life is only one aspect of the performance of a formulation. Persistent effect after application and ease of application have also been issues with biopesticides that have reduced uptake. Ideally, for a leaf-eating insect, 3 weeks of residual activity is the target (Glare et al. 2012) although this is rarely achieved.

6 Issues holding back new biocontrol products: field efficacy - perception and fact

In a recent survey by Fera Science (Fera 2023), nearly 58% of respondents cited efficacy concerns as the biggest barrier to using biopesticides. Cost and uncertainty about integrating biocontrols with other products were also mentioned, but efficacy in all situations has long been considered an issue for biopesticides. This is partly due to many products being launched onto the market before sufficient testing was completed, and also the strong effect the biotic and abiotic environments can have on biopesticide success. Memories of early failures may have also influenced some farmers to be wary of biopesticides that did not live up to the hype. Biological approaches to pest control are not new, so it is unsurprising there is a mixed history of effectiveness in the field. Many early biopesticides, for example, had variable efficacy in the field and were often very susceptible to environmental conditions. It does not take many failures to taint growers' perception of an entire approach. There have been several eras of misplaced optimism about the next generation of biopesticides, such as during the 1980s when the leading pesticide manufacturers produced what were intended as mainstream products but (other than *B. thuringiensis*-based products) generally performed poorly in the market.

However, biopesticides have improved in recent years and can often compete with traditional pesticides in terms of efficacy. An example of a demonstration at the farm level was the UK Crop Health North project (https://www.crophealthnorth.co.uk/, accessed November 2023), which set up on-farm, 2-year comparative trials with wheat, looking at biological crop protection products vs. conventional chemical pesticide approaches. They found that the biocontrol technologies, alone or in combination with conventional crop chemistry, obtained similar yields and grain quality.

There is a need for more field research to assess the effects on specific pest problems under diverse cropping systems (Damalas and Koutroubas 2018). However, large-scale field trials are often beyond the financial ability of research institutions or start-up biopesticide companies. This problem may have a solution. As summarised by Marrone (2023a), large companies have been acquiring smaller biopesticide companies. This approach means the early research and development is done by innovative start-ups, then the technology is acquired by the larger, often multinationals who have greater market access and research budgets (Marrone 2019).

It is interesting that in many countries, it is only recently that consumer, grower and regulation push has led to more interest in biopesticides than at any other time. For example, the European Union proposed regulations to decrease the use of chemical pesticides by 50% by 2030 (Regulation (EU) 2021/2115), discussed above, in response to the lack of impact of a previous Sustainable Use of Pesticides Directive. These new measures include a focus on IPM to be considered before chemical pesticides are used.

With the increase in synthetic pesticide resistance and residue concerns on marketable crops (especially those exported), growers need alternatives and are interested in the new generation of biopesticides, among other green technologies. There is a new wave of regenerative approaches to farming which, like organics, covers a very broad area of farming practice. Biopesticides fit well into such approaches.

However, around the world, grower attitudes to biopesticides drives use. Generally, growers who have knowledge and use biopesticides have a good opinion of them, but many growers around the world are unaware of biopesticides. In the United States, there appears to be a growing acceptance that biologicals are equivalent to conventional products or have unique attributes for specific applications (Marrone 2023b). However, lack of trust in product performance was listed by 41% of respondents in a US survey (Sfiligoj 2022, reported in Marrone 2023b). A study of pear and walnut growers in the western USA adopting biological control approaches (Goldberger and Lehrer 2016) found that 54% of walnut growers and 74% of pear growers used some biological control practices. Of the multiple factors that influenced grower decisions on pest control approaches, perceived benefits included less reliance on chemicals, reduced production costs, environmental protection and improve human health while perceived barriers included lack of knowledge, questionable effectiveness, high cost and lack of recommendation by pest management consultants. In Kenya and Uganda, willingness to pay for biopesticides among 1300 farmers surveyed was higher for risk taking, educated, higher-income farmers who were aware of the negative effects of chemical pesticide use, while older farmers, large family size and having trust of other community members decreased the willingness to pay for biopesticides (Nyangau et al. 2022).

An interesting question when considering the uptake of new technologies, including biopesticides, is whom do growers trust as an information source. It is a big decision for growers to try new approaches, which may require a complete overhaul of their crop protection strategies, especially when it is their livelihoods are at stake. Biologicals may be incompatible with some synthetic pesticides, such as fungicides and fungal-based biopesticides, so growing practices may need to adapt. This is not always the case but must be considered when advocating biopesticides. Growers want to see a new product working in a real-world situation and often get their recommendations from other growers. Some studies have shown that education and training can be a major influence on attitudes toward more environmentally sound pest control approaches (Khan and Damalas 2014; Thao et al. 2019; Bakhtawer 2021).

7 Issues holding back new biocontrol products: meeting market needs

It is basic economics that products will sell when they meet a need and a price point. For many biopesticide development projects, a fundamental question is, 'Is there a need for a new product against the particular target(s)?' While there is widespread support for increasing the use of biopesticides in crop protection, biopesticides still have to meet price expectations and competition. Synthetic chemicals still make up the majority of pest control products used. Synthetic pesticides are often broad-spectrum, less expensive, easier to apply and in some cases more efficacious than alternatives. Broad-spectrum activity is generally not compatible with an IPM system. Overuse of synthetic pesticides has led to problems with resistance, but surprisingly these pesticides often continue to be used even where they are clearly not controlling pest problems. The reasons are multiple. Hu (2020) classified the dependence on agrochemicals into five interacting themes: (1) existing agricultural regime, (2) social process of pesticide application, (3) economic reasons, (4) politics and governance and (5) promotional failure of alternatives. Hu (2020) and others have classified the indiscriminate use of pesticides as mainly a human-made issue, with deep-seated socio-economic and political reasons behind it. But clearly, agrochemical overuse has become an issue.

The cost of biopesticides as an alternative to synthetic chemicals has often been cited as a limitation on uptake, so parity with existing products, even where these products no longer work or are unavailable, is of benefit. If the cost of the solution is more than the benefit the grower receives, the biopesticide will obviously not be successful. This fact is sometimes left out of development thinking. Biopesticides can offer additional benefits beyond just direct action against an arthropod pest. As discussed previously, benefits can include plant growth promotion, compatibility with other approaches forming an IPM solution and benefits in farm systems, such as no withholding period after applications. Use of biopesticides can result in the reestablishment of native beneficials and opportunities to conduct augmentative releases, for example, *Trichogramma* wasps that are now released on thousands of acres of leafy greens and other crops in the United States to control caterpillars, a practice made possible since these growers only spray *B. thuringiensis* and no longer use broad-spectrum insecticides (Steven Arthurs, pers. comm.). These additional benefits can be hard to quantify but increase the ability of meeting the market needs by biopesticides.

Costs are also comparative, so if competing products change or can no longer be used, markets will shift to other options including biopesticides. In addition, other hurdles in commercialisation include quality control problems, shelf-life, low awareness of products and higher expense (Arthurs and Dara, 2018).

8 Issues holding back new biocontrol products: lack of novel actives and approaches

One of the notable aspects of the currently registered microbial biopesticides used against arthropods is the limited range of active agents represented. While there are a range of products available, most are based on a few organisms, dominated by *Bacillus* spp. and several entomopathogenic Ascomycetes. There are obvious reasons for this, as these are the organisms that are consistently virulent against a range of hosts, can be mass-produced economically and have good shelf life after formulation. But there is a need for more diversity in the active agents used in biopesticides to help reach new markets and overcome constraints in some situations. There are some recently developed microbials and bioactives that show the promise of unexplored entomopathogens, such as *Chromobacterium subtsugae* (Martin et al. 2007), developed as the product Grandevo®. Smagghe et al. (2023) listed microbial entomopathogens *Paenibacillus elgii*, *Pseudomonas* spp., *Aspergillus* spp. and *Cladosporium* spp. as some of the novel actives potentially reaching the market in Australia, for example. Other novel microbial insecticides commercialised include a new species *Burkholderia rinojensis* (Venerate®) (Cordova-Kreylos et al. 2013; Marrone 2019) for a range of insects, and *Cordyceps javanica* (Lalguard Java WP) for *Bemisia* control (AgNews 2022). Venerate® is used as a seed treatment in approximately 15 million acres in the United States.

In addition, there are new areas of research and development that may offer some potential to overcome constraints on microbial biopesticides, such as the use of microbial consortia (Maciąg et al. 2023; Batista and Singh 2021; Nunes et al. 2024) or synthetic biology approaches to produce bioactives cheaply (Pham et al. 2019). In Brazil, there are a number of products that are based on mixtures of active microbes (Nunes et al. 2024). For some of these approaches and those of more traditional biopesticides to work, especially in the soil environment, more understanding is needed of the ecology and modes of action of inoculant strains and their interaction with other biotic and abiotic factors (O'Callaghan et al. 2022).

There are also exciting developments in the area of genetic modification, which can lead to novel strains with enhanced capabilities (Chen et al. 2017; Wang and Coleman 2019; Song et al. 2022). Such approaches will require regulatory changes in many countries, but the general acceptance of CRISPR-Cas knockouts in many countries (Zhang et al. 2020) points to a future where such modifications are common.

9 Conclusion

Summarising the chapter, what is likely to improve the success of biopesticides?

- Production and formulation improvements: improved shelf life, and wider target ranges that include more pest species affected by one product.
- Increasing market recognition for the multiple benefits: economic and environmental.
- New actives, including those developed through biotechnological means, that overcome current biological limitations.
- Embracing new approaches, such as synthetic biology, to produce bioactives more cheaply.
- Reducing costs: whether it is through production, application or formulation induced efficiencies, costs in the marketplace matter.
- Field efficacy – really an outcome of improvement in the other areas, but demonstrable efficacy is key.
- Regulation that balances the need with ensuring products.
- Focus on areas of market need.
- Increased grower awareness and education.

10 Where to look for further information

Useful websites:

- The British Crop Protection Council maintains a subscription-based service that offers a comprehensive dataset on pesticides: https://www.bcpc.org/product/bcpc-online-pesticide-manual-latest-version.
- CABI maintains the Bioprotection Portal https://bioprotectionportal.com/.

- The Biological Products Industry Alliance (BPIA) promotes the responsible development of safe and effective biological products, including biopesticides, biostimulants and biofertilisers. https://www.bpia.org/.
- The Society for Invertebrate Pathology is a source of information on insect pathogens. https://www.siponline.org/.

More general edited books:

- Birch, N. and Glare, T.R. (Eds) (2020). *Biopesticides for Sustainable Agriculture*. Burleigh Dodds Series in Agricultural Science.
- Nollet, L.M.L. and Rathore, H.S. (Eds.) (2015). *Biopesticides Handbook*. CRC Press, Boca Raton.

11 References

AgNews (2022). First biopesticide with *Cordyceps javanica* fungus registered against *Bemisia tabaci* fly in Brazil. https://news.agropages.com/News/NewsDetail---44364.htm

Arthurs, S. and Dara, S.K. (2018). Microbial biopesticides for invertebrate pests and their markets in the United States. *J. Invertebrate Pathol.*, 165, 13–21. https://doi.org/10.1016/j.jip.2018.01.008

Astudillo, A., et al. (2023). Advances in agroindustrial waste as a substrate for obtaining eco-friendly microbial products. *Sustainability*, 15, 3467. https://doi.org/10.3390/su15043467

Ayilara, M.S., et al. (2023). Biopesticides as a promising alternative to synthetic pesticides: a case for microbial pesticides, phytopesticides, and nanobiopesticides. *Front. Microbiol.*, 14, 1040901. https://doi.org/10.3389/fmicb.2023.1040901

Bakhtawer, A.S. (2021). A cross sectional survey of knowledge, attitude and practices related to the use of insecticides among farmers in industrial triangle of Punjab, Pakistan. *PLoS One*, 16(8), e0255454. https://doi.org/10.1371/journal.pone.0255454

Batista, B.D. and Singh, B.K. (2021). Realities and hopes in the application of microbial tools in agriculture. *Microb. Biotechnol.*, 14(4), 1258–1268. https://doi.org/10.1111/1751-7915.13866

Biggel, M., et al. (2022). Recent paradigm shifts in the perception of the role of *Bacillus thuringiensis* in foodborne disease. *Food Microbiol.*, 105, 104025. https://doi.org/10.1016/j.fm.2022.104025

Chen, J., et al. (2017). CRISPR/Cas9-mediated efficient genome editing via blastospore-based transformation in entomopathogenic fungus *Beauveria bassiana*. *Sci. Rep.*, *7*, 45763. https://doi.org/10.1038/srep45763

Cordova-Kreylos, A.L., et al. (2013). Isolation and characterization of *Burkholderia rinojensis* sp. nov., a non-*Burkholderia cepacia* complex soil bacterium with insecticidal and miticidal activities. *Appl. Environ. Microbiol.*, 24, 669–678. https://doi.org/10.1128/AEM.02365-13

Croplife (2022). Unlocking the potential of biopesticides. https://croplifeeurope.eu/wp-content/uploads/2022/05/CLE_Unlocking-the-Potential-of-Biopesticides_2022.pdf

Damalas, C.A. and Koutroubas, S.D. (2018). Current status and recent developments in biopesticide use. *Agriculture*, 8, 13. https://doi.org/10.3390/agriculture8010013

Fenibo, E.O., et al. (2021) Biopesticides in sustainable agriculture: a critical sustainable development driver governed bygreen chemistry principles. *Front. Sustain. Food Syst.*, 5, 619058. https://doi.org/10.3389/fsufs.2021.619058

Fera (2023). Biopesticides: current challenges and future opportunities. https://www.fera.co.uk/news/biopesticides-current-challenges-and-future-opportunities-webinar/ 25 April 2023

Glare, T.R., et al. (2012). Have biopesticides come of age? *Trends Biotechn.*, 30, 250–258.

Glare, T.R. and Nollet, L.M.T. (2023). Types of biopesticides. In L.M.L. Nollet and H.S. Rathore (Eds.), *Handbook of biopesticides, second edition*. CRC Press, in press.

Glare, T.R. and O'Callaghan, M. (2000) *Bacillus thuringiensis; biology, ecology and safety*. John Wiley and Sons, Chichester, UK. 350 pp.

Goldberger, J.R. and Lehrer, N. (2016). Biological control adoption in western U.S. orchard systems: results from grower surveys. *Biological Control*, 102, 101–111. https://doi.org/10.1016/j.biocontrol.2015.09.004

Hu, Z. (2020). What socio-economic and political factors lead to global pesticide dependence? A critical review from a social science perspective. *Int. J. Environ. Res. Public Health*, 17, 8119. https://doi.org/10.3390/ijerph17218119

Iwanicki, N.S.A., et al. (2021). Development of novel spray-dried and air-dried formulations of *Metarhizium robertsii* blastospores and their virulence against *Dalbulus maidis*. *Appl. Microbiol. Biotechnol.*, 105(20), 7913–7933. https://doi.org/10.1007/s00253-021-11576-5

Jackson, M.A. and Mascarin, G.M. (2016). Stable fungal blastospores and methods for their production, stabilization and use. US Patent 2015049673.

Jackson, T.A., et al. (1992). Pathogen to product - development of *Serratia entomophila* (Enterobacteriaceae) as a commercial biological control agent for the New Zealand grass grub (*Costelytra zealandica*). In T.A. Jackson and T.R. Glare (Eds.), *Use of pathogens in scarab pest management*. Intercept, Andover, UK. pp. 191–198.

Jaronski, S.T. and Mascarin, G.M. (2017). Mass production of fungal entomopathogens. In L.A. Lacey (Ed.), *Microbial* control insect mite *pests*, Academic Press, New York, pp. 141–155. https://doi.org/10.1016/B978-0-12-803527-6.00009-3

Jiang, Y. and Wang, J. (2023). The registration situation and use of mycopesticides in the world. *J. Fungi (Basel).*, 9(9), 940. https://doi.org/10.3390/jof9090940

Johnson, V.W., et al. (2001). Formulation of *Serratia entomophila* for biological control of grass grub. *NZ Plant Prot.*, 54, 125–127.

Khan, M. and Damalas, C.A. (2015). Factors preventing the adoption of alternatives to chemical pest control among Pakistani cotton farmers. *Int. J. Pest Manag.*, 61, 9–16. http://dx.doi.org/10.1080/09670874.2014.984257

Kassa, A., et al. (2004) Production and processing of *Metarhizium anisopliae* var. *acridum* submerged conidia for locust and grasshopper control. *Mycol. Res.*, 108, 93–100. https://doi.org/10.1017/S0953756203008827

Koul, O. (2019). Nanobiopesticides: an introduction. In O. Koul (Ed.), *Nano-biopesticides. Today and future perspectives*. Academic Press, New York, pp. 1–15.

Koul, O. (Ed.) (2023). *Development and commercialization of biopesticides*. Academic Press, New York. https://doi.org/10.1016/C2021-0-02105-2

Kumar, S. and Singh, A. (2015). Biopesticides: present status and the future prospects. *J. Fertil. Pestic.*, 6, e129.

Maciag, T., et al. (2023). Microbial consortia for plant protection against diseases: more than the sum of its parts. *Int. J. Mol. Sci.*, 24(15), 12227. https://doi.org/10.3390/ijms241512227

Marrone, P.G. (2019). Pesticidal natural products – status and future potential. *Pest Manag. Sci.*, 75, 2325–2340.

Marrone, P.G. (2023a). Status of the biopesticide market and prospects for new bioherbicides. *Pest Manag. Sci.*, 80(1), 81–86.

Marrone, P.G. (2023b). Biopesticide commercialization in North America: state of the art and future opportunities. In O. Koul (Ed.), *Development and commercialization of biopesticides*. Academic Press, New York, pp. 173–202.

Martin, P.A., et al. (2007). Toxicity of *Chromobacterium subtsugae* to southern green stink bug (Heteroptera: Pentatomidae) and corn rootworm (Coleoptera: Chrysomelidae). *J. Econ. Entomol.*, 100(3), 680–684. https://doi.org/10.1603/0022-0493

Mascarin, G.M., et al. (2015). Liquid culture fermentation for rapid production of desiccation tolerant blastospores of *Beauveria bassiana* and *Isaria fumosorosea* strains. *J. Invertebr. Pathol.*, 127, 11–20.

Pham, J.V., et al. (2019). A review of the microbial production of bioactive natural products and biologics. *Front. Microbiol.*, 10, 1404. https://doi.org/10.3389/fmicb.2019.01404

Qu, M., et al. (2022). Bioinsecticides as future mainstream pest control agents: opportunities and challenges. *Front. Agr. Sci. Eng.*, 9(1), 82–97. https://doi.org/10.15302/J-FASE-2021404

Muskat, L.C., et al. (2022). Encapsulation of the psyllid-pathogenic fungus *Pandora* sp. nov. inedit. and experimental infection of target insects. *Pest Manag Sci.*, 78(3), 991–999. https://doi.org/10.1002/ps.6710

Muskat, L.C., et al. (2023). Development of attract and kill formulations for biological psyllid pest control. *IOBC-WPRS Bulletin*, 166, 59–60.

Nunes, P.S.O., et al. (2024). Microbial consortia of biological products: do they have a future? *Biol. Cont.*, 188, 105439. https://doi.org/10.1016/j.biocontrol.2024.105439

Nyangau, P., et al. (2022). Farmers' knowledge and management practices of cereal, legume, and vegetable insect pests, and willingness to pay for biopesticides. *Intern. J. Pest Manag.*, 68(3), 204–216. https://doi.org/10.1080/09670874.2020.1817621

O'Callaghan, M., et al. (2022). Soil microbial inoculants for sustainable agriculture: limitations and opportunities. *Soil Use Manag.*, 38, 1340–1369. https://doi.org/10.1111/sum.12811

Reportlinker (2024). Biopesticides Global Market Report 2023. https://www.reportlinker.com/p06282631/Biopesticides-Global-Market-Report.html

Samada, L.H. and Tambunan, U.S.F. (2020). Biopesticides as promising alternatives to chemical pesticides: a review of their current and future status. *OnLine J. Biol. Sci.*, 20(2), 66–76. https://doi.org/10.3844/ojbsci.2020.66.76

Smagghe, F., et al. (2023). Biological control of arthropod pests in protected cropping by employing entomopathogens: efficiency, production and safety. *Biol. Cont.*, 186, 105337. https://doi.org/10.1016/j.biocontrol.2023.105337

Song, Y., et al. (2022). Development and application of CRISPR-based genetic tools in *Bacillus* species and *Bacillus* phages. *J. Appl. Microbiol.*, 133(4), 2280–2298. https://doi.org/10.1111/jam.15704

Stokstad, E. (2018). France's decade-old effort to slash pesticide use failed. Will a new attempt succeed? *Science*. https://www.science.org/content/article/france-s-decade-old-effort-slash-pesticide-use-failed-will-new-attempt-succeed

Thao, C., et al. (2019). Pesticide knowledge, attitudes, and practices among small-scale hmong farmers in the San Joaquin Valley of California. *J. Integr. Pest Manag.*, 10, 32. https://doi.org/10.1093/jipm/pmz030

Verma, D.K., et al. (2021). Recent trends in plant-and microbe-based biopesticide for sustainable crop production and environmental security. In R. Prasad, et al. (Eds.), *Recent developments in microbial technologies*. Springer, Singapore, pp. 1–37.

Wang, Q. and Coleman, J.J. (2019). Progress and challenges: development and implementation of CRISPR/Cas9 technology in filamentous fungi. *Comput. Struct. Biotechnol. J.*, 17, 761–769. https://doi.org/10.1016/j.csbj.2019.06.007

Zhang, D., et al. (2020). Genome editing with the CRISPR-Cas system: an art, ethics and global regulatory perspective. *Plant Biotech. J.*, 18, 1651–1669. https://doi.org/10.1111/pbi.13383

Zimmermann, G., et al. (1995). Elias Metschnikoff, Elie Metchnikoff or Ilya Ilich Mechnikov (1845–1916): a pioneer in insect pathology, the first describer of the entomopathogenic fungus *Metarhizium anisopliae* and how to translate a Russian name. *Biocon. Sci. Technol.*, 5, 527–530.

Chapter 9

Developments in application technologies for biocontrol agents for insect pest control

Claudia Preininger and Viktoria Stagl, AIT Austrian Institute of Technology, Austria

1 Introduction

The incidence of plant pests and the transmission of plant diseases through insects is increasing at a steady pace, with the promotion of this phenomenon occurring as a result of global warming and fluctuations in climate change. Plant pests and diseases are responsible for an estimated 20–40% of crop losses, with the losses of staple cereal crops (rice, wheat, maize) and tuber crops (potatoes and sweet potatoes) directly impacting food security and nutrition (see https://www.cabidigitallibrary.org/journal/cabicompendium). The majority of countries have identified a priority list of the most serious economically and environmentally important and/or invasive pests, including *Xylella fastidiosa* (Morelli, 2021), the Japanese beetle, the Asian long-horned

http://dx.doi.org/10.19103/AS.2024.0149.11

beetle, citrus greening and citrus black spot (COMMISSION DELEGATED REGULATION (EU) 2019/1702). In order to combat plant pests and ensure the continued production of food for an ever-increasing population, the use of pesticides is on the rise.

Concurrently, the European Commission has proposed a new Regulation on the Sustainable Use of Plant Protection Products, which includes European Union (EU)-wide targets to reduce the use and risk of chemical pesticides by 50% by 2030. The Regulation gives priority to alternative pest control methods in integrative pest management (IPM) practice, including biocontrol methods through various active substances, such as microorganisms, natural enemies and others. Bejarano and Puopolo (2020) list microorganisms approved as active substances in the European Union.

A variety of active ingredients can be used to biocontrol insect pests, with parasites, pathogens and predators being the main groups. Predators (Biondi, 2018; Choudhary, 2022) kill and feed on target pests and include insects such as ladybirds, green lacewings, rove beetles, hoverflies and predatory mites. Parasites, mostly flies and wasps, live and feed in or on the body of the host or parasitise the insect eggs. Insect pathogens (Sabbahi, 2022) are microorganisms including certain fungi (Poveda, 2021), bacteria and viruses. Examples of entomopathogenic fungi are *Beauveria bassiana* and *Metarhizium anisopliae*. Both have an extremely wide host range from flies, thrips to beetles and attack insect hosts in various crops from coffee to beans, cabbage, maize, potatoes and tomatoes by attaching to the cuticular substrate and producing enzymes that break down and penetrate the insect cuticle (Lacey, 2015).

The most widely used bacterial agent for biocontrol of mosquitoes, flies, moth caterpillars and beetle larvae is *Bacillus thuringiensis* (Bt) with several subspecies (Lacey, 2015). However, some insects (e.g. corn earworm, diamondback moth) have developed resistance to Bt toxins (Sabbahi, 2022). Viruses for pest control are centred on baculoviruses, which are used to control major lepidopteran pests. Entomopathogenic nematodes (Beck, 2013; Vashisth, 2013; Shapiro-Ilan, 2006; Morton, 2017) of the genera *Steinernema* and *Heterorhabditis* are also used, moreover botanicals (Choudhary, 2022), pheromones (Biondi, 2018; Choudhary, 2022; Mateos Fernández et al., 2022), peptides (Tran, 2024), nanoparticles (Choudhary, 2022) and RNA/dsRNA (Niu, 2024).

As the characteristics, including the nature, specificity and size of the active ingredients for insect control vary widely, the methods of application in the field also differ for different species. In fact, the efficacy of a biocontrol agent depends on a number of factors, including the activity of the biocontrol agent *per se* and at the site of action, the effective concentration and composition of the formulation and the delivery of the biocontrol agent to achieve the best possible coverage of the target plant or insect. However, coverage is also influenced by environmental conditions, crop type and spray systems used.

This chapter addresses several key topics related to the use of biocontrol agents in foliar spray applications for sustainable pest management. These include the use of predators, nematodes, microorganisms, double-stranded RNA (dsRNA)/RNA interference (RNAi) and botanicals. Additionally, it considers the challenges associated with foliar spray applications, including equipment, spraying parameters, formulation and their impact on treatment efficacy.

This chapter presents an overview of biocontrol agents used in spray applications against insect pests. It describes their working principle and the results obtained in the lab, greenhouse and field trials by giving examples from the literature. It discusses the various factors that impact effective foliar treatment and gives examples on how to optimize them. These factors include the persistence of biocontrol agents, which primarily concerns the stability and efficacy of the agent at the point of action and the protection of the agent from degradation due to heat and UV radiation; secondly, the impact of spraying systems, ranging from handheld sprayers and ground sprayers to unmanned aerial vehicles (UAVs) like drones on surface coverage and its optimization through digital image analysis of water-sensitive papers; thirdly the potential of equipment and adjuvants to develop more targeted foliar spray applications and finally, rainfastness of biocontrol agents including methods for lab testing.

2 Issues in spraying biocontrol agents: predator insects

Predatory insects include beetles, true bugs, lacewings, wasps and dragonflies. They hunt and feed on many pest insects. For example, predatory bugs play an important role in controlling whiteflies, thrips, caterpillars and leaf miner larvae, while the small (~0.3 mm) *Trichogramma* spp. wasps (Kumari, 2022) are used to control lepidopteran (moth) pests such as cornstalk borer, carpenterworm and American plum borer. These wasps have been widely used as parasitized eggs glued onto cardboard cards or as female *Trichogramma* larvae in perforated capsules distributed either manually on the field or by using ground sprayers or airplanes.

Once the first moths lay eggs, the *Trichogramma* larvae must be sprayed. Because of the short life span of *Trichogramma* larvae (only a few days), it is important to apply them at the right time. In this case, aerial spraying by airplane or drone is more time-effective and energy-saving than manual ground spraying. In contrast to the plane, larvae can be positioned quickly (1 ha in just 2 min, Drone dropping of *Trichogramma* against the European corn borer (Skyinnov, accessed on 1 August 2024) with extreme precision by drone, which flies a few meters above the plots.

3 Issues in spraying biocontrol agents: enthomopathogenic nematodes

Nematodes are simple roundworms, colorless, unsegmented and lacking appendages. They can be classified as either free-living, predaceous or parasitic (Vashisth, 2013). The most commonly used entomopathogenic nematodes are members of the genera *Steinernema* and *Heterorhabditis*, which exist in a symbiotic relationship with bacteria of the genus *Xenorhabdus* and *Photorhabdus*. Upon entering the insect, they release the symbiotic bacterium, which produces a toxin that eventually kills the insect larvae. Entomogenic nematodes are highly lethal to many important insect pests while being safe for the surrounding environment. In contrast to other biological control agents, such as microbial agents, the nematodes are highly efficacious, with the larvae being killed within 1 or 2 days. Furthermore, they do not leave residues and are stable in mixtures with a multitude of pesticides. However, entomopathogenic nematodes are susceptible to heat and sunlight/UV radiation, which presents a challenge in their application and necessitates the development of appropriate formulations to maintain nematode stability. To maintain a moist environment, nematodes can be produced within the insects in which they reproduce or entrapped in a moist and porous matrix, such as alginate.

It is recommended that spraying takes place in the evening when the air temperature is below 30°C. The activity of the nematodes is enhanced in cool, wet weather. The successful delivery of the nematodes to the target plant is of critical importance to the success of the treatment. The efficacy of the delivery process is contingent upon the specific crop in question and the type of spraying apparatus employed, that is to say, the ease with which the spray can reach the intended target crop. Brusselman (2012) demonstrated that plants with limited accessibility, such as Savoy cabbage and cauliflower, necessitate modified spray application techniques to ensure uniform deposition on all plant parts (base, underside of leaves and stem). Otherwise, deposition is less than 40% on average. The type of nozzle used had a negligible effect on the extent of coverage achieved. However, the deployment of an air support system or a row application system led to an enhancement in the deposition of nematodes at the base of savoy cabbage.

Furthermore, while entomopathogenic nematodes can be applied with a wide range of spray systems, a few key considerations must be taken into account. Operating pressure limits for different nematode species vary considerably and can be as low as 1380 kPa for *Heterorhabditis megidis* (Shapiro-Ilan, 2006). Indeed, the spray system may impact nematode viability. Several online sources that provide guidance on integrating nematodes into an insect control programme recommend pump pressures below 300 psi (2068×10^3 kPa), nozzle orifices greater than 0.5 mm, and filters greater than 50 mesh.

4 Issues in spraying biocontrol agents: microorganisms

The number of different genera of microorganisms used as biocontrol agents is rather small. Of the 100 microorganisms currently approved and pending as active substances in the European Union (EU Pesticides Database – Active substances (europa.eu), accessed on 29 July 2024), most bacteria are *Bacillus* strains, mainly *Bacillus amyloliquefaciens* and *B. thuringiensis*, while the majority of fungal strains belong to the *B. bassiana*, *Trichoderma* and *Metarhizium* strain families. Very few viruses (granulovirus, pepino mosaic virus) have been approved for biocontrol purposes. Table 1 gives examples of microbial spray applications against insect pests in F(ield), G(reen house) or L(ab).

Bacteria from the genus *Bacillus* are capable of forming spores which protect them against different abiotic and biotic stresses, they form biofilm by becoming aggregated and attached to a surface and can tolerate long storage periods (Fessia et al., 2022).

Bacillus thuringiensis Berliner (Bt) is a Gram-positive, aerobic, spore-forming and endotoxin-producing bacterium. When the bacterium produces spores, it also produces unique crystalline proteins that are toxic to certain insects when ingested, killing them within a few days. Bt has been reported for the treatment of grubs (Xin, 2018), *Helicoverpa armigera* (Marzban, 2021; Malinga, 2022; Vimala Devi, 2015), bagworm *Metisa plana* (Masri, 2022) and cotton bollworm and corn borer larvae (Li, 2023) amongst others.

Bt is sensitive to sunlight and heat and therefore has limited persistence on leaves (up to a week). As a result, several spray cycles may be required to achieve the desired level of control. In fact, in a study by Malinga and Laing (2022) weekly spray applications of the biopesticides Eco-Bb® (*B. bassiana*) (2×10^9 spores/g), Bb endophyte (*B. bassiana*) (2×10^9 spores/g), Bolldex® (Nucleopolyhedrovirus) (7.5×10^{12} spores/g) and Delfin® (*B. thuringiensis*) (32 000 IU/mg) had to be applied against the African bollworm (*H. armigera*) up to 23 weeks after planting. High doses were used in the field to achieve good efficacy. Bb strains are highly temperature sensitive, and as a general recommendation, microbial sprays need to be applied in the late afternoon/evening under calm and humid conditions (Wraight et al., 2021).

Beauveria bassiana is a fungus that is employed to control the infestation of crops by aphids, thrips and whitefly. It is also effective against subterranean termite pests (Ambele et al., 2020). The conidia of this fungus are applied to plants as an emulsion or a wettable powder. *Metarhizium anisoplae* was reported as biocontrol agent against *Tuta absoluta* on tomato (Kabaale et al., 2022), for the biocontrol of locusts (Matthews, 2021), against subterranean termite pests (Ambele et al., 2020), to control *Phyllotreta striolata* in Chinese Flowering Cabbage (Chen et al., 2023).

Table 1 Figures of merit for microbial sprays against insect pests

Target plant	Pathogen/pest	Type of formulation	F(ield), G(reen house) or L(ab)	Spraying equipment	Application/ plant stage	Volume	Concentration/ amount of active agent	Adjuvants	References
				Beauveria bassiana					
Bean	*Bemisia* whitefly	S	G	Dual-action gravity-feed artist's airbrush	One application	0.4 mL per leaf	5×10^6 conidia/mL	Rhamnolipid	do Nascimento Silva et al. (2019)
Cucumber, tomato, eggplant, green bean	Two spotted spider mite *Tetranychus urticae*	WP	G	Hand-held sprayer	Twice at an interval of 1 week	Sprayed to run-off	$1.05–4.2 \times 10^6$ conidia/mL	Mineral oil (paraffin), Break-thru	Gatarayiha et al. (2010)
Eggplant	Two spotted spider mite *Tetranychus urticae*	WP	F	Hand-held sprayer	Intervals of 1–4 weeks	Sprayed to run-off	$0.7–1.6 \times 10^{12}$ conidia/ha	Break-thru	Gatarayiha et al. (2011)
Onion	Onion thrips	OBF	F	Aspee sprayers, knapsack sprayer, controlled droplet applicator	One or two applications	N/A	10^8 spores/mL	Paraffin oil, other adjuvants	Murugasridevi et al. (2022)
Chili, tomato	Chilli thrips and tomato thrips	OBF	F	Aspee sparyers, knapsack sprayer, controlled droplet applicator	One to three applications	N/A	10^8 spores/mL	N/A	Parveen et al. (2020)

Grass	*Dichroplus maculipennis*	S	F (cages)	Manual 2-l sprinkler	N/A	N/A	1 × 10^8 conidia/mL	Canola oil, wheat bran	Pelizza et al. (2019)
Potato	Colorado potato beetle *Leptinotarsa decemlineata*	OBF, WP	F	Backpack, single-row, CO_2 –powered, hydraulic sprayer	Two to four applications	280 L/ha	10^{13} conidia/ha	N/A	Wraight and Ramos (2002)
Coffee	Coffee berry borer *Hypothenemus hampei*	OBF	F	Backpack sprayer or tractor-powered, air-blast orchard sprayers	Up to nine spray applications, 1–4 weeks interval	2.34 L/ha suspension (in 300–935 L/ha water)	N/A	Surfactant	Wraight et al. (2021)
Tomato	Aphid *Aphis gossypii*	OBF, TF, CF	F	Aspee sprayers, knapsack sprayer, controlled droplet applicator	Two applications in 14 days interval	N/A	10^8 spores/mL	N/A	Prithiva et al. (2019)
Bhindi plants	Two spotted spider mite *Tetranychus urticae*	WP	G	Hand sprayer	Two applications in 14 days interval	N/A	1 × 10^8 conidia/mL	Tween 80, talc, CMC, calcium carbonate, citric acid	Abarna et al. (2022)
Sugar beet, soybean	*Helicoverpa armigera*	OBF, WP	G	N/A	N/A	N/A	2.1 × 10^4 spores/mL	Talc, gycerol, gum, corn oil, bentonite	Ritu et al. (2012)
Tomato	Whitefly *Bemisia tabaci*	OBF, TF, CF	F	Hand sprayer	Two applications in 14 days interval	N/A	10^8 conidia/mL	N/A	Prithiva et al. (2017)

(Continued)

Table 1 (*Continued*)

Target plant	Pathogen/pest	Type of formulation	F(ield), G(reen house) or L(ab)	Spraying equipment	Application/ plant stage	Volume	Concentration/ amount of active agent	Adjuvants	References
Corn	Cotton bollworm and corn borer larvae	WP	F	Electric sprayer	One application	N/A	6000 g/ha	N/A	Li et al. (2023)
Cabbage	Peach-potato, cabbage and currant-lettuce aphid	S	L, F	N/A	Two applications with 7 days interval	N/A	2.44 mL/L (Botanigard ES)	Codacide and Addit	Prince and Chandler (2020)
Cotton	*Helicoverpa armigera*	WP	F	Knapsack sprayer	Weekly applications for 10 weeks	N/A	6×10^{11} spores/ha	N/A	Malinga and Laing (2022)
Cocoa	Termites *Odontotermes spp.*	S	G	Hand sprayer	One application	25 mL per plant	1×10^{8} conidia/mL	Triton X-100	Ambele et al. (2020)
Tomato	Tomato leafminer	S	L	Hand-held plastic sprayer	One application	N/A	10^{8} conidia/mL	N/A	Zheng et al. (2023)
				Bacillus thuringiensis					
Sunflower	*Helicoverpa armigera*	S	F	High volume knapsack sprayer	Star bud stage	N/A	N/A	Boric acid	Vimala Devi and Vineela (2015)
Oil palm	Bagworm *Metisa plana*	S	F	JMR V1650 drone: VP110 nozzle, 0.2 Mpa, 2.8 m/s permanent speed in 2 m height	Two applications	1.5 L/ha Bt in 30 L/ha	N/A	N/A	Masri et al. (2022)

Corn	Cotton bollworm and corn borer larvae	S	F	Electric sprayer	One application	N/A	900 g/ha	N/A	Li et al. (2023)
Cotton	*Helicoverpa armigera*	WP	F	Knapsack sprayer	Weekly applications for 10 weeks	N/A	3×10^{10} IU/ha	N/A	Malinga and Laing (2022)
				Metarhizium acridum, anisopliae, brunneum					
N/A	Desert locust *Schistocerca gregaria*	N/A	F	DJI Agras T16 drone	N/A	4.8 L/min	N/A	N/A	Matthews (2021)
N/A	Desert locust *Schistocerca gregaria*	OBF	F (cages)	Drone: large micro-light aircraft fitted with Micronair AU7000 atomizers	N/A	1–2.5 L/ha	2.0×10^{12} conidia/ha	50% Ondina EL and 50% 'Jet Al'	Price et al. (1997)
N/A	Desert locust *Schistocerca gregaria*	N/A	F (cages)	DJI Agras T20 drone fitted with an ultra-low volume atomizer	N/A	N/A	500 g Novacrid® in 20 l	Diesel	Ochieng' et al. (2023)
Chinese flowering cabbage	Cabbage flea beetle *Phyllotreta striolata*	WP	G, F	N/A	N/A	1 L/plot (20 m^2)	400 mg wettable powder/L	Chlorfenapyr	Chen et al. (2023)
Tomato	*Tuta absoluta*	CP, OBF	F	Hand-operated knapsack sprayers	weekly interval	400 mL/ha, applied in 20 L per 10 mL	1×10^9 conidia/mL	Nutrient agar, glycerine, molasses, canola oil	Kabaale et al. (2022)

(Continued)

Table 1 (*Continued*)

Target plant	Pathogen/pest	Type of formulation	F(ield), G(reen house) or L(ab)	Spraying equipment	Application/ plant stage	Volume	Concentration/ amount of active agent	Adjuvants	References
Cocoa	Termites *Odontotermes spp.*	S	G	Hand sprayer	One application	25 mL per plant	1 × 10^8 conidia/mL	Triton X-100	Ambele et al. (2020)
Tomato	Tomato leafminer	S	L	Hand-held plastic sprayer	One application	N/A	10^8 conidia/mL	N/A	Zheng et al. (2023)
					Mixtures				
Castor, sunflower	Lepidoperan pests	S	F	Knapsack sprayer	One application	500 L/ha	N/A	Tween 80, Span 20, oleic acid	Vimala Devi and Vineela (2015)
Soybean	*Anticarsia gemmatalis, Chrysodeixis includens*	OBF	L, F	CO_2-powered backpack sprayer	One application	200 L/ha of 1% formulation in water	10^9 conidia or OB/mL, 10^{11}–10^{12} OB or conidia/ha	Sunflower oil and emulsifying agents	Lopes et al. (2020)
Maize	*Spodoptera frugiperda*	OBF	L, G, F	Tractor-mounted bar sprayer, CO_2-powered backpack sprayer	One application	60–300 L/ha	10^8–10^{12} OB or conidia/ha	Sunflower oil and emulsifying agents	Faria et al. (2022)
					Virus				
Cotton	*Helicoverpa armigera*	S	F	Knapsack sprayer	Weekly applications for 10 weeks	N/A	1.5 × 10^{12} OB/ha	N/A	Malinga and Laing (2022)

				Cydia pomonella					
Apple	Codling moth	WP	F	Motorized backpack airblast sprayer	Three or four applications with 14 days interval	935 l/ha	6.57 × 10^{12} OB/ha or 2.2 × 10^{12} OB/ha	Lignin, TiO_2, sugar	Arthurs et al. (2006)
				Pseudozyma flocculosa					
Strawberry	Two-spotted spider mite *Tetranychus urticae*	WP	G	Mini manual sprayer	One application	N/A	WP diluted 50-, 100-, and 200-fold;	Talc, SDS, hydrated silica, diatomaceous earth, KH_2PO_4	Zhang et al. (2016)
				Cordyceps javanica					
Bean	*Bemisia* whitefly	S	G	Artist's airbrush	One application	0.4 mL per leaf	5 × 10^6 conidia/mL	Rhamnolipid	Do Nascimento Silva et al. (2019)

S, suspension; WP, wettable powder; CP, conidia powder; OBF, oil-based formulation; TF, talc formulation; CF, crude formulation; G, granulate; MC, microcapsules; SRF, slow release formulation; H, hydrogel; SOL, solution; EM, emulsion.

Zheng et al. (2023) showed for the five entomopathogenic fungal isolates *Metarhizium flavoviride* (root dipping), *M. anisopliae* (seed coating), *M. rileyi* (root dipping), *Cordyceps fumosorosea* (seed coating) and *B. bassiana* (foliar spray) that the most effective application method for control of tomato leafminer *Phthorimaea absoluta* depends on the type of fungal species (Zheng et al., 2023).

Field efficacy is also influenced by microbial formulation. Prithiva et al. (2019) showed that the use of an oil dispersion (OD) of *B. bassiana* against the aphid *Aphis gossypii* (Aphididae: Hemiptera) in tomato was superior to crude and talc-based formulations, with an average aphid population reduction of 45.50% compared to 34.04% and 25.22%, respectively. The superior performance of OD was attributed to better suspension of dry conidia in the oil and better distribution on the leaves, facilitating conidia contact and adhesion to the insect cuticle and subsequent insect infection. The importance of emulsifiable formulations was also highlighted by Inglis et al. (2000) who studied the effect of artificial rain on the persistence of *B. bassiana* on both leaves and the pest insect (Colorado potato beetle). The choice of oil may also contribute to increased larval mortality, as reported for *B. bassiana* and *M. anisopliae* in the control of *Spodoptera litura* larvae (Ummidi and Vadlamani, 2014). As suggested by a number of authors (Pelizza et al., 2019; Prithiva et al., 2017; Ritu et al., 2012; Kaiser, 2019), formulated *B. bassiana* was more effective than unformulated crude conidia, for example against the grasshopper pest *Dichroplus maculipennis*, larvae of *H. armigera* and *Bemisia tabaci* on tomato.

Microbe dosage plays a critical role in the treatment success. It was found that the effective concentration in the field was generally higher than in the lab and/or green house. Frequently, infection rates are dose-dependent as reported for *B. bassiana* against larvae of *H. armigera* (Ritu et al., 2012).

Efficacy of microbial pesticides may also be enhanced, if used in combination with chemical additives, for example, Bt with boric acid (Opisa, 2020) or with other active ingredients as reported by Lopes et al. (2020), for *Metarhizium rileyi* and nucleopolyhedrovirus (NPV) in soybean fields infested by both *A. gemmatalis* and *C. includens*. The combined use of *M. rileyi* and NPV resulted in 50.8% of overall mortality 6 days after spraying, whereas for treatments with the fungus or the virus alone the percentages were 41.2% and 8.8%, respectively. This is in contrast to the findings of Faria et al. (2022), who observed no improvement in the efficacy of control measures for the fall armyworm in maize. Furthermore, the combination of NPV and the fungus *M. rileyi* did not result in increased larval mortality in the field. The co-formulation of *B. bassiana* and colza oil resulted in significantly higher mortality of pollen beetles (*Brassicogethes* spp.) featuring a strong synergistic effect (Kaiser et al., 2019). Indeed, the deployment of two or more microbial species, which must necessarily be compatible with one another, can enhance the efficacy of biological control through a variety of mechanisms, e.g. competition, production of toxins, volatile compounds or secondary metabolites (Sindhu et al., 2017) and interactions with the targeted pest. In a recent publication, Nunes et al. (2024) presented a list of microbial consortia that demonstrated superior efficacy in controlling insect pests compared to bioagents when deployed individually. The majority of these consortia were found to be combinations of *B. bassiana*. The combination of microbials with diverse modes of action may also prove an effective method of slowing down the process of resistance development, as has been suggested for *B. thuringiensis* (Bt) (Vimala Devi, 2020).

Baculoviruses, including nucleopolyhedroviruses (NPVs) and granuloviruses (GVs), are naturally occurring living organisms that act in a species-specific manner without affecting non-target organisms. NPVs and GVs differ in the viral protein that constitutes the crystalline matrix of the occlusion bodies (OBs). This protein is known as polyhedrin for NPVs and granulin for GVs. They are large, circular double-stranded DNA viruses with a genome size ranging from 80 to 180 kilobases, which infect the larval stages of mainly lepidopteran species, e.g. the fall armyworm (FAW), Spodoptera frugiperda (Hussain et al., 2021). To ensure the optimal timing of baculovirus spraying, the use of monitoring traps, weather stations, or computer modelling is crucial. The stability of baculoviruses in a field environment, particularly with regard to protection from UV radiation, may be enhanced through encapsulation. For example, Wilson et al. (2020) employed the use of an ENTOSTAT wax combined with minute quantities of the UV absorber TiO_2 for the encapsulation of the virus. The TiO_2 was not simply added to the NPV suspension; rather, it was closely bound to the wax-encapsulated OB. This allows for the use of only low quantities of TiO_2 – one hundred times less than would be required in a suspension. Behle et al. (2010) employed lignin as a UV protectant, either incorporated as an adjuvant or spray-dried to encapsulate the baculovirus. The researchers demonstrated on field-grown cabbage plants that the original insecticidal activity remaining (OAR) was more than three times higher for virus formulations with lignin than without.

5 Issues in spraying biocontrol agents: RNA interference/double-stranded RNA interference

RNA interference (RNAi) has been extensively researched in insects to develop effective strategies for pest control. RNAi is a form of gene silencing that is triggered by dsRNA (Niu et al., 2024). The dsRNA can be expressed directly in transgenic plants, allowing insects to uptake it while feeding on the plant. Alternatively, the dsRNA can be expressed in microbes or chemically synthesized and applied through spraying at high concentrations to the plant. In general, the use of dsRNA offers several advantages over conventional insecticides.

One such advantage is the ability to control species specificity by targeting gene regions that vary between species (Mateos Fernández et al., 2022).

However, up to date only a few stable gene transformation protocols are available and successful spray applications are compromised by the rapid degradation of RNA in the environment (Mann et al., 2023). This lability due to nuclease attack in the leaves, the low pH of many plant leaf surfaces and UV radiation (Lucena-Leandro et al., 2022) makes it difficult to provide a sufficiently high concentration on the plant surface (where the insects feed) to effectively combat the insect pest.

dsRNA delivery and stability on the plant is a significant challenge that is being addressed through different formulation approaches, such as the use of macromolecular polymers like chitosan (Lucena-Leandro et al., 2022) or nanomaterials through RNA linkage and complexation (Arjunan, 2024; Christiaens et al., 2020). Once delivered to the plant, dsRNA must enter the plant system, which may be impeded by leaf morphology and the leaf's physico-chemical properties (e.g. hydrophobicity), as well as the permeability of cuticles or the absence of stomata (Hoang, 2022). In their 2021 review, Rank and Koch summarize the findings of various plant studies that have employed RNA sprays. Notwithstanding, the successful foliar application of exogenous dsRNAs or siRNAs has been documented for *B. cinerea*, *F. graminearum*, *S. sclerotiorum*, *F. asiaticum*, and more recently against *Fusarium oxysporum*, *Phakopsora pachyrhizi*, *Aspergillus niger*, *Rhizoctonia solani* and *Austropuccinia psidii* (Mann, 2023).

6 Issues in spraying biocontrol agents: botanicals

Botanical pesticides are derived from plants and act on insect pests either by contact (e.g. pyrethrins) or ingestion (e.g. neem oil). As with other biological agents, botanical pesticides are sensitive to sunlight, moisture and air, requiring precise timing and a higher frequency of application. Indeed, Kobenan (2022) applied *Ocimum gratissimum* and *Cymbopogon citratus* essential oils to cotton plants every 2 weeks from day 45 to 115 after the emergence of the whitefly (*B. tabaci*) and jassid (*J. fascialis*). The quality of botanicals can vary from batch to batch, and as broad-spectrum insecticides, they will also kill non-target insects. A number of essential oils have been subjected to investigation with regard to their potential insecticidal properties: *O. gratissimum* and *C. citratus* have been demonstrated to be effective against whitefly (*B. tabaci*) and jassid (*J. fascialis*) in cotton (Kobenan et al., 2022) and neem leaf extract in eggplant (Abubakar, 2023). Orange wax (OW) extracts containing terpineol, limonene and eremophilene have been shown to be effective against *F. oxysporum* and *Spodoptera frigiperda* larvae – spraying OW extract at 50 mg/mL for 10 s caused the highest larval mortality (García-Cruz, 2024). A combination of *Rosmarinus officinalis* extract and auxiliary agents has been demonstrated to effectively control aphids in cucumber greenhouses (Homayoonzadeh, 2022). Linalool sprays have been shown to be effective against two species of thrips (Gharbi and Tay, 2022). Safflower and cottonseed oils have been identified as effective treatments for the spider mite *Tetranychus urticae* Koch (Acari: Tetranychidae) (Takeda et al., 2020). Selected botanical oils (citronella (*Cymbopogon nardus* (L.) Rendle), fennel (*Foeniculum vulgare* Mill.), garlic (*Allium sativum* L.) lemon grass, patchouli and peppermint (*Mentha X piperita* L., geranium (*Pelargonium* spp.), neem, patchouli and thyme were tested against the cassava whitefly (Amour et al., 2023) being neem oil the most effective in reducing the number of eggs, nymphs and adults, while patchouli was identified as the most effective repellent for whiteflies. Additionally, nano-emulsions of essential oils derived from mandarin, marigold and camphor exhibited promising results in controlling leaf miner (*T. absoluta*) larvae in tomato (Marouf and Harras, 2022).

The concentration of essential oil must be low (approximately 1%) and adjusted to avoid phytotoxicity to the plants. Phytotoxicity can be caused by disturbances in osmoregulation, mitochondrial respiration or the generation of reactive oxygen species.

7 Challenges in foliar spraying: persistence of spray on plants and insects

In comparison to chemical pesticides, biocontrol agents demonstrate superior sustainability, with no residual contamination following their use and a minimal risk of insect resistance. However, the rate of killing is relatively slow, and the agents are more susceptible to environmental conditions (temperature, humidity, sunlight) that may impair their efficacy. To circumvent the detrimental effects of elevated temperatures and UV radiation, the application of pesticides to crops is typically conducted in the late afternoon or evening. Nevertheless, the efficacy of the agent may be enhanced by the addition of adjuvants (Holka, 2023), which protect the active ingredient from UV damage, regardless of wind and rain. The optimal selection of adjuvants is contingent upon a number of factors, including the physico-chemical properties of the active ingredient (a.i.), the mode of action of the a.i., the targeted crop and pathogen, the formulation type and the interaction with other co-formulants. It is not possible to identify a single adjuvant that fulfils all the criteria, and the choice of adjuvant is always made on a case-by-case basis (Holka, 2023).

The application of formulation technology serves to enhance the biological efficacy of the active ingredient, while simultaneously facilitating its handling and application. The use of adjuvants, such as humectants, wetters, UV protectants and stabilizers, is beneficial in maintaining microbial stability and creating a protective environment (Preininger et al., 2018; Berninger, 2018). In order to protect delicate active ingredients from sunlight and UV radiation, carrier particles such as wheat bran, talc and bentonite (Abarna, 2022; Nithya, 2017; Ritu, 2012) are often incorporated into the formulation; furthermore, UV absorbers such as Congo red and lignin (Arthurs et al., 2006; Tamez-Guerra et al., 2000). Additionally, modified soybean or sunflower oil (Behle, 2010; Nithya, 2017), graphene oxide (Maghsoudi, 2017) and encapsulated active ingredients, such as polydopamine-nucleopolyhedrovirus microcapsules (Yan, 2022), Bt in alginate with hollow glass beads (Zhang, 2016) or baculovirus in wax/TiO2 (Wilson, 2020) or lignin (Behle, 2010) have been explored. For an overview of solar protectants, see Tamez-Guerra et al. (2005), for encapsulation of biocontrol agents see Chapter 7.

8 Challenges in foliar spraying: spraying systems and their capabilities

The type of equipment (Gan-Mor, 2014) and spraying parameters used with biocontrol agents are critical to treatment performance. It is obvious that ground and aerial sprayers, which apply pesticides from greater heights and at higher speeds, have different requirements, and that hand-held knapsack sprayers, although cheap and easy to use, are significantly more prone to inconsistent spray patterns and spray drift than tractor-towed or larger mechanical equipment. This is due to pressure fluctuations and variable flow rates as a result of the lack of pressure control. On the other hand, the use of a knapsack sprayer allows for efficient spot treatment of infested fields as opposed to mechanical equipment that requires treatment of the entire field. The use of UAVs as agricultural sprayers is an emerging technological trend that offers numerous advantages, particularly in challenging terrain. It enables more precise treatment and is less labour-intensive.

Indeed, there are diverse types of sprayers with varying features and technological capabilities, different active ingredients and formulations, and diverse crops, leaf morphologies, and growth stages. To optimize biocontrol efficacy, it is essential to examine the interactions between the various components and parameters.

Six types of sprayers, including a power sprayer, a battery sprayer, two hand sprayers, a controlled droplet applicator (CDA) and an ultra-low volume (ULV) sprayer, were tested using an oil-based formulation of *B. bassiana*. The CDA sprayer was the most effective sprayer in reducing *A. gossypii* in tomato by 37–46% compared to the control in field trials, slightly better than the ULV sprayer (Prithiva, 2019). Similar results have been reported for tomato thrips (Parveen, 2020) and onion thrips (Murugasridevi, 2022).

Treatment efficacy can be affected not only by the type of sprayer but also by the spraying conditions as shown by Grella (2023) for commercial *Trichoderma harzianum* (Trianum-P, Koppert) and *B. amyloliquefaciens* (Serenade ASO, Bayer) using two different airblast sprayers. Mechanical and thermal stress assessment by simulating spray application conditions showed that sprayer type, initial spray mixture temperature and temperature increase during the trials significantly reduced microbial activity (CFU) (Grella, 2023).

Attention should be paid to the mechanical effect of pump recirculation on the temperature rise of the spray mixture, and thus the mechanical and temperature stress on the biocontrol agents, as shown for *T. harzianum* T 22 and *B. amyloliquefaciens* QST713 (Grella, 2023). Mechanical stress due to different nozzle sizes, at different operating pressures and pumping times for *B. thuringiensis* showed that its viability decreased with increasing operating pressure, decreasing nozzle size and increasing pumping time. The study recommended that the application of biopesticides may be delivered through a nozzle with a larger orifice size (3.56 mm) at low operating pressure (145 kPa) for getting maximum viability of the biopesticide (Kumar, 2023).

It is also crucial to agitate the spray tank to prevent particulate matter from settling at the bottom of the tank and to guarantee a homogeneous spray mixture, thereby ensuring that the same effective dose is delivered throughout the spraying process. Alternatively, or in addition, a dispersing agent, such as xanthan or CMC, may be incorporated. For example, Gauthier (2019) utilized a combination of guar gum and xanthan to disperse pupae in the spray tank of a field prototype boom sprayer and facilitate their adhesion to maize plants.

Droplet size and droplet size distribution impact plant- and insect coverage, furthermore, spray drift which all are affected by the spray parameters (e.g. flow rate, pressure) and choice of equipment tools (e.g. nozzles). (Droplet size and coverage are also impacted by plant morphology, stage of plant growth and leaf area index). The main factors affecting droplet size distribution using aerial spraying in combination with microbial biocontrol agents (Bt, *B. bassiana*) were rotation speed > nozzle diameter > media > flow rate (Ru, 2020). In that work, the rotational speed correlated negatively with the retention rate of the biocontrol agents, while the flow rate and nozzle diameter had no significant effect on the activity retention rate of the two biocontrol agents. According to the authors, this is due to the fact that the decrease in nozzle diameter or the increase in rotational speed leads to the instantaneous pressure and cutting force of the biological pesticides in vivo, resulting in the destruction of the bacterial structure and the decrease in germination rate. For best performance, the authors suggest a working speed between 3000 rpm and 5000 rpm.

9 Challenges in foliar spraying: more targeted foliar application

The delivery of the biopesticide spray is of critical importance with regard to the efficacy of the treatment applied. In fact, Warneke (2021) reported around 40% chemical wastage and 15% drift of the applied spray volume when using a radial air blast sprayer. While biocontrol agents are rarely harmful to non-target organisms and the environment, to avoid wasting spray volume and to be cost- and energy-efficient, the foliar spray should reach the target crop 100%. It is imperative that the target area be adequately covered in order to achieve the desired efficacy.

The rate of uptake depends on the type of crop and adjuvants. Additionally, the application conditions and the bioactive's intrinsic properties, such as hydrophobicity, as well as its mode of action, influence the rate of uptake. Jin (2013) found that conidial density had a significant effect on droplet size distribution and deposition coverage, probably because of conidial surface hydrophobicity which is different for different strains. Better adhesion by film formation (important for contact biologicals) or stomatal leaf penetration (important for systemic biologicals) reduces spray volumes and spraying frequency. This also helps to reduce the waste of the sprayed biopesticides while at the same time increases treatment efficacy.

Spray coverage can be estimated through the counting of droplets per unit area on water-sensitive paper, which, due to the incorporation of an indicator (bromophenol blue), exhibits a blue color in the presence of a liquid. The water-sensitive papers are positioned between plant rows and underplant leaves. Software for the real-time assessment of droplets deposited on water-sensitive papers coupled with image analysis software has been developed. Examples of such software include freely available smartphone apps, like SnapCard (Nansen, 2015), Depositscan or Drop Leaf (Brandoli, 2021). Snapcard can be used to optimize spray coverage, reduce spray drift and minimize the risk of resistance development in target pest populations. Additionally, Drop Leaf provides information necessary for calibrating system pressure, flow rate and travel speeds for sprayers. Alternatively, Image J can be employed to calculate counts, density, and coverage. Our experience indicates that it produces comparable values to those obtained with Depositscan. The values obtained are contingent upon the measurement method. For instance, droplets can be fitted to thresholds with Depositscan, but this is not feasible with Dropleaf.

Droplet deposits are a measure for the treatment success and help to optimize spraying parameters, for example flight velocity and spraying height in case of UAVs (drones) (Masri, 2022).

Reasons for drift are manifold, but most important is the weather including wind speed and direction, temperature and relative humidity. It impacts the effectiveness of spray applications and the likelihood of wastage through run-off and drift. Spraying is recommended for the cooler time of day (morning or evening) at consistent and low wind to prevent transport of smaller drops (<200 μm), moderate temperature to prevent thermal drift, and medium air humidity (above 50%) to prevent water evaporation causing increased number of fine drops prone to drift. Drift should be minimized by not spraying into the wind (especially important for direct spraying, e.g. air blast spraying) and using proper application technique (precision sprayers), equipment (nozzle tips with a larger orifice) and device settings (low spray pressure, low boom height). For optimum spray and treatment performance, online spray advisory forecasts based on prediction decision models or a fixed or handheld weather station are recommended.

The use of adjuvants (Wang, 2024) allows for the control of the spray droplet spectrum, the reduction of drift and the enhancement of the adhesion and retention of spray droplets on plants. Additionally, adjuvants can facilitate the coverage of active ingredients on plants and support their penetration into the plant. Adjuvants such as rapeseed oil may also contribute to the reduction of drift.

Spray booms may be adapted to the shape of the crop simply by changing the distances between the nozzles suffices. It was shown that an automated tunnel sprayer equipped with vertical spray booms in a vertical crop like bay laurel could increase the uniformity of spray distribution as compared with the traditional spray lances and knapsack sprayers.

Good coverage of the underside of cauliflower leaves with entomopathogenic nematodes (*Steinernema carpocapsae* Weiser) against *Mamestra brassicae* was achieved using different boom configurations with vertical extensions carrying underleaf spray nozzles (Brusselmann et al., 2012). In a field experiment, it was shown that cabbage head damage was effectively reduced (Beck, 2013). In a separate study (Faria et al., 2022), it was demonstrated that microorganisms, regardless of formulation, were ineffective when sprayed with bar or CO_2 backpack sprayers with fixed spray nozzles oriented horizontally. However, when a manual sprayer was employed, directing the bioactive solution solely to the maize whorls, successful results were achieved.

In case of helicopter or drone spraying the spray height strongly affects droplet density and treatment efficacy. For example, Ochieng' (2023) has demonstrated that spraying *Metarhizium acridum* on desert locusts using a drone at any height below 10 m may lead to over-deposition of the pesticide, while heights above 10 m may lead to under-application, which may limit exposure of the locusts to *Metarhizium spores* or pesticide molecules.

10 Challenges in foliar spraying: rainfastness

Biocontrol agents represent sustainable alternatives to complement or substitute chemical pesticides. However, they frequently suffer from wash-off by rain, which has a detrimental impact on treatment efficacy. This occurs either because the bioactive is completely washed off or diluted to a concentration that is less effective, or because it is redistributed to surfaces that are not the intended target. Rainfastness describes the ability of a bioactive to resist wash-off by rain. A pesticide is considered rainfast after application if it has adequately dried or has been absorbed by plant tissues so that it will still be effective after rainfall or irrigation (Efficacy guideline 119, 2021).

Rainfastness is highly complex and variable and depends on the type, quantity and intensity of the rain, the physical and chemical characteristics of the biocontrol agent, dose, formulation, temperature and relative humidity, and leaf surface morphology. Leaf morphologies which are directly related to surface structure, roughness and wettability exhibit considerable variation not only between different plant species but also between different growth stages of the same plant (Efficacy guideline 119, 2021). Leaf surface characteristics clearly affect spray deposition, as shown for two rice cultivars, one with a smooth and one with a rough, hairy leaf surface (Ji, 2021). Spray deposition per unit leaf area was higher for the hairy surface than for the smooth surface. However, there were more small droplets on the smooth surface and larger droplets on the hairy, rough surface. To increase the deposition of biocontrol droplets on the smooth surface, nozzles with lower spray pressure and higher flow rates were recommended. Proper wetting of the leaf surface is usually achieved by the addition of non-ionic surfactants, which are evaluated in spreadability tests of formulations on parafilm (Fig. 1). The performance of a pesticide in terms of rainfastness can be enhanced through the addition of so-called sticker adjuvants, which serve to increase the retention and adhesion of the active ingredient to the target surface. These are of particular importance for contact pesticides. Typical rainfastness adjuvants that prevent the spray from bouncing or rolling off and promote adhesion to the leaf include block co-polymers like polyether trisiloxanes (Gatarayiha, 2010), waxes, celluloses, oils (Ummidi and Vadiamani, 2014; Kabaale, 2022), poly(vinyl alcohol) (Symonds et al., 2016b), chitosan (Symonds, 2016a), xanthan gum (Berninger, 2021), rhamnolipid (do Nascimento Silva, 2019), Triton X-100 (Homayoonzadeh, 2022), gelatin and gum Arabic (Xin, 2018), Tween (Vimala Devi, 2015) to name a view. It was demonstrated that the rainfastness of PVAs increases with increasing molecular weight. A linear correlation between the rainfastness of PVA and its molecular weight was observed when the threshold molecular weight of 33–52 kDa was exceeded (Symonds, 2016b). The addition of oils can render water-soluble biocontrol agents more hydrophobic, forming a water-resistant barrier layer. Surfactants may be regarded as rainfastness adjuvants, given that

Figure 1 Left: droplets on a hairy potato leaf surface. Right: testing different surfactants to improve spreadability of the spray.

in conjunction with systemic bioactives, they facilitate enhanced penetration into the leaves prior to any precipitation. Moreover, spreading agents ensure rapid drying, thereby contributing to rainfastness. It is essential to maintain a low surfactant concentration, as excessive levels may result in phytotoxicity due to the up concentration of the surfactant upon evaporation following spraying. Additionally, high concentrations may lead to runoff of the spray.

The type of formulation was also demonstrated to be a significant factor. In conditions of precipitation, the foliar treatment of the oil formulation of *B. bassiana* was found to be considerably more effective than the respective wettable powder (WP) formulation (65% vs 8% control), as reported by Wraight (2002).

A number of techniques has been devised to assess the rainfastness of adjuvants in biological formulations. These employ parafilm or microscopic slides as an artificial plant surface, with water from tips (Symonds et al., 2016b) or spray/rain towers (Symonds et al., 2016b) as an artificial rain source. The deposition and subsequent drying of formulation droplets on these surfaces, followed by exposure to an artificial rain event, provides the basis for calculating the retention of the deposited bioactive. This is achieved through the use of imaging technology and digital software analysis. While these methods offer a valuable approach to adjuvant screening, they are not a substitute for testing on plants. This is particularly relevant in cases where plants are difficult to wet or reach, due to factors such as plant geometry, insect attack, or the presence of epicuticular wax structures and low surface areas.

11 Future trends

The field of pest management will undergo a significant transformation, evolving into a multifaceted approach that integrates a wide range of biocontrol agents and cutting-edge data-driven technologies, such as IPM packages consisting of several different types of biocontrol agents, e.g. *B. thuringiensis* var. kurstaki, *B. bassiana*, neem and chlorantraniliprole, combined with the installation of pheromone traps, as reported by Buragohain (2021), or the development of microbial consortia (Nunes, 2024), which are considered more versatile and efficient than individual strains due to their different modes of action. Unmanned aerial sprayers, equipped with real-time sensors, will allow for a reduction in spray volumes while allowing for faster and more site-specific pest control. Technologies used in precision agriculture are constantly evolving and include the Internet of Things (IoT), big data analytics, artificial intelligence (AI) and machine learning. In addition, high-resolution satellite imagery and remote sensing are being utilized to facilitate informed management decisions (https://ec.europa.eu/eip/agriculture/sites/default/files/minipaper_1_220908_finaldraft.pdf, accessed on 1 August 2024)

12 Where to look for further information

This chapter is based on a synthesis of multiple review articles, including Arjunan (2024), Berninger (2018), Choudhary (2022), Fessia (2022), Hoang (2022), Hussain (2021), Kumar (2021), Lucena-Leandro (2022), Mateos Fernández (2022), Morelli (2021), Niu (2024), Nunes (2024), Preininger et al. (2018), Rank (2021), Tran (2024), Vashisth (2013) and Zabot (2022).

Moreover, it is recommended to consult the websites of current and past European projects such as:

- *BeXyl* – Xylella, from emergency to management, Integrated Management Strategies for Mitigating *Xylella fastidiosa* impact in Europe, https://bexylproject.org/
- *Mosquarel* – Drones to help spread of sterile mosquitos, https://cordis.europa.eu/project/id/899888/de
- *Smart Droplets* – Accelerating the achievement of EU Green Deal Goals for pesticide and fertilizer reduction through AI, data and robotic technologies, Smart Droplets – Smart Droplets, https://smartdroplets.eu
- *SiNanoCrop* – Bio-Based Polyelectrolyte Nanoformulations for Foliar Delivery of Multitarget dsRNA: An Approach towards novel BioControl Agents for Crop Protection, https://cordis.europa.eu/project/id/101154872
- *Cerberus* – A sustainable approach to crop protection based on early detection and smart actuation, https://cordis.europa.eu/project/id/101134878

A significant gathering of professionals engaged in the field of biocontrol is the Annual Biocontrol Industry Meeting (ABIM), which features a comprehensive array of exhibitions and research presentations. The event is held annually in Basel, Switzerland, and can be accessed via the following link: https://www.abim.ch/.

At https://www.bvl.bund.de/DE/Arbeitsbereiche/04_Pflanzenschutzmittel/01_Aufgaben/02_ZulassungPSM/05_psm_EUPAF/psm_EUPAF_node.html the European Precision Application Task Force (EUPAF) provides a scientific platform for the exchange of information on precision applications in crop protection and promotes dialogue between experts in the development of precision application equipment, farmers, regulatory scientists and risk assessors from public and private organizations and research institutions.

13 References

Abarna, V.P., et al., 2022. Efficacy of talc-based formulation of *Beauveria bassiana* (Bals.) Vuill. (MZ749636) against two spotted spider mite, *Tetranychus urticae* Koch. *J Appl Nat Sci* 14, 232–237. https://doi.org/10.31018/jans.v14iSI.3707

Abubakar, M., et al., 2023. Efficacy of Eco-friendly Bio-pesticides Against The Whitefly *Bemisia tabaci* (Gennadius) for sustainable eggplant cultivation in Kebbi State, Nigeria. *Agronomy* 13, 3083. https://doi.org/10.3390/agronomy13123083

Ambele, C.F., et al., 2020. Entomopathogenic fungi as endophytes for biological control of subterranean termite pests attacking cocoa seedlings. *J Fungi* 6, 1–18. https://doi.org/10.3390/jof6030126

Amour, M., et al., 2023. Efficacy of selected botanical oils against the cassava whitefly (*Bemisia tabaci*) and their effects on its feeding behaviour. *J Appl Entomol* 147, 473–485. https://doi.org/10.1111/jen.13128

Arjunan, N., et al., 2024. Nanoparticle-mediated dsRNA delivery for precision insect pest control: a comprehensive review. *Mol Biol Rep* 51. https://doi.org/10.1007/s11033-023-09187-6

Arthurs, S.P., et al., 2006. Evaluation of spray-dried lignin-based formulations and adjuvants as solar protectants for the granulovirus of the codling moth, *Cydia pomonella* (L). *J Invertebr Pathol* 93, 88–95. https://doi.org/10.1016/j.jip.2006.04.008

Beck, B., 2013. *Sustainable insect control in vegetables through optimized applications of entomopathogenic nematodes*. Ghent University, Ghent, Belgium.

Behle, R.W., et al., 2010. Improving formulations for biopesticides: enhanced UV protection for beneficial microbes. *J ASTM Int* 8, JAI102793. https://doi.org/10.1520/JAI102793

Bejarano, A. and Puopolo, G., 2020. Bioformulation of Microbial Biocontrol Agents for a Sustainable Agriculture, in: De Cal, A., et al. (eds.), *How Research Can Stimulate the Development of Commercial Biological Control Against Plant Diseases*. Springer International Publishing, Cham, pp. 275–293. https://doi.org/10.1007/978-3-030-53238-3_16

Berninger, T., et al., 2018. Maintenance and assessment of cell viability in formulation of non-sporulating bacterial inoculants. *Microb Biotechnol* 11, 277–301. https://doi.org/10.1111/1751-7915.12880

Berninger, T., et al., 2021. Water-soluble polymers in agriculture: xanthan gum as eco-friendly alternative to synthetics. *Microb Biotechnol.* 14, 1881–1896. https://doi.org/10.1111/1751-7915.13867

Biondi, A., et al., 2018. Ecology, worldwide spread, and management of the Invasive South American Tomato Pinworm, *Tuta absoluta*: Past, present, and future. *Annu Rev Entomol* 63, 239–58. https://doi.org/10.1146/annurev-ento-031616

Brandoli, B., et al., 2021. DropLeaf: a precision farming smartphone tool for real-time quantification of pesticide application coverage. *Comput Electron Agric* 180, 105906. https://doi.org/https://doi.org/10.1016/j.compag.2020.105906

Brusselman, E., et al., 2012. Effect of the spray application technique on the deposition of entomopathogenic nematodes in vegetables. *Pest Manag Sci* 68, 444–453. https://doi.org/10.1002/ps.2290

Buragohain, P., et al., 2021. Development and validation of an integrated pest management strategy against the invasive South American tomato leaf miner, *Tuta absoluta* in South India. *Crop Protection* 139. https://doi.org/10.1016/j.cropro.2020.105348

Chen, W., et al., 2023. Screening of fungal strains and formulations of *Metarhizium anisopliae* to control *Phyllotreta striolata* in Chinese Flowering Cabbage. *Insects* 14. https://doi.org/10.3390/insects14060567

Choudhary, K., et al., 2022. A review on destructive tomato pest, *Phthorimaea absoluta* (Lepidoptera: Gelechiidae) and its management. *J Biol Control* 36, 84–93. https://doi.org/10.18311/jbc/2022/28704

Christiaens, O., et al., 2020. The Use of Nanocarriers to Improve the Efficiency of RNAi-Based Pesticides in Agriculture, in: Fraceto, L.F., et al. (eds.), *Nanopesticides*. pp. 49–68. https://doi.org/10.1007/978-3-030-44873-8_3

do Nascimento Silva, J., et al., 2019. Novel combination of a biosurfactant with entomopathogenic fungi enhances efficacy against *Bemisia* whitefly. *Pest Manag Sci* 75, 2882–2891. https://doi.org/10.1002/ps.5458

EU Pesticides Database - Active substances [WWW Document]. https://ec.europa.eu/food/plant/pesticides/eu-pesticides-database/start/screen/active-substances (accessed 7.29.24).

Faria, M., et al., 2022. Evaluation of key parameters for developing a Metarhizium rileyi-based biopesticide against *Spodoptera frugiperda* (Lepidoptera: Noctuidae) in maize: laboratory, greenhouse, and field trials. *Pest Manag Sci* 78, 1146–1154. https://doi.org/10.1002/ps.6729

Fessia, A., et al., 2022. Could *Bacillus* biofilms enhance the effectivity of biocontrol strategies in the phyllosphere? *J Appl Microbiol* 133, 2148–2166. https://doi.org/10.1111/jam.15596

Gan-Mor, S., et al., 2014. Equipment for Application of Biopesticides, in: Sahayaraj, K. (ed.), *Basic and Applied Aspects of Biopesticides*. Springer, India, pp. 333–343. https://doi.org/10.1007/978-81-322-1877-7_19

García-Cruz, A., et al., 2024. Control of phytopathogen organisms using bioactive compounds contained in orange wax. *Biofuel Bioprod Bioref* 18, 510–523. https://doi.org/10.1002/bbb.2582

Gatarayiha, M.C., et al., 2010. Effects of adjuvant and conidial concentration on the efficacy of *Beauveria bassiana* for the control of the two spotted spider mite, *Tetranychus urticae*. *Exp Appl Acarol* 50, 217–229. https://doi.org/10.1007/s10493-009-9307-6

Gatarayiha, M.C., et al., 2011. Field evaluation of *Beauveria bassiana* efficacy for the control of *Tetranychus urticae* Koch (Acari: Tetranychidae). *J Appl Entomol* 135, 582–592. https://doi.org/10.1111/j.1439-0418.2010.01569.x

Gauthier, P., et al., 2019. Technical Feasibility of Spraying *Trichogramma ostriniae* Pupae to Control the European Corn Borer in Sweet Corn Crops. *Appl Eng Agric* 35, 185–192. https://doi.org/10.13031/aea.12833

Gharbi, K. and Tay, J.W., 2022. Fumigant toxicity of essential oils against *Frankliniella occidentalis* and *F. insularis* (Thysanoptera: Thripidae) as affected by polymer release and adjuvants. *Insects* 13. https://doi.org/10.3390/insects13060493

Grella, M., et al., 2023. Assessment of microbial biocontrol agent (BCA) viability to mechanical and thermal stress by simulating spray application conditions. *Pest Manag Sci* 79, 4423–4438. https://doi.org/10.1002/ps.7643

Hoang, B.T.L., et al., 2022. RNAi as a foliar spray: efficiency and challenges to field applications. *Int J Mol Sci* 23, 6639. https://doi.org/10.3390/ijms23126639

Holka, M. and Kowalska, J., 2023. The potential of adjuvants used with microbiological control of insect pests with emphasis on organic farming. *Agriculture* 13, 1659. https://doi.org/10.3390/agriculture13091659

Homayoonzadeh, M., et al., 2022. A novel biopesticide formulation for organic management of *Aphis gossypii* in cucumber greenhouses. *Chem Proc* 10, 55. https://doi.org/10.3390/iocag2022-12325

Hussain, A.G., et al., 2021. Viruses of the fall armyworm *Spodoptera frugiperda*: a review with prospects for biological control. *Viruses* 13, 2220. https://doi.org/10.3390/v13112220

Inglis, G.D., et al., 2000. Influence of rain and conidial formulation on persistence of *Beauveria bassiana* on potato leaves and Colorado potato beetle larvae. *Biol Control* 18, 55–64. https://doi.org/10.1006/bcon.1999.0806

Ji, G., et al., 2021. Leaf surface characteristics affect the deposition and distribution of droplets in rice (*Oryza sativa* L.). *Sci Rep* 11. https://doi.org/10.1038/s41598-021-97061-5

Jin, X., et al., 2013. Effects of conidial densities and spray volume of *Metarhizium anisopliae* and *Beauveria bassiana* fungal suspensions on conidial viability, droplet size and deposition coverage in bioassay using a novel bioassay spray system. *Biocontrol Sci Technol* 23, 362–366. https://doi.org/10.1080/09583157.2012.758687

Kabaale, F.P., et al., 2022. First report of field efficacy and economic viability of *Metarhizium anisopliae*-ICIPE 20 for *Tuta absoluta* (Lepidoptera: Gelechiidae) management on tomato. *Sustainability (Switzerland)* 14. https://doi.org/10.3390/su142214846

Kaiser, D., et al., 2019. Co-formulation of *Beauveria bassiana* with natural substances to control pollen beetles-Synergy between fungal spores and colza oil. *Biol Control* 140, 144106. https://doi.org/10.3929/ethz-b-000374460

Kobenan, K.C., et al., 2022. Application of essential oils of *Ocimum gratissimum* and *Cymbopogon citratus* as Bioinsecticides for the Management of Two Major Biting-Sucking Insects (*Bemisia tabaci* Gennadius and *Jacobiella fascialis* Jacobi) and the improvement of seed and fiber quality of cotton plants in Ivory Coast. *Chem Biodivers* 19. https://doi.org/10.1002/cbdv.202100801

Kumar, M., et al., 2023. Effect of spray operating parameters on viability of *Bacillus thuringiensis* based biopesticide under laboratory condition. *Agric Res* 12, 189–196. https://doi.org/10.1007/s40003-022-00639-2

Kumar, P. et al., 2021. *Bacillus thuringiensis* as microbial biopesticide: uses and application for sustainable agriculture. *Egypt J Biol Pest Control* 31, 95. https://doi.org/10.1186/s41938-021-00440-3

Kumari, M., et al., 2022. Biological Control of Agricultural Insect Pests. Insecticides - Impact and Benefits of Its Use for Humanity. IntechOpen. https://doi.org/http://dx.doi.org/10.5772/intechopen.104464

Lacey, L.A., et al., 2015. Insect pathogens as biological control agents: Back to the future. *J Invertebr Pathol* 132, 1–41. https://doi.org/10.1016/j.jip.2015.07.009

Li, X., et al., 2023. The efficiency of pest control options against two major sweet corn ear pests in China. *Insects* 14. https://doi.org/10.3390/insects14120929

Lopes, R.B., et al., 2020. Efficacy of an oil-based formulation combining *Metarhizium rileyi* and nucleopolyhedroviruses against lepidopteran pests of soybean. *J Appl Entomol* 144, 678–689. https://doi.org/10.1111/jen.12787

Lucena-Leandro, V.S., et al., 2022. Current scenario of exogenously induced RNAi for lepidopteran agricultural pest control: from dsRNA design to topical application. *Int J Mol Sci* 23, 15836. https://doi.org/10.3390/ijms232415836

Maghsoudi, S. and Jalali, E., 2017. Noble UV protective agent for *Bacillus thuringiensis* based on a combination of graphene oxide and olive oil. *Sci Rep* 7, 11019. https://doi.org/10.1038/s41598-017-11080-9

Malinga, L.N. and Laing, M.D., 2022. Efficacy of biopesticides in the management of the cotton bollworm, *Helicoverpa armigera* (Noctuidae), under field conditions. *Insects* 13, 673. https://doi.org/10.3390/insects13080673

Mann, C.W.G., et al., 2023. RNA-based control of fungal pathogens in plants. *Int J Mol Sci* 24, 12391. https://doi.org/10.3390/ijms

Marouf, A. and Harras, F., 2022. Efficacy of three plant oil extracts and their nano-emulsions on the leaf miner, *Tuta absoluta* Larvae (Meyrick) (Lepidoptera: Gelechiidae). *J Plant Prot Pathol* 13, 207–213. https://doi.org/10.21608/jppp.2022.153798.1090

Marzban, R., et al., 2021. Preparation of wettable powder formulation of *Bacillus thuringiensis* KD2. *J Appl Biol Scie* 15, 285–293. https://www.jabsonline.org/index.php/jabs/article/view/830

Masri, M.M.M., et al., 2022. Effectiveness of *Bacillus thuringiensis* aerial spraying against the bagworm, *Metisa plana* Walker (Lepidoptera: Psychidae) outbreak in oil palm using drone. *J Oil Palm Res* 34, 276–288. https://doi.org/10.21894/jopr.2021.0045

Mateos Fernández, R., et al., 2022. Insect pest management in the age of synthetic biology. *Plant Biotechnol J* 20, 25–36. https://doi.org/10.1111/pbi.13685

Matthews, G.A., 2021. New technology for desert locust control. *Agronomy* 11, 1052. https://doi.org/10.3390/agronomy11061052

Morelli, M., et al., 2021. *Xylella fastidiosa* in olive: a review of control attempts and current management. *Microorganisms* 9,1771. https://doi.org/10.3390/microorganisms9081771

Morton, A. and Garcia-del-Pino, F., 2017. Laboratory and field evaluation of entomopathogenic nematodes for control of *Agriotes obscurus* (L.) (Coleoptera: Elateridae). *J Appl Entomol* 141, 241–246. https://doi.org/10.1111/jen.12343

Murugasridevi, K., et al., 2022. Oil based formulation of *Beauveria bassiana* (BB 112) against onion thrips *Thrips tabaci* Lindeman. *Indian J Entomol* 84, 268–273. https://doi.org/10.55446/IJE.2021.20

Nansen, C., et al., 2015. Optimizing pesticide spray coverage using a novel web and smartphone tool, SnapCard. *Agron Sustain Dev* 35, 1075–1085. https://doi.org/10.1007/s13593-015-0309-y

Nithya, P.R. and Reji Rani, O.P., 2017. Evaluation of carrier materials for formulating entomopathogenic fungus *Lecanicillium Lecanii* (Zimmermann) Zare and Gams. *J Biol Control* 31, 50–55. https://doi.org/10.18311/jbc/2017/15556

Niu, J., et al., 2024. RNA interference in insects: the link between antiviral defense and pest control. *Insect Sci* 31, 2–12. https://doi.org/10.1111/1744-7917.13208

Nunes, P.S.O., et al., 2024. Microbial consortia of biological products: -do they have a future? *Biol Control* 188. https://doi.org/10.1016/j.biocontrol.2024.105439

Ochieng', V., et al., 2023. Optimum flight height for the control of desert locusts using unmanned aerial vehicles (UAV). *Drones* 7, 233. https://doi.org/10.3390/drones7040233

Opisa, S., et al., 2020. Chemical additives enhance the activity of a Bt-based biopesticide targeting the beet webworm larvae. *J Appl Entomol* 144, 26–32. https://doi.org/10.1111/jen.12706

Parveen, S.S., et al., 2020. Influence of spraying techniques on the efficacy of *Beauveria bassiana*-based mycoinsecticide against chilli and tomato thrips. *Vegetos* 33, 345–351. https://doi.org/10.1007/s42535-020-00112-x

Pelizza, S.A., et al., 2019. Application of *Beauveria bassiana* using different baits for the control of grasshopper pest *Dichroplus maculipennis* under field cage conditions. *J King Saud Univ Sci* 31, 1511–1515. https://doi.org/10.1016/j.jksus.2019.02.014

Poveda, J., 2021. *Trichoderma* as biocontrol agent against pests: new uses for a mycoparasite. *Biol Control* 159. https://doi.org/10.1016/j.biocontrol.2021.104634

Preininger, C., et al., 2018. Concepts and applications of foliar spray for microbial inoculants. *Appl Microbiol Biotechnol* 102, 7265–7282. https://doi.org/10.1007/s00253-018-9173-4

Price, R.E., et al., 1997. Aerial spray trials against brown locust (*Locustana pardalina*, Walker) nymphs in South Africa using oil-based formulations of *Metarhizium flavoviride*. *Crop Protection* 16, 345–351.

Prince, G. and Chandler, D., 2020. Susceptibility of *Myzus persicae*, *Brevicoryne brassicae* and *Nasonovia ribisnigri* to fungal biopesticides in laboratory and field experiments. *Insects* 11, 55. https://doi.org/10.3390/insects11010055

Prithiva, J.N., et al., 2017. Efficacy of different formulations of *Beauveria bassiana* (Bb 112) against *Bemisia tabaci* on tomato. *J Entomol Zool Stud* 5, 1239–1243.

Prithiva, J.N., et al., 2019. Study on efficacy of *Beauveria bassiana* (Bb 112) formulations and validation of delivery equipments against *Aphis gossypii* on tomato. *J App Hortic* 21, 231–236. https://doi.org/10.37855/jah.2019.v21i03.40

Rank, A.P. and Koch, A., 2021. Lab-to-field transition of RNA spray applications – How far are we? *Front Plant Sci* 12. https://doi.org/10.3389/fpls.2021.755203

Ritu, A., et al., 2012. Biopesticidal formulation of *Beauveria bassiana* effective against larvae of *Helicoverpa armigera*. *J Biofertil Biopestic* 3. https://doi.org/10.4172/2155-6202.1000120

Ru, Y., et al., 2020. Experimental study on spraying performance of biological pesticides in aerial rotary cage nozzle. *Int J Agric Biol Eng* 13, 1–6. https://doi.org/10.25165/j.ijabe.20201306.5511

Sabbahi, R., et al., 2022. A global perspective of entomopathogens as microbial biocontrol agents of insect pests. *J Agric Food Res* 10, 100376. https://doi.org/10.1016/j.jafr.2022.100376

Shapiro-Ilan, D.I., et al., 2006. Application technology and environmental considerations for use of entomopathogenic nematodes in biological control. *Biol Control* 38, 124–133. https://doi.org/10.1016/j.biocontrol.2005.09.005

Sindhu, S.S., et al., 2017. Biological Control of Insect Pests for Sustainable Agriculture, in: Adhya, T.K., et al. (eds.), *Advances in Soil Microbiology: Recent Trends and Future Prospects: Volume 2: Soil-Microbe-Plant Interaction*. Springer Singapore, Singapore, pp. 189–218. https://doi.org/10.1007/978-981-10-7380-9_9

Skyinnov. Drone dropping of *Trichogramma* against the European corn borer [WWW Document]. https://skyinnov.fr/en/2021/12/02/the-application-of-trichogramma-by-drone-to-control-the-corn-borer/ (accessed 8.9.24).

Symonds, B.L., et al., 2016a. Chitosan as a rainfastness adjuvant for agrochemicals. *RSC Adv* 6, 102206–102213. https://doi.org/10.1039/c6ra23485b

Symonds, B.L., et al., 2016b. Rainfastness of Poly(vinyl alcohol) Deposits on Vicia faba Leaf Surfaces: From Laboratory-Scale Washing to Simulated Rain. *ACS Appl Mater Interfaces* 8, 14220–14230. https://doi.org/10.1021/acsami.6b01682

Takeda, N., et al., 2020. A vegetable oil–based biopesticide with ovicidal activity against the two-spotted spider mite, *Tetranychus urticae* Koch. *Eng Life Sci* 20, 525–534. https://doi.org/10.1002/elsc.202000042

Tamez-Guerra, P., et al., 2000. Sunlight Persistence and Rainfastness of Spray-Dried Formulations of Baculovirus Isolated From *Anagrapha falcifera* (Lepidoptera: Noctuidae). *J Econ Entomol* 93, 210–218. https://doi.org/10.1603/0022-0493-93.2.210

Tamez-Guerra, P., et al., 2005. Solar Radiation (UV) Protectants for Microbial Insecticides, in: *Semiochemicals in Pest and Weed Control, ACS Symposium Series*. American Chemical Society, pp. 127–147. https://doi.org/doi:10.1021/bk-2005-0906.ch010

Tran, G.H., et al., 2024. Cyclotides: the next generation in biopesticide development for eco-friendly agriculture. *J Pept Sci* 30, 3570. https://doi.org/10.1002/psc.3570

Ummidi, V.R.S. and Vadlamani, P., 2014. Preparation and use of oil formulations of *Beauveria bassiana* and *Metarhizium anisopliae* against *Spodoptera litura* larvae. *Afr J Microbiol Res* 8, 1638–1644. https://doi.org/10.5897/ajmr2013.6593

Vashisth, S., et al., 2013. Entomopathogenic nematodes: a review. *Agric Rev* 34, 163. https://doi.org/10.5958/j.0976-0741.34.3.001

Vimala Devi, P.S., et al., 2020. Novel formulations of *Bacillus thuringiensis* var. kurstaki: an eco-friendly approach for management of lepidopteran pests. *World J Microbiol Biotechnol* 36, 78. https://doi.org/10.1007/s11274-020-02849-8

Vimala Devi, P.S. and Vineela, V., 2015. Suspension concentrate formulation of *Bacillus thuringiensis* var. kurstaki for effective management of *Helicoverpa armigera* on sunflower (*Helianthus annuus*). *Biocontrol Sci Technol* 25, 329–336. https://doi.org/10.1080/09583157.2014.977846

Wang, L., et al., 2024. Tank-mix adjuvants improved spray performance and biological efficacy in rice insecticide application with unmanned aerial vehicle sprayer. *Pest Manag Sci* 80, 4371–4385. https://doi.org/10.1002/ps.8143

Warneke, B.W., et al., 2021. Canopy spray application technology in specialty crops: a slowly evolving landscape. *Pest Manag Sci* 77, 2157–2164. https://doi.org/10.1002/ps.6167

Wilson, K., et al., 2020. A novel formulation technology for baculoviruses protects biopesticide from degradation by ultraviolet radiation. *Sci Rep* 10, 13301. https://doi.org/10.1038/s41598-020-70293-7

Wraight, S.P., et al., 2021. Efficacy of *Beauveria bassiana* strain GHA spray applications against coffee berry borer *Hypothenemus hampei* on Hawai'i Island. *Biol Control* 161, 104587. https://doi.org/10.1016/j.biocontrol.2021.104587

Wraight, S.P. and Ramos, M.E., 2002. Application parameters affecting field efficacy of *Beauveria bassiana* foliar treatments against Colorado potato beetle *Leptinotarsa decemlineata*. *Biol Control* 23, 164–178. https://doi.org/10.1006/bcon.2001.1004

Xin, L.L., et al., 2018. Preparation and bioassay of *Bacillus thuringiensis* microcapsules by complex coacervation. *Dig J Nanomater Biostruct* 13, 1239–1247.

Yan, G., et al., 2022. Surface modification of nucleopolyhedrovirus with polydopamine to improve its properties. *Pest Manag Sci* 78, 456–466. https://doi.org/10.1002/ps.6640

Zabot, G.L. et al., 2022. Encapsulation of Bioactive Compounds for Food and Agricultural Applications. Polymers 2022, 14(19), 4194. https://doi.org/10.3390/polym14194194

Zhang, L., et al., 2016. A new formulation of *Bacillus thuringiensis*: UV protection and sustained release mosquito larvae studies. *Sci Rep* 6, 39425. https://doi.org/10.1038/srep39425

Zheng, Y., et al., 2023. Dual role of endophytic entomopathogenic fungi: induce plant growth and control tomato leafminer *Phthorimaea absoluta*. *Pest Manag Sci* 79, 4557–4568. https://doi.org/10.1002/ps.7657

Chapter 10

Challenges in risk assessment of biopesticides for insect pests

Shannon Borges and Mike Mendelsohn, Office of Pesticide Programs – U.S. Environmental Protection Agency, USA

1 Introduction

Conventional chemical pesticides currently have the largest portion of the market for externally applied insect pest control worldwide (Damalas and Koutroubas, 2018). However, the growth of the world's population, climate change, and globalization place increasing demands on agriculture and other sectors for innovative and agile pest control strategies (Calicioglu et al., 2019; Charles et al., 2010; Bertelsmeier, 2021; Hulme, 2009). Recognition of these needs is reflected in government-driven initiatives, such as sustainable development and agriculture concepts and goals of the United Nations 2030

http://dx.doi.org/10.19103/AS.2024.0149.12

agenda (Fenibo et al., 2021), the European Green Deal (EC, 2019), efforts to support the global bioeconomy (e.g. EC, 2018, 2019; Exec. Order No. 14081, 2022), and the State of California's Sustainable Pest Management Roadmap (CalDPR, 2023). It is also seen in the growing awareness of the impacts of chemicals on food safety and the environment and the desire among consumers for alternatives to conventional chemical pesticides (Damalas and Koutroubas, 2018; Fenibo et al., 2021). Biopesticides present alternatives to conventional chemical pesticides and can play a key role in meeting these needs. With the growing demand for sustainability, biopesticides are and will be important tools for insect pest control in the future.

Biopesticides are pesticides that are naturally occurring or derived from natural materials, such as animals, plants, microorganisms, and certain minerals. They may provide means of more sustainable pest control since they generally have lower human health and nontarget organism toxicity profiles, do not persist in the environment, and are more specific to their target pests compared to conventional chemical pesticides (Leahy et al., 2014). In the United States, biopesticides are categorized into three classes:

- Biochemical pesticides are naturally occurring substances (or if synthetic, they must be structurally similar and functionally identical) that control pests through a non-toxic mode of action and have a history of exposure to humans and the environment demonstrating minimal toxicity (USEPA, 2007a). Examples of biochemicals that are used to control insects include mating disruptors, such as arthropod pheromones; chemicals that inhibit insect development; antifeedant chemicals; or others that repel or kill insects, such as essential or other oils and mineral clays; among many others.
- Microbial pesticides, which consist of living or dead microorganisms as active pesticidal ingredients, include eukaryotic microorganisms (e.g. protozoa, algae, and fungi) and prokaryotic microorganisms (e.g. Eubacteria, Archaebacteria) or parasitically replicating microscopic elements (e.g. viruses) (USEPA, 2007b). Microbial pesticides may act upon insects through toxicity, pathogenicity, or by other means, such as anti-feeding effects. Commonly known microbial active ingredients used to control insects include the bacterium *Bacillus thuringiensis* (Bt), the fungus *Beauveria bassiana*, and baculoviruses. Genetically engineered (GE) microorganisms can be registered as pesticides; however, they are generally handled as emerging technology pesticides. For the purposes of this chapter, 'microbial pesticides' refers only to those that are naturally occurring.
- Emerging technology biopesticides.

This last group includes the following:

1 plant-incorporated protectants (comprised of pesticidal substance(s) that are intended to be produced and used in a living plant, or in the produce thereof, and the genetic material necessary for their production (e.g. the *cry1Ab* gene and Cry1Ab protein in a Bt corn));
2 genetically engineered (GE) microbial pesticides (e.g. bacteria, viruses);
3 microbial pesticides introduced into mosquitoes (e.g. *Wolbachia* mosquito population suppression products);
4 genetic modifications in pest animals (e.g. GE mosquito population suppression products);
5 externally applied double-stranded RNA (dsRNA) pesticide products; and
6 certain peptides and proteins (e.g. harpin and spider toxins).

Other means of control of insect pests include biological control or 'biocontrol' through unmodified natural enemies such as predatory or parasitic insects and other invertebrates. However, unlike modified insects (e.g. GE mosquitoes), these organisms are regulated and assessed separately from biopesticides in the United States and are not included in the scope of this chapter.

The U.S. Environmental Protection Agency (EPA) has established a regulatory framework for biopesticides that facilitates their development. This approach includes lower registration fees, shorter timeframes for processing biopesticide registrations, and tiered data requirements that focus testing where it is needed. While this framework has worked well as intended over several decades of implementation, risk assessments for biopesticides can be challenging due to their unique properties, modes of action, and rapid advances in technology and development. Often, these testing challenges impact the resolution of the risk assessments, which may lead to more conservative assumptions when evaluating the risk of the biopesticide and more restrictive use patterns than would be the case with improved testing methods. With rising global interest in biopesticides and their promise in meeting future sustainable pest control needs, it is important to address these challenges and minimize barriers to bringing these viable alternatives to chemical pesticides into the marketplace. An exhaustive account of the challenges encountered with biopesticide risk assessment is beyond the scope of this chapter; however, we highlight major and emerging challenges and information needs with the three main areas affecting risk conclusions, including product identity and characterization, exposure analyses, and hazard testing.

2 Biopesticide data requirements and problem formulation for risk assessment

Specific challenges with biopesticide testing are described later in this chapter; however, a discussion of biopesticide risk assessment must begin with hazard testing, since it lies at the center of most risk assessments for biopesticides. Biopesticides are generally understood to be of minimal risk, so hazard tests that demonstrate a lack of adverse effects under worst-case exposure conditions are usually considered sufficient to confirm this understanding. For this reason, data to support biopesticide registrations are required on a tiered basis, where Tier I hazard tests are conceptually simple and represent worst-case exposures, and higher tiers progress through more complex designs that mimic increasingly realistic conditions. For example, nontarget organism testing data requirements for microbial pesticides begin at Tier I with high-dose testing on birds, wild mammals, freshwater and marine/estuarine fish and invertebrates, nontarget plants, nontarget insects, and honey bees. Mortality is often the endpoint observed, and if no effects are evident, testing is concluded. Otherwise, testing advances as needed through Tiers II, III, and IV, which gather data about microbial environmental expression, chronic and range testing, and semi-field and field testing, respectively. When minimal persistence or effects are observed within a tier, testing ceases.

Full descriptions of the data requirements for biochemical and microbial pesticides are described in the *Code of Federal Regulations* at 40 CFR Subparts U and V, respectively (USEPA, 2007a, b). Some data are conditionally required for certain uses or under specific conditions, which are described in the regulations. The EPA also has the flexibility to require, or not require, on a case-by-case basis data and information to support pesticide regulatory decisions and can establish or modify data on an individual basis to fully characterize risks associated with pesticides (USEPA, 2007c). Testing conducted according to established guidelines is the primary means by which data are generated to support biopesticide risk assessments but other approaches may also be employed. If similar data are available in the public literature, these data may simply be cited in lieu of testing. Additionally, applicants may submit a scientifically based rationale describing minimal exposure, hazard, or both. These arguments are typically supported with data from the public literature to allow regulators to conclude unequivocally that testing is not needed because the risk is minimal. The use of these alternatives offers multiple benefits by lowering the costs of biopesticide development and reducing the use of animals for testing.

Whether guideline testing is needed or other information can suffice in lieu of this testing is determined through problem formulation, which is a process in which assessment goals are set, risk hypotheses are generated, and the scope of the risk assessment is defined (USEPA, 1998; 2014). This step is particularly important for biopesticides because the range of potential biopesticide active ingredients is wide, even within each biopesticide category, their properties can vary significantly, and data sets supporting registrations are often limited. Product identity and characterization data are central to this process since this information describes the properties of the biopesticide active ingredient and product, distinguishing what may already be known and what additional information may be needed. This information, along with literature search results and consideration of how, when, and where the biopesticide is used, can identify uncertainties and areas where more information is needed. This process may be iterative and dependent on the results of tiered testing described earlier, and it is critical for defining the expected data set and the direction of the biopesticide risk assessment.

3 Characterizing the active ingredient: biochemical pesticides

Characterizing the biopesticide active ingredient involves identifying the substance and its defining characteristics, including physical and chemical properties and other qualities relevant to understanding its nature and disposition in the environment. As noted above, this information is key to defining data needs and setting the scope of any analyses, particularly since biopesticides are often naturally occuring and information significant to the risk assessment may already be known. This information is also important to show that the pesticide to be applied is the same as the pesticide used in testing. In many cases, especially for well-researched substances, characterization is relatively straightforward. However, advances in technology or increased knowledge of biopesticides have raised new questions that regulators must address, as described further in this chapter.

Biochemical pesticides include substances that range widely in their composition from purified compounds to complex chemical mixtures, so while there are standard data requirements for product identity, these data must capture important nuances where necessary. Basic data requirements involve identification by chemical name, Chemical Abstracts Service (CAS) registry number, structural and empirical formula, amount present, specificity, and mode of action, and if derived from a microorganism, the genus and species of the microorganism (USEPA, 1996b). However, the complexity of some biochemicals can pose additional challenges that may influence risk assessment hypotheses.

Many biochemicals exist as complex mixtures containing numerous active and non-active constituents. Traditional means of chemical analysis (e.g. high-performance liquid chromatography) may not be applicable to some biochemicals that are complex substances or mixtures, and in such cases, other approaches to analysis may need to be employed, such as fingerprint chromatograms or analysis for certain chemical groups (e.g. proteins). It is not feasible or necessary to identify each constituent in a mixture, but identifying active constituents aids in characterizing hazard and at a minimum some constituent (i.e. a marker compound) must be identified for purposes of quantifying the active ingredient. In these cases, the link between the activity of the biochemical and the amount of the constituent must be clear. Confusion can arise when the active constituent is named as the active ingredient but is actually part of a complex mixture. In cases where the active constituent comprises a high percentage (e.g. 90%) of the mixture, designating it as the active ingredient may be functionally accurate and acceptable. However, where it is contained in lower amounts, the activity of other constituents may need to be identified to fully understand the identity of the active substance. Since the mixture can contain other constituents that may have effects on humans or other nontarget organisms, it is important to fully identify what constitutes the active ingredient.

Biochemicals present other unique challenges in identification, particularly where two are claimed or assumed to be similar when they may not be. This can result from differences in identification among sources of information, wherein substances that are not compositionally identical can have the same common name, CAS number, be considered synonyms (e.g. acetic acid and vinegar), or from situations where the common name is very broad but encompasses many substances that vary in composition (e.g. plant oils). Additionally, extracts from different plant parts (e.g. leaves, stems, roots, or bark) may involve unique mixtures of constituents (Cesa et al., 2019; Vukovic et al., 2009) and may not be equivalent. Similarly, different extraction methods could produce substances of differing composition and physical and chemical properties (e.g. see Richter and Schellenberg (2007); Aziz et al. (2018)). In these cases, the active substance must be examined closely to confirm any claimed equivalence. Where equivalence cannot be shown, separate data sets would be required to support registration. Additionally, composition of some oils and plant extracts can vary by season and growth stage (Hudaib et al., 2002; Lemos et al., 2017; Adams, 1987), requiring a focus on quality control. In all cases, proper nomenclature must be assigned to each active ingredient to correctly inform users of the active pesticide content.

Biochemicals produced and extracted from microorganisms can be straightforward in their identification; however, these chemicals may not be fully extracted or purified, and the EPA has encountered cases where cellular and fermentation products are still present, sometimes to a large degree. In such cases, the classification of the active ingredient as a biochemical or microbial may be unclear, triggering differing data requirements, or a hybrid set of identity and hazard data may be required.

4 Characterizing the active ingredient: microbial pesticides

Research into new microbial pesticides often involves the discovery of new strains of known species but is increasingly leading to proposals for new species that have not been previously identified. In the United States, each new isolate is considered to be a unique active ingredient and should be identified to the greatest extent possible by taxonomic position, serotype, strain, or any other appropriate designation. Additional characteristics requiring description are described in the EPA's guideline for microbial pesticide product identity (USEPA, 1996a). Identifying the taxonomic classification of the microorganism as accurately and completely as possible is necessary, since this information is important in determining a microorganism's likelihood of producing toxic metabolites or possessing virulence factors associated with adverse effects. Comprehensive microbial identification can also help provide information on a microorganism's biological properties with respect to target pest, host range, life cycle, and mode of action.

One of the challenges with microbial pesticide identification is the rapid advancement of genetic analysis and how best to employ it in the identification of microbial pesticides. Classic phenotypic approaches to identification are often used but must be supplemented by molecular analyses such as those evaluating traits within the genome like multi-locus sequence typing and 16s rRNA for bacteria, internal transcribed spacer sequence for fungi, digital DNA–DNA hybridization, and whole genome sequencing (average nucleotide identity). As sequencing ability increases, the number and types of traits analyzed can increase; however, these additional data will have limited meaning without adequate data quality endpoints. Significant factors include the depth and coverage of sequencing analysis, quality and consistency of sequence annotation, quality of database curation, and the strength and consistency of database acceptance criteria for sequence data. Of importance also is the level of information that is available for well-studied microorganisms (e.g. human or plant pathogens) versus those with fewer data and a combination of identification methods used over time that inhibits comparison where similarities may be proposed. EPA guidelines are flexible enough to allow the employment of advancing technologies in microbial identification; however, the above limitations should be considered for data-supporting registrations.

There are currently available approaches that may have the potential to be used with genetic information to inform problem formulation or make risk assessment predictions. For instance, whole genome sequencing and bioinformatic analyses are regularly used in monitoring programs of foodborne pathogens (e.g. Köser et al., 2012, Brown et al., 2019) and in characterizing microorganisms used in food additives (EFSA, 2018; 2021).

Additionally, research into pathogenicity and virulence factors expressed in insecticidal microorganisms has contributed to the understanding of infection processes and toxin production (e.g. Koduru et al., 2014; Fedhila et al., 2004; Valero-Jiménez, 2016), which can assist in identifying candidate biopesticide strains and possible effects. However, in the above examples, the microorganisms and associated systems are well-studied and utilize comprehensive databases. The use of similar types of analyses to make risk assessment predictions for microbial pesticides may be possible, and supportive in the vein of developing new approach methodologies (NAMs) for microbials but such efforts need more work before they can be incorporated into a regulatory evaluation scheme that is applicable to a broad range of microbes. Phenotypic function would need to be better predicted from genetic sequence or transcriptomic data and would require focused, hypothesis-driven analyses and robust, well-curated databases for a range of microbial taxa to reliably link genetic information to hazard. While much of this information may be available for the more extensively researched microbial groups, this approach would not likely be an option for many types of microbial active ingredients without more research, and bioassay data are still needed at this time.

Risk assessment can also be influenced by materials produced by microorganisms. Microbial active ingredients that include fermentation products are complex mixtures consisting of hundreds or thousands of large and small molecules. As with biochemicals, identifying all constituents is not needed, but may be critical where secondary metabolites of concern may be produced or where identification of these metabolites can better characterize hazards. For well-studied microorganisms known to produce chemicals that can be toxic to humans or other nontarget organisms (e.g. mycotoxins produced by fungi), accurate identification of that microorganism can indicate the needs for additional higher-tier toxicity testing or further characterization of toxin production in the microbial pesticide where necessary. However, challenges arise for newly identified species, where properties may be less known depending on taxonomy. In such cases, characterizing any secondary metabolites of concern will aid in problem formulation and risk assessment (OECD, 2018) and hazard testing will be critical to understanding the potential for effects. Other cases exist where the microorganism may be intentionally rendered non-viable, and the pesticide functions strictly through chemical means. In such cases, at a minimum, a marker compound, preferably one that is an active constituent, is necessary to represent the active ingredient, since typical methods of microbial enumeration cannot be used. These active ingredients also pose a challenge in determining the appropriate hazard and product characterization data requirements, since their production methods include aspects of both microbial and biochemical pesticides.

5 Challenges with exposure analysis: human health

In most cases, exposure analyses for biopesticides are qualitative and do not require residue or environmental fate data or extensive analyses to support risk assessment. If the hazard can be shown to be negligible, then exposure is of minimal concern for humans and nontarget organisms. Residue testing, which measures the potential exposure to humans on food, is a specific set of data that is required under certain conditions (USEPA, 2007a, b), and they are generally needed when hazard is of concern or residue data are needed to inform a quantitative exposure analysis. They may also be used to support claims that no residues are present on food commodities. Exposure and environmental fate and residue testing for nontarget organisms are Tier II tests and are not triggered unless adverse effects are observed at Tier I or if there are other uncertainties, such as where the biopesticide is a novel active ingredient and little is known about its fate and persistence in the environment.

For all biopesticides, the EPA assesses exposure to humans for oral/dietary (including drinking water), dermal, and inhalation routes and the assessment includes occupational and non-occupational exposure scenarios. Where potential risks of concern are identified, this information is used to inform the need for exposure mitigation, such as restrictions on uses (e.g. no residential uses), application methods or rates, personal protective equipment for applicators, and label warnings. Pesticides with food uses require a tolerance, identical to a maximum residue limit, which is a level of the pesticide in food that the EPA finds is 'safe,' meaning that there is a 'reasonable certainty that no harm will result from aggregate exposure to the pesticide residue' (USEPA, 2023a). Where hazard is minimal, an exemption from the requirement of a tolerance can be issued, meaning that any level of exposure is considered to meet this standard. In these cases, data must show unequivocally that a hazard is not anticipated.

While quantitative human health risk assessments are rarely performed for microbial active ingredients, they are, on occasion, necessary for biochemical active ingredients. When quantitative analyses are needed for human health risk assessment of biochemical pesticides, the EPA can utilize exposure models developed for conventional chemical pesticides. Difficulties with these analyses can exist; however, if there is uncertainty as to the correct quantification of the active ingredient, such as when the active ingredient is a complex mixture without a reliable marker compound when multiple active ingredients are tested at once, or when exposures from non-pesticidal uses are significant (e.g. when the active ingredient is found in food or cosmetics) and limit the safe level of pesticidal uses. Greater challenges exist, though these cases are rare, where no safe level is known; for example, when the active ingredient is a potential allergen, or in cases where an identified impurity, such as a metal of concern, may be present. In such cases, exposure generally must be shown to be negligible. More commonly, however, exposure is difficult to evaluate when the pesticide label is imprecise with respect to how the pesticide is applied, as discussed later in this chapter.

For microbial pesticides, human health risk assessments rely primarily on what is known from characterization of the microorganism and hazard testing performed on animals. This testing includes single high-dose toxicity and pathogenicity testing that is performed with the active ingredient and also limit-dose acute toxicity testing with each product to be registered. Cell culture testing with mammalian cell lines is required for viruses, and hypersensitivity incidents must also be reported if they occurred during research and development of the active ingredient. As described earlier, where concerns are identified in Tier I tests, additional toxicity testing may be required at higher tiers. However, most microbial pesticides showing hazards in Tier I tests are highly scrutinized and may not be supported for registration, particularly where an exemption from the requirement of a tolerance cannot be issued. Numerical tolerances are difficult to establish for live microorganisms because their fate in the environment is driven by numerous biotic and abiotic factors, and modeling to determine residue levels above zero must capture all these possible factors for all or a representative subset of application methods and commodity groups. For almost all microbial pesticide registrations, thorough characterization of the microorganism and mammalian hazard testing is sufficient to determine appropriate exposure mitigation and is key for human health safety assessments.

6 Challenges with exposure analysis: nontarget organisms

As with human exposure, if testing or other information indicates that hazards to nontarget organisms is minimal, then exposure analyses can be qualitative. Where needed, developing quantitative exposure analyses for nontarget organisms can be challenging because of the number and variety of environmental factors involved, the nature of the biopesticide, and the lack of environmental expression or fate data. At a minimum, conservative estimates can be calculated, assuming maximum application rates are applied to the environment, but refinements are possible where data are available. Additionally, the EPA is obligated under the Endangered Species Act to consult with relevant agencies when effects to federally listed threatened and endangered ('listed') species are possible (USEPA, 2023b). Where screening level assessments are based on population-level effects, biological evaluations for listed species assessments must consider effects on individuals, so exposure analyses should be rigorous and as realistic as possible.

When quantitative estimates are needed, most biopesticide nontarget organism risk assessments can be performed with conservative estimates based on initial exposure immediately after application, when environmental concentrations are expected to be highest. Exposure analyses for biochemicals can rely on models that exist for conventional chemical pesticides for further refinements to determine initial estimated environmental concentrations (EECs) in dietary items for birds, mammals, and honey bees; and exposures to nontarget plants and aquatic environments via drift and runoff (USEPA, 2023c). Most of these models provide conservative EEC estimates. Persistence can be determined where data are available or where default assumptions can be applied, though this information is usually applied qualitatively to show rapid degradation for biochemicals, rather than to estimate numeric chronic exposure levels.

One issue that is somewhat unique to biochemical pesticides is the ambiguity in the levels of exposure that may produce effects on nontarget organisms. Biochemical pesticides are, by definition, non-toxic to the target pests, and their pesticidal properties are exerted through other means, such as desiccation, smothering, repelling, adherence to foliage, or other effects. Since exposure decreases with distance from the area treated, understanding the level at which effects can occur relative to the amount applied is important in determining the potential geographic footprint of effects. For example, an LC_{50} may be known from toxicity testing, and if the EEC can be estimated for a specific distance from the treated area, then risk assessors can estimate the area where effects could be expected to occur. This information would inform on mitigation measures that may be employed, such as setting spray buffers at a certain distance. However, for other effects, little data are available to understand the exposure levels at which such effects do or do not occur, and they do not necessarily have a dose–response relationship where such a level can be estimated. In these cases, conservative assumptions must be used, which often results in the overestimation of the footprint of effects. A better understanding of the levels and conditions at which these effects occur would help to refine whether and at what distance effects may be expected to occur off-site.

Exposure estimation tools have been developed for chemical pesticides, but these models are not easily adapted to microbials except in a few cases, such as spray drift modeling that calculates drift deposition based on the fraction of the amount of the pesticide that is applied (see AgDrift in USEPA (2023c)). Where the initial level of exposure must be known, the simplest approach is to calculate the concentration that exits the application equipment and is applied in the environment; however, this amount could result in a high estimate of exposure if the pesticide is diluted by environmental matrices, which is likely in many situations. More research in this area to link application rates of microbial pesticides to initial concentrations expected in the environment, particularly for bees and other arthropods, would help not only to refine risk assessments but also to support hazard testing at more realistic levels, as discussed in subsequent sections. This research should also focus on exposure via relevant routes (e.g. dietary, contact), particularly defining various sources of exposure in the environment for all nontarget taxa.

Microbial pesticides may be applied through 'inundative' applications, wherein the pesticide is applied at high levels, typically above natural levels of the microorganism in the environment. Due to limitations on resources (e.g. sufficient nutrients and substrates for growth), the microorganism is not expected to persist in the environment at the level applied. This is anticipated especially when the microorganism is applied to an environmental compartment in which it is not normally found, for example, where a soil fungus sensitive to ultraviolet light is applied to foliage (Kaiser et al., 2018). Environmental expression testing is a Tier II data requirement that is specific to single active ingredients and can shed some light on persistence. However, a better understanding of the general persistence of microbial pesticides across multiple environmental compartments (e.g. soil, foliage, and water) and in food sources for nontarget organisms is needed. Information on the movement of microbial pesticides in the environment, particularly in or across soil, could provide refinement to microbial assessments but is not typically available. For example, microbial pesticides applied as soil applications are assumed not to move far; however, microbes can move through and over the soil though this can depend on many conditions (Abu-Ashour et al., 1994; Alegbeleye and Sant'Ana, 2020). While in-depth investigation into all of the mechanisms driving soil fate may not be needed for microbial pesticides, a basic understanding of the relevance of this movement to exposure off-site exposure, particularly to water, could help to inform problem formulation and decisions on whether hazard testing is warranted or could assist in refining risk conclusions.

7 Improving guidance on use and labeling

Pesticide uses or use sites describe specifically where pesticides may be applied and they are listed on the label (see USEPA, 2022). Refining biopesticide labels by making it clear where, how, and when they are used most effectively, particularly identifying their ideal application conditions, assists risk assessors in producing the most accurate risk assessments. Labels may be ambiguous with respect to the maximum single and total application rates, application equipment, or application methods, or they may not cover all the uses described. These factors influence routes, levels, and durations of exposure, particularly to pesticide applicators, and when there is uncertainty resulting from ambiguous information on the label, conservative assumptions must be used to assess exposure and the impact of mitigations (e.g. equipment controls and personal protective equipment). In such cases, these conservative assumptions lead to overestimates of exposure that may require more restrictive use patterns, more complex label instructions, and/or erroneously give the impression a pesticide product is associated with greater risk in comparison to other products.

Many biochemical and microbial pesticide products have labels that include a wide variety of uses. However, in practice, they may have narrower or even very specific or niche uses. The EPA must assess all proposed uses on a label, regardless of whether the pesticide product is applied in actuality to only a subset of them, which particularly impacts nontarget organism risk assessment. More uses on the label broaden the required scope of the risk assessment, requiring predictions of risks to a wider set of organisms or a larger geographic area than may be realized and the incorporation of more mitigations to minimize any risks when necessary. The result may also be unnecessary process delays if effects to fewer nontarget species will occur based on the actual uses. Narrowing the list of uses to a more realistic set would help to streamline risk assessments.

Further defining the use conditions under which biopesticides are most effective simultaneously reduces the scenarios in which adverse effects could occur and helps pesticide users employ products in a more targeted and effective manner. For example, certain strains of *Pseudomonas fluorescens* used in control of invasive grasses have optimum activity at temperatures of 15–20°C (Dooley and Beckstead 2010), which narrows the scope of nontarget exposure and can be incorporated into problem formulation. Similarly, usage information (the amounts and locations of pesticide applications) would help inform and refine ecological risk assessments during reevaluation or registration review (these data would not be available during initial registration). These data are very limited for biopesticides though. Most usage data for biopesticides is limited to reporting broadly on biopesticides' share of the marketplace (e.g. Marrone et al., 2014, Bailey et al., 2010). However, actual usage of active ingredients by commodity, geography, and sector would assist in further refining where effects to nontarget organisms may be expected to occur. These are areas where more research could not only be put into appropriate uses and markets for biopesticides, possibly helping to further their adoption, but also help to refine and streamline risk assessments.

8 Challenges with hazard testing: mammals

As discussed previously, hazard testing forms the basis of many biopesticide risk assessments. While biopesticides are often naturally occurring, this quality alone is not sufficient to justify minimal hazard, so some testing is often required. Where testing can show minimal hazard by established standards, minimal risk can be concluded and risk assessments simplified if the information obtained from the biochemical or microbial characterization confirms these findings. However, a robust testing scheme and informative guidelines are key components to identifying concerns.

Biopesticide testing for mammals to inform human health risk assessment generally requires testing with the active ingredient to determine acute, subchronic, developmental toxicity and genotoxicity for biochemicals and acute toxicity and toxicity/pathogenicity for microbials (USEPA, 2007a). These tests are straightforward, and guidelines supporting them have a long history of successfully supporting pesticide registrations. Therefore, these studies generally do not present many novel issues.

However, recent efforts to reduce animal testing and improve assessments using more species-relevant information offer viable alternatives to certain biopesticide mammalian toxicology tests. These include advances in NAMs, such as *in vitro* tests to replace whole animal testing (e.g. for primary eye irritation testing with biochemicals, see USEPA (2015)), waiving data requirements based on other scientific evidence (USEPA, 2020), read-across, *in chemico* testing, and use of *in silico* tools. To date, exploration of *in vitro* testing and other NAMs has not focused on biopesticides or taken into consideration the unique properties associated with biopesticides, especially microbials, which can present unique challenges. Consequently, continued research in this area is worthwhile in order to remain competitive with other types of pesticides where alternative testing also has the potential to reduce development costs.

New challenges have arisen with microorganisms that are intentionally rendered non-viable. In these cases, where effects arise because of chemicals produced by the microorganism, testing based on biochemical data requirements and testing guidelines may be most useful in understanding hazard as long as the microorganism is truly non-viable and pathogenicity is not possible. Additionally, where all active constituents cannot be identified, data requirements may need to be addressed with a full suite of Tier I testing to understand potential hazards, as opposed to addressing data requirements with scientific rationale.

9 Challenges with hazard testing: nontarget organisms

Hazard testing for nontarget organisms involves acute testing at Tier I on birds, wild mammals (where necessary), fish and invertebrates, nontarget plants, and nontarget insects, including honey bees (*Apis mellifera*). Hazard testing for nontarget organisms with biochemical pesticides utilizes well-established and reproducible guidelines developed for chemical pesticides. However, as with human health testing, similar challenges have arisen with microorganisms that have been rendered non-viable and, in these cases, full Tier I testing with biochemical data requirements may be required, as well as any other testing if there are remaining uncertainties.

Challenges exist with refining nontarget hazard testing for live microbial pesticides, and the remainder of this section addresses challenges specifically with microbial pesticides. This testing was recently the subject of an OECD Cooperative Research Program-sponsored conference on 'Innovating Microbial Pesticide Testing' (OECD, 2023a), since it is recognized that stronger guidance is needed for nontarget organism hazard testing for microbial pesticides. The EPA's guidelines specific to microbial pesticides (Series 885; see USEPA (2023d)) are meant to be flexible, and certain ones could be adapted to be more fit-for-purpose (see Borges et al., 2021). Otherwise, chemical guidelines are available, such as EPA 850 Series guidelines (USEPA, 2023e), OECD guidelines (OECD, 2023b), and International Organization for Biological Control (IOBC) guidelines for testing on other insects and arthropods (Candolfi et al., 2000); however, guidelines developed for chemicals are unable to account for the unique needs of testing with microorganisms. If inconsistent methods are used for testing across different laboratories, the reliability and reproducibility of the results may be impacted. If so, additional testing may be required, which can increase the cost of development and cause delays in registration. Additionally, if tests are not reliable, the true hazard may not be defined, which could lead to impacts to nontarget organisms in the field if risks are underestimated, or unnecessary limitations on use if risks are overestimated.

Since this book focuses on the control of insect pests, specific challenges here focus on testing with arthropods. In the United States, arthropod tests are conducted most frequently for nontarget hazard evaluation and their challenges stem from the wide range of both microbial test material and nontarget arthropods that may be used in testing and the specific conditions and test methods required to accommodate each combination. Interest is growing in resolving these issues, and Borges et al. (2021) extensively cover common issues encountered with microbial testing with bees. Many of these apply to testing with other nontarget arthropods, and each main area of needed research is discussed in the next sections. Ideally, the outcome of research on these topics would inform the development of more detailed testing guidance for bees and other arthropods.

9.1 Species, life stage, and source

For bee testing, the EPA guideline specifically indicates the honey bee as the species required for microbial pesticide testing. Testing is typically done on adult bees, but larvae are included if they may be affected (USEPA, 1996c). OECD guidelines for testing chemical pesticides with other species and life stages of bees are available (OECD, 2023b) but must be adapted for microbial pesticides. However, given that one of the challenges with testing microbial pesticides on bees is test duration (described further below), other species and life stages may be more suited for microbial testing (Borges et al., 2021).

For other arthropods, EPA's guideline suggests three tests with at least two species represented by parasitic dipterans or hymenopterans; or predaceous hemipterans, coleopterans, mites, or neuropterans. Particularly where arthropods are the target of the biopesticide, host range is an important consideration (USEPA, 1996d), and testing on a variety of beneficial species is combined with knowledge of host range to determine effects on insects. Justification of test species should be described by the testing laboratory, and the guideline provides further suggestions for the selection of predatory or parasitic species associated with the ecosystem where the pesticide will be applied or specific predators or parasites of pests targeted by the microbial pesticide (USEPA, 1996c). However, in practice, the same few species are used repeatedly in these studies. This narrow range likely is the result of the limited number of species that can survive in a laboratory setting and their availability on the commercial market, or their use may simply be routine within a laboratory. A guidance on arthropod species that are viable in a laboratory setting, including testing conditions and husbandry practices that may assure better survival in the laboratory, would be useful toward expanding the range of species tested or focusing testing on the most appropriate species. This includes other species that could be valuable indicators of effects, such as pollinators other than bees.

Some of the species used for nontarget arthropod testing are sourced commercially, rather than raised in colonies at the testing facility, which could be why most tests are performed on adults. However, immature animals are expected to be more susceptible to infection or sensitive to toxins produced by microorganisms and may produce more meaningful results (USEPA, 1996d), so further investigation into their optimal use would support the development of larval tests. Testing on commercially sourced adults might also influence test results in that the age of the adults may be unknown and mixed, such that some test insects are near the normal end of their lives. Additionally, their previous exposure to pesticides, other contaminants, and microorganisms is not known and may also affect results.

9.2 Study duration

The potential to cause pathogenic effects is a unique property of microbial pesticides that is not captured in testing guidelines designed for chemicals, and a primary adaptation for microbials requires a study duration that is long enough to observe pathogenicity. Requirements to observe test animals for long periods (e.g. 30 days post dosing) should capture pathogenic effects if they occur; however, certain species such as honey bees are generally unable to survive outside of the hive for that duration, and effects resulting from exposure to the test material can be confounded with other factors that limit survival. Shorter durations (e.g. 96 h as required by USEPA, 2012) are not long enough to observe pathogenic effects of most microbial pesticides, and the optimal time could vary by test species and the microorganism (Borges et al., 2021). While the addition of a 20% control effect cutoff helps to prevent confounding of other factors, poor husbandry and other uncertainties as described in this section can result in early mortality and suboptimal observation periods. Additionally, it is unlikely that specific guidelines can be developed to capture optimal durations for every combination of test insect and individual or class of microbial pesticide; however, further research into this subject would support the development of robust guidance for laboratories and regulators.

9.3 Exposure considerations

The exposure level, route, and duration are also major questions for honey bee and arthropod testing. These issues are covered for bees by Borges et al. (2021), and they also apply to other arthropods. The EPA utilizes a maximum hazard approach to testing, which is meant to produce testing with a high level of confidence that no unreasonable adverse effects will result from the use of the microbial pesticide. Given that pathogenic effects are typically unlikely to follow a dose–response relationship, and because environmental levels of microbial pesticides cannot be feasibly predicted, testing at very high levels that incorporate safety factors is expected to provide a high degree of confidence (USEPA, 1996d). Dosages of up to 100 times an endpoint level (e.g. LD_{50} or LC_{50} in its natural host), or 10–100 times the recommended field dosage are recommended (e.g. USEPA (1996e)). Since the LD_{50} or LC_{50} is often not known, most testing levels instead involve exposure levels of 10–100 times the EEC. However, as noted above, guidance for estimating the EEC in various environmental matrices and arthropod diets is needed. Without a standard method for estimating the EEC, exposure levels across studies vary and may not reliably demonstrate a lack of hazard at environmental levels.

Testing may be more appropriately done with multiple doses instead of a single high/maximum hazard dose. Single doses are easier conceptually and functionally and work well when there are no effects; however, in some cases effects are observed that are not easily interpreted. For example, low levels of mortality (e.g. 20–30%) may not indicate risk to populations at the tested exposure level; however, testing at multiple levels would be more useful toward establishing some endpoint value, particularly a no observed adverse effect concentration (NOAEC), and understanding the environmental levels at which these effects could be expected to occur (USEPA, 1996d).

More investigation into the optimum exposure duration is needed and could include the combination of exposure duration and level that sufficiently replicates a maximum hazard level or potential exposure in the field. Available guidelines do not specify the duration of exposure, and in practice, both honey bee and other nontarget arthropod tests range from a few hours of exposure to continuous exposure throughout the study. Continuous exposure reduces uncertainties around whether animals are dosed at the maximum level and helps to ensure that all animals have consumed the test item. However, microbial pesticides may have palatability or other issues that can affect feeding and produce effects that are not necessarily related to exposure to the test item. Exposure durations of a few hours may not be sufficient to deliver a maximum dose that would capture both pathogenicity and toxicity.

Many microbial pesticides must be consumed to have targeted effects on arthropods, but several, such as certain fungal entomopathogens, are known to infect insects through contact exposure. In such cases, contact tests may be more appropriate. However, unknown effects may also be important considerations in choosing an exposure route. For example, microorganisms that may produce biosurfactants or other enzymes may be disruptive to the arthropod cuticle, and effects may be better observed through contact rather than oral tests. At the same time, effects may occur through consumption and toxicity of other unknown compounds produced by the microorganism. Therefore, the development of guidance on decision points that optimize the exposure route would assist test designers and likely would rely on robust microbial identity and characterization data as described earlier in this chapter. Additionally, exposure methods for contact tests would need to be explored, since it is unlikely that passive methods, such as those described in Candolfi et al. (2000), will be sufficient in all cases. However, contact exposure where a known amount is placed on the arthropod could be more precise than oral exposure, and developing methods could advance testing techniques.

9.4 Endpoints observed and controls

Guidelines are flexible with respect to toxicity or pathogenicity endpoints, and mortality is often the effect that is measured. Other endpoints as described in several bee testing guidelines may be reasonable for testing with microbial pesticides (Borges et al., 2021), though mortality is usually sufficient. It may be important in some cases to distinguish between toxicity and pathogenicity, in which case inactivated control groups can be informative. EPA guidelines recommend a concurrent control group containing either microbe-free or nonviable microbe material from the same culture system used to produce the test material (USEPA, 1996c, e). These control groups should contain any toxic materials and not produce pathogenic effects; therefore, theoretically, if effects are observed in the microbe-free control and in the group with the live microorganism, then effects may be concluded to result from toxicity of chemicals produced by the microorganism. Microbe-free preparations may be produced by filtering the microbial test substance to remove cellular material; however, often these controls are produced through autoclaving, which may produce substances that can cause detrimental effects (Borges et al., 2021). Further research into simple methods to produce microbe-free materials for testing would reduce uncertainties around effects observed in these groups.

9.5 Environmental conditions

One of the major challenges in testing with live microorganisms and arthropods is balancing conditions to optimize both the survival of the test insect and the growth of the microorganism. Environmental conditions for bee testing are not described in EPA's microbial guideline, but OECD guidelines offer possibilities for research into optimum conditions for honey bees and other bee species. The EPA guideline for testing with nontarget insects suggests that environmental conditions are important at the time of exposure and only provides an example of conditions for testing with fungi at >90% relative humidity (USEPA, 1996e). However, while testing at this level would help to ensure the growth of the fungus, this level may be beyond the tolerance of the test animal (e.g. 50–60% is recommended by OECD Test Guideline 245 for honey bees [OECD, 2017]). Therefore, significant research is needed to balance testing conditions between test animals and microorganisms and to inform recommendations that can assist in improving the design of these tests.

Similarly, diet must also be optimized to ensure that test animals are receiving sufficient nutrition particularly in cases where studies must be conducted for long durations. Borges et al. (2021) covers this issue and related research needs for honey bee adults and larvae. Similar considerations should be made for testing with other arthropods. A specific difficulty in many cases is that most of these species are predatory or parasitic and therefore have particular dietary needs which may complicate exposure to the test material. The guidance provided within Candolfi et al. (2000) includes relatively more detail regarding diet during testing compared to other guidance. However, for microbial pesticides, the test substance must often be delivered by the diet, which is typically a sucrose solution. If this diet is not sufficient to ensure a viable test over extended durations, further information into the dietary needs of typical arthropod test species could be helpful.

10 Risk assessment of double-stranded RNA pesticides

Externally applied dsRNA presents some unique considerations for risk assessment. Externally applied dsRNA pesticides are emerging technology biopesticides that are similar to biochemical active ingredients and whose pesticidal function relies on the natural mechanism of gene silencing through RNA interference (RNAi). RNAs are present in all living organisms, where they serve essential functions of gene regulation (e.g. transfer RNA, messenger RNA, and micro RNA), and RNAi is a mechanism present in many eukaryotic organisms where it regulates gene expression and provides a defense against transposable elements and RNA-based viruses by targeting them for degradation (Torri et al., 2022).

Externally applied dsRNA products typically exploit the target pest's natural RNAi machinery to downregulate a protein that is essential for the organism's survival. Once the product is ingested by the target pest, it is processed by the pest's cellular RNAi machinery into small interfering (siRNA) molecules. These short nucleotide length sequences are then selected and incorporated into the pest's RNA-induced silencing complex (RISC), resulting in the downregulation of the protein that is essential for the pest's survival. An overview of externally applied dsRNA pesticide risk assessment considerations can be found in the OECD Working Papers titled, 'Considerations for the Environmental Risk Assessment of the Application of Sprayed or Externally Applied ds-RNA-Based Pesticides' (OECD, 2020) and 'Considerations for the Human Health Risk Assessment of Externally Applied ds-RNA-Based Pesticides' (OECD, 2023c). Information on EPA's assessment of Ledprona, the first externally applied dsRNA pesticide registered in the United States, can be found in EPA's Pesticide News Story and associated docket cited therein (USEPA, 2023f).

EPA primarily bases its data needs for externally applied dsRNA pesticides on its biochemical pesticide data requirements. Some of the data listed there may be appropriate to waive, and additional data described earlier in this chapter will likely be required to address the unique nature of these biopesticides. EPA's microbial pesticide data requirements are also considered, particularly with regard to the product's manufacturing process and the duration of nontarget organism studies. Exposure to humans and nontarget organisms and sequence matches are important considerations in the design of the dsRNA and product formulation and impact the risk assessments and data needed to support them. Human bioinformatics analyses should consider potential mismatches and are useful in determining whether there are any unintended sequence matches to human gene transcripts. However, sequence data alone are not a stand-alone predictor of such gene silencing off-target effects, as there are numerous physiological barriers to the uptake of RNA in mammals.

Exposure to humans and nontarget organisms can be impacted by product formulation and chemical modifications to the RNA. These can impact environmental stability and uptake in target pests, nontarget organisms and humans. Because of the impact formulation can have on exposure, mammalian and nontarget organism toxicity testing may need to be conducted using the formulated product. In addition to evaluating environmental fate in soil and water, testing the stability of the dsRNA in simulated gastric fluid in the formulated product together with representative tank mix ingredients further informs the potential for human exposure.

The formulation and chemical alteration of the dsRNA and sequence matches are important considerations in the data needed to support nontarget organism risk assessments. Bioinformatics analyses for nontarget organisms can help determine whether there are any unintended sequence matches to nontarget organism gene transcripts and can also inform the selection and prioritization of nontarget species to be tested. However, sequence data alone are not a stand-alone predictor of these unintended off-target gene silencing effects caused by RNAi of nontarget organisms RNA transcripts (i.e. the ability to detect matches is limited by genome quality and a detected match is not necessarily indicative of a hazard). Testing the activity spectrum of the dsRNA against closely related and less-related species to the pest can add to the bioinformatics in refining the known activity of the dsRNA. Product formulation and chemical modifications of the RNA can impact environmental stability and uptake in nontarget organisms. Evaluation of these as well as nontarget organism toxicity testing using the formulated product will typically be needed. Given the longer timeframe than typical chemical toxicants that it takes for externally applied dsRNA to take effect, nontarget organism toxicity testing duration should follow the microbial pesticide test guidelines.

11 Case study: nontarget organism testing

The case study focuses on nontarget organism testing and illustrates some of the challenges faced by regulators in evaluating studies and developing accurate risk assessments. In this case, a strain of bacteria, ubiquitous in the environment and not known widely to be toxic or pathogenic to arthropods, was tested in several insect species. In studies with adult mealybug destroyer beetles (*Cryptolaemus montrouzieri*) and lady beetles (*Hippodamia convergens*), mortality of <50% was observed with continuous exposure to the active ingredient but was significantly higher than control mortality. In both studies, the maximum hazard dose of ten times the EEC was tested. The LC_{50} was established as >10 times the EEC and a NOAEC could not be defined. Without any additional information, it was concluded that the effects resulted from exposure to the test item. Similarly, mortality was observed in maximum hazard dose testing at 100 times the EEC with honey bees but exposure only occurred for 19 h, food consumption was reduced, and it was not clear whether reduced feeding may have produced the observed results. Additionally, green lacewing (*Chrysoperla rufilabris*) adults were tested under continuous exposure at ten times the EEC, and no adverse effects were observed. Therefore, testing on four species of nontarget insects provided inconsistent results indicating some level of effect in some species, but without establishing any toxicity or pathogenicity endpoint values. As a result, the totality of the testing is difficult to interpret in terms of actual hazard to nontarget arthropods.

Information on the studies does not indicate any problems with testing except for food avoidance by the honey bees. Typically, follow-up inquiries about individual studies occur more than a year after study completion, and details to evaluate confounding factors are difficult to obtain at that time. However, several of the issues described in the previous sections, such as the diet and age of the test organisms, could have influenced the results. Concerning diet, mealybug destroyer and ladybird beetle were fed 50% sucrose solution, whereas green lacewings were fed a powdered diet incorporated into the 50% sucrose solution. Predatory species normally consume other higher-protein foods as part of their diets, which primarily support growth, reproduction, and development. Sugars are important to the diets of predatory insects (Lundgren, 2009), and 50% sucrose likely was sufficient for adult survival. However, the diets among the non-bee species differed, and it is not known whether this aspect may have impacted the results among the three species. The diet fed to juvenile bees, however, may have been inadequate (Borges et al., 2021), but its influence is confounded by reduction in food consumption in that study. The sources of these three insect species were also not described, but if they were sourced commercially or captured, their ages would likely be unknown and mixed, and observed mortality could have resulted from natural deaths due to the expected life span. The studies producing adverse effects also illustrate the utility of testing at multiple exposure levels, particularly when effects are observed. In this case, a multiple-exposure level study conducted below the maximum hazard dose should have defined a NOAEC, which could inform risk characterization or mitigations if needed. In practice, range-finding studies conducted initially would set appropriate test levels prior to conducting the full study.

The low level of mortality (<50%) that was observed in three species, with no significant mortality observed in one species, indicates that there may be some kind of effect due to exposure to this microorganism and its related materials (e.g. secondary metabolites) in the diet. If effects were observed early in the studies, these may be more likely the result of toxicity, since pathogenic effects would be expected to take longer to appear. This bacterium is not known as an insect pathogen, so observed effects are more likely the result of toxicity. A microbe-free control was only included in the bee study, but inclusion in all studies would help to determine the role of potential toxins, as well as immune effects (Borges et al., 2021). Further characterization of the potential metabolites produced by the microorganism would also be helpful.

It is also possible that these studies indicated true effects to be expected from this particular biopesticide. In a case like this, risk assessments could reasonably conclude that adverse effects are not anticipated under field conditions because testing only resulted in 20–40% mortality at very high exposure that is unlikely to occur in the environment. Other factors, such as application sites and methods, could also limit exposure spatially, and the presence of alternate untreated food sources would further reduce exposure and risk to insect populations. However, in a case like this, it is difficult to conclude that effects would not occur to individuals of sensitive species if exposed, which is a required consideration in the United States. In cases where mortality occurs at higher levels, particularly if results are mixed above and below 50%, or where high mortality occurs in some studies while none occurs in others, risk determination and characterization are much more difficult. For these cases, studies may need to be repeated, adding to the cost and length of time to registration. Consequently, with better guidance in conducting these tests to address the challenges discussed in this chapter, more consistent and reliable results could have been obtained leading to better characterization of the risks.

12 Conclusion

Biopesticides are important tools for controlling insect pests, and demand for them is growing globally. Addressing testing questions and challenges will help better refine environmental risk assessments by limiting the need for conservative assumptions and increasing our understanding and real-world relevance of test results. Certain aspects of risk assessment for biopesticides are well-established, such as human health hazard assessment for microbial and biochemical pesticides. However, uncertainties present in other aspects, such as questions of pesticide identity or low-resolution hazard testing for nontarget organisms, can introduce difficulties into risk assessments. Additionally, as new assessment tools become available for chemicals and conventional pesticides, such as advancements in NAMs, adapting these tools to biopesticides or developing tools taking into consideration biopesticides can help maintain their advantage.

This chapter has identified some of the common issues faced by regulators in assessing risks of microbial and biochemical pesticides and certain dsRNA emerging technology pesticides. Pesticide identity sets the stage for risk assessment, so it is critical to fully identify and characterize biopesticide active ingredients not only to identify anticipated concerns but also to potentially reduce testing. Common problems related to the identification of the biopesticide have been highlighted, which, if resolved, can enhance understanding within the risk assessment. Common issues related to exposure analysis have also been identified, including uncertainties raised by ambiguities on pesticide labels. Additionally, hazard testing issues, especially those related to nontarget organism testing with microbial pesticides, have also been highlighted. While biopesticides are generally considered to be of low risk, robust risk assessments are still needed to support their registration. Advancement in clarifying these issues will strengthen biopesticide risk assessments by increasing their resolution and decreasing the need for conservative assumptions.

13 Disclaimer

This manuscript does not necessarily represent US Environmental Protection Agency (US EPA) policy. This manuscript has been subjected to review and approved for publication by the US EPA's Office of Chemical Safety and Pollution Prevention. Mention of trade names or commercial products does not indicate endorsement by the US EPA.

14 Where to look for further information

The following articles provide a good overview of this subject:

- Wend, K., Zorrilla, L., Freimoser, F.M. et al. Microbial pesticides – challenges and future perspectives for testing and safety assessment with respect to human health. Environ Health 23, 49 (2024). https://doi.org/10.1186/s12940-024-01090-2
- Karaoğlan, B., Alkassab, A.T., Borges, S. et al. Microbial pesticides: challenges and future perspectives for non-target organism testing. Environ Sci Eur 36, 205 (2024). https://doi.org/10.1186/s12302-024-01017-1
- OECD 2023. OECD Series on Pesticides No. 110, Considerations for the Human Health Risk Assessment of Externally Applied ds-RNA-Based Pesticides. (ENV/CBC/MONO (2023) 26, OECD Environment Directorate, Paris, France.
- OECD 2023. OECD Series on Pesticides No. 104, Considerations for the Environmental Risk Assessment of the Application of Sprayed or Externally Applied dsRNA-based Pesticides. ENV/JM/MONO (2020) 26, OECD Environment Directorate, Paris, France.

15 References

Abu-Ashour, J., et al. (1994) 'Transport of microorganisms through soil', *Water, Air and Soil Poll.*, 75, 141–158.

Adams, R.P. (1987) 'Yields and seasonal variation of phytochemicals from *Juniperus* species of the United States', *Biomass*, 12: 129–139.

Alegbeleye, O.O. and Sant'Ana, A.S. (2020) 'Manure-borne pathogens as an important source of water contamination: an update on the dynamics of pathogen survival/ transport as well as practical risk mitigation strategies', *Int. J. Hygiene and Environ. Health*, 227, 113524. https://doi.org/10.1016/j.ijheh.2020.113524

Aziz, Z.A.A., et al. (2018) 'Essential oils: extraction techniques, pharmaceutical, and therapeutic potential – a review', *Curr. Drug Metab.* 19(13), 1100–1110. https://doi.org/10.2174/1389200219666180723144850

Bailey, K.L., et al. (2010) 'Social and economic drivers shaping the future of biological control: a Canadian perspective on the factors affecting the development and use of microbial biopesticides', *Biol. Control*, 52, 221–229. https://doi.org/10.1016/j.biocontrol.2009.05.003

Bertelsmeier, C. (2021) 'Globalization and the anthropogenic spread of invasive social insects', *Curr. Opin. Insect Sci.*, 46, 16–23. https://doi.org/10.1016/j.cois.2021.01.006

Borges, S., et al. (2021) 'Overview of the testing and assessment of effects of microbial pesticides on bees: strengths, challenges, and perspectives', *Apidologie*, 52, 1256–1277. https://doi.org/10.1007/s13592-021-00900-7

Brown, E., et al. (2019) 'Use of whole-genome sequencing for food safety and public health in the United States', *Foodborne Path. and Dis.*, 16(7), 441–450. https://doi.org/10.1089/fpd.2019.2662

Calicioglu, O., et al. (2019) 'The future challenges of food and agriculture: an integrated analysis of trends and solutions', *Sustainability*, 11(1), 222–243. https://doi.org/10.3390/su11010222

California Department of Pesticide Regulation (CalDPR) (2023) *Accelerating Sustainable Pest Management: A Roadmap for California*, Retrieved from https://www.cdpr.ca.gov/docs/sustainable_pest_management_roadmap/spm_roadmap.pdf, January 13, 2024.

Candolfi, M.P., et al. (2000) *Guidelines to Evaluate Side-Effects of Plant Protection Products to Non-Target Arthropods*, The Publication Commission of the International Organization for Biological and Integrated Control of Noxious Animals and Plants, West Palearctic Regional Section (IOBC/WPRS), Ghent, Belgium.

Cesa, S., et al. (2019) 'Phytochemical analyses and pharmacological screening of Neem oil', *S. African J. Botany*, 120, 331–337. https://doi.org/10.1016/j.sajb.2018.10.019

Charles, H., et al. (2010) 'Food security: the challenge of feeding 9 billion people, *Science*, 327, 812–818. https://doi.org/10.1126/science.1185383

Damalas, C.A. and Koutroubas, S.D. (2018) 'Current status and recent developments in biopesticide use', *Agriculture*, 8(1), 13–18. https://doi.org/10.3390/agriculture8010013

Dooley, S.R. and Beckstead, J. (2010) 'Characterizing the interaction between a fungal seed pathogen and a deleterious rhizobacterium for biological control of cheatgrass', *Biol. Control*, 53: 197–203. https://doi.org/10.1016/j.biocontrol.2009.12.007

European Commission (EC), Directorate-General for Research and Innovation (2018) *A Sustainable Bioeconomy for Europe – Strengthening the Connection between Economy, Society and the Environment – Updated Bioeconomy Strategy*, 107 p, Retrieved from https://data.europa.eu/doi/10.2777/792130 on August 18, 2023.

EC (2019) *The European Green Deal. Communication from the Commission to the European Parliament, the European Council, the European Economic and Social Committee and the Committee of the Regions*, 24 p, Retrieved from https://eur-lex.europa.eu/legal-content/EN/TXT/?uri=COM%3A2019%3A640%3AFIN on August 18, 2023.

European Food Safety Authority (EFSA) (2018) 'Guidance on the characterization of microorganisms used as feed additives or as production organisms' *EFSA J.*, 16(3), e05206. https://doi.org/10.2903/j.efsa.2018.5206

EFSA (2021) 'EFSA statement on the requirements for whole genome sequence analysis of microorganisms intentionally used in the food chain', *EFSA J.*, 19(7), 6506. https://doi.org/10.2903/j.efsa.2021.6506

Executive Order No. 14081 (2022) Executive Office of the President, 87 U.S. Federal Register No. 178, September 15, 2022.

Fedhila, S., et al. (2004) 'Characterization of two *Bacillus thuringiensis* genes identified by *in vivo* screening of virulence factors', *Appl. Environ. Micro.*, 70(8), 4784–4791. https://doi.org/10.1128/AEM.70.8.4784-4791.2004

Fenibo, E.O., et al. (2021), 'Biopesticides in sustainable agriculture: a critical sustainable development driver governed by green chemistry principles', *Front. Sustain. Food Syst.*, 5, 619058. https://doi.org/10.3389/fsufs.2021.619058

Hudaib, M., et al. (2002) 'GC/MS evaluation of thyme (*Thymus vulgaris* L.) oil composition and variations during the vegetative cycle', *J. Pharm. Biomed. Analysis*, 29(4), 691–700. https://doi.org/10.1016/S0731-7085(02)00119-X

Hulme, P.E. (2009), 'Trade, transport and trouble: managing invasive species pathways in an era of globalization', *J. Appl. Ecol.*, 46, 10–18. https://doi.org/10.1111/j.1365-2664.2008.01600.x

Kaiser, D., et al. (2018) 'Efficiency of natural substances to protect *Beauveria bassiana* conidia from UV radiation', *Pest Manage. Sci.*, 75(2), 556–563. https://doi.org/10.1002/ps.5209

Koduru, U.D., et al. (2014) 'Bioinformatic tools in the analysis of determinants of pathogenicity and ecology of entomopathogenic fungi used as microbial insecticides in crop protection'. In K. Kishor et al. (Eds), *Agricultural Bioinformatics*, Springer, New Delhi, India, pp. 215–234.

Köser, C.U., et al. (2012) 'Routine use of microbial whole genome sequencing in diagnostic and public health microbiology', *PloS Pathog.*, 8(8), e1002824. https://doi.org/10.1371/journal.ppat.1002824

Lemos, M.F., et al. (2017) 'Seasonal variation affects the composition and antibacterial and antioxidant activities of *Thymus vulgari*', *Indust. Crops Products*, 95: 543–548. http://dx.doi.org/10.1016/j.indcrop.2016.11.008

Leahy, J., et al. (2014). 'Biopesticide oversight and registration at the U.S. Environmental Protection Agency in Biopesticides: state of the art and future opportunities'. In Coats et al. (Eds.), *ACS Symposium Series*. Washington, DC: American Chemical Society. https://doi.org/10.1021/bk-2014-1172.ch001

Lundgren, J.G. (2009) *Relationships of Natural Enemies and Non-Prey Foods*, Springer, New York, 452 pp.

Marrone, P.G. 2014. The Market and Potential for Biopesticides. In Coats et al. (Eds.), *ACS Symposium Series*. Washington, DC: American Chemical Society. https://pubs.acs.org/doi/abs/10.1021/bk-2014-1172.ch016

Organization for Economic Cooperation and Development (OECD) (2017) *OECD Guideline for the Testing of Chemicals, Honey Bee (Apis mellifera), Chronic Oral*

Toxicity Test (10-day Feeding), OECD/OCDE 245, Paris, France, Adopted October 9, 2017.

OECD (2018) *OECD Series on Pesticides No. 98, Working Document on the Risk Assessment of Secondary Metabolites of Microbial Biocontrol Agents*, ENV/JM/MONO(2018)33, OECD Environment Directorate, Paris, France.

OECD (2020). OECD Series on Pesticides No. 104, *Working Document on Considerations for the Environmental Risk Assessment of the Application of Sprayed or Externally Applied dsRNA-based Pesticides.* ENV/JM/MONO (2020) 26, OECD Environment Directorate, Paris, France.

OECD (2023a) *OECD Series on Pesticides No. 109, Innovating Microbial Pesticides Testing: Conference Proceedings*, ENV/CBC/MONO(2023)10, OECD Environment Directorate, Paris, France.

OECD (2023b) *OECD Guidelines for the Testing of Chemicals, Section 2*, Retrieved from https://www.oecd-ilibrary.org/environment/oecd-guidelines-for-the-testing-of-chemicals-section-2-effects-on-biotic-systems_20745761, on September 25, 2023.

OECD (2023c). OECD Series on Pesticides No. 110, *Working Document on Considerations for the Human Health Risk Assessment of Externally Applied ds-RNA-Based Pesticides.* (ENV/CBC/MONO (2023) 26, OECD Environment Directorate, Paris, France.

Richter, J. and Schellenberg, I. (2007) 'Comparison of different extraction methods for the determination of essential oils and related compounds from aromatic plants and optimization of solid-phase microextraction/gas chromatography', *Anal. Bioanal. Chem.*, 387, 2207–2217. https://doi.org/10.1007/s00216-006-1045-6

Torri, A., et al. (2022) The origin of RNA interference: Adaptive or neutral evolution? *PLoS Biol.*, 20(6), e3001715.

U.S. Environmental Protection Agency (USEPA) (1996a) *Microbial Pesticide Test Guidelines: OPPTS 885.1100 Product Identity*, U.S. Environmental Protection Agency, Washington, DC, February 1996.

USEPA (1996b) *Biochemical Pesticide Test Guidelines: OPPTS 880.1100 Product Identity and Composition*, U.S. Environmental Protection Agency, Washington, DC, February 1996.

USEPA (1996c) *Microbial Pesticide Test Guidelines: OPPTS 885.4380 Honey Bee Testing, Tier I*, U.S. Environmental Protection Agency, Washington, DC, February 1996.

USEPA (1996d) *Microbial Pesticide Test Guidelines: OPPTS 885.4000 Background for Nontarget Organism Testing of Microbial Pest Control Agents*, U.S. Environmental Protection Agency, Washington, DC, February 1996.

USEPA (1996e) *Microbial Pesticide Test Guidelines OPPTS 885.4340 Nontarget Insect Testing, Tier I*, U.S. Environmental Protection Agency, Washington, DC, February 1996.

USEPA (1998) *Guidelines for Ecological Risk Assessment*, U.S. Environmental Protection Agency, Risk Assessment Forum, EPA/630/R-95/002F, Washington, DC, April, 1998.

USEPA (2007a) "Biochemical Pesticides", *Code of Federal Regulations*, Title 40, Part 158, Subpart U.

USEPA (2007b) "Microbial Pesticides", *Code of Federal Regulations*, Title 40, Part 158, Subpart V.

USEPA (2007c) "Flexibility", *Code of Federal Regulations*, Title 40, Part 158, Subpart A, Section 158.30.

USEPA (2012) *Ecological Effects Test Guidelines OCSPP850.3020: Honey Bee Acute Contact Toxicity Test*, U.S. Environmental Protection Agency, Washington, DC, June 2012.

USEPA (2014) *Framework for Human Health Risk Assessment to Inform Decision Making*, U.S. Environmental Protection Agency, Risk Assessment Forum, Washington, DC, April, 2014.

USEPA (2015) Use of an Alternate Testing Framework for Classification of Eye Irritation Potential of EPA Pesticide Products, U.S. Environmental Protection Agency, Office of Pesticide Programs, March 2, 2015.

USEPA (2020) *Guidance for Waiving Acute Dermal Toxicity Tests for Pesticide Technical Chemicals & Supporting Retrospective Analysis*, U.S. Environmental Protection Agency, Office of Pesticide Programs, Docket Document ID EPA-HQ-OPP-2016-0093-0181, Washington, DC, December 31, 2020.

USEPA (2022) *Pesticide Use Site Index*, Retrieved from https://www.epa.gov/pesticide-registration/pesticide-use-site-index on November 3, 2023.

USEPA (2023a) *Regulation of Pesticide Residues on Food*, Retrieved from https://www.epa.gov/pesticide-tolerances on October 27, 2023.

USEPA (2023b) *Protecting Endangered Species from Pesticides*, Retrieved from https://www.epa.gov/endangered-species on October 30, 2023.

USEPA (2023c) *Models for Pesticide Risk Assessment*, Retrieved from https://www.epa.gov/pesticide-science-and-assessing-pesticide-risks/models-pesticide-risk-assessment on October 16, 2023.

USEPA (2023d) Series 885 – Microbial Pesticide Test Guidelines, Retrieved from https://www.epa.gov/test-guidelines-pesticides-and-toxic-substances/series-885-microbial-pesticide-test-guidelines on September 25, 2023.

USEPA (2023e) Series 850 – Ecological Effects Test Guidelines, Retrieved from https://www.epa.gov/test-guidelines-pesticides-and-toxic-substances/series-850-ecological-effects-test-guidelines on September 25, 2023.

USEPA (2023f) *EPA Pesticides News Stories - EPA Registers Novel Pesticide Technology for Potato Crops December 22, 2023*, Retrieved from https://www.epa.gov/pesticides/epa-registers-novel-pesticide-technology-potato-crops on January 13, 2024.

Valero-Jiménez, C.A., et al. (2016) 'Genes involved in virulence of the entomopathogenic fungus *Beauveria bassiana*', *J. Invertebrate Pathol.*, 133, 41–49. http://dx.doi.org/10.1016/j.jip.2015.11.011

Vukovic, N., et al. (2009) 'Antimicrobial activity of the essential oil obtained from roots and chemical composition of the volatile constituents from the roots, stems, and leaves of *Ballota nigra* from Serbia', *J. Medicinal Food*, 12(2), 435–441. https://doi.org/10.1089/jmf.2008.0164

Index

www.ingramcontent.com/pod-product-compliance
Ingram Content Group UK Ltd.
Pitfield, Milton Keynes, MK11 3LW, UK
UKHW021826100926
8556IPUK00003B/184

9 781801 468398